TRIGONOMETRY

HARCOURT BRACE COLLEGE OUTLINE SERIES

TRIGONOMETRY

James Gehrmann and Thomas Lester

Department of Mathematics
California State University at Sacramento

Harcourt Brace College Publishers

Fort Worth Philadelphia San Diego New York Orlando Austin San Antonio
Toronto Montreal London Sydney Tokyo

Printed in the United States of America

Library of Congress Cataloging in Publication Data

Gehrmann, James.
  Trigonometry.

  (Harcourt Brace college outline series)
  Includes index.
  1. Trigonometry. I. Lester,Thomas. II. Title.
III.Series. IV.Series: Books for professionals.
QA531.G44 1984   516.2'4   83-22702

ISBN 0-15-601693-1

  8 9 0 1 2 3 4 5 074 13 12 11 10

First edition

CONTENTS

PREFACE

This outline has three purposes: (1) to offer practical exercise in problem-solving to students interested in sharpening their trigonometric skills, (2) to provide supplementary explanations, additional examples, and extra solved problems for students experiencing difficulties in formal trigonometry classes, and (3) to present a self-contained outline of trigonometry for those pursuing an independent study or review of trigonometry.

If you plan to use this outline as a supplement to your trigonometry textbook, your best strategy is to follow the exposition in your class textbook until you experience difficulty, or want practice with a method explained there. Then study the examples and work the solved problems of the appropriate type in this outline. When you think that you have mastered the method illustrated by our examples and solved problems, attempt to solve supplementary exercises of the same type. If you can correctly solve the first few supplementary exercises, you have overcome the difficulty; if you cannot solve the supplementary exercises, reread the examples and work through the solved problems before attempting other supplementary exercises of the same type.

If you plan to use this outline in an independent study or review of trigonometry, you must be familiar with the material covered in the first year of high school algebra (including the quadratic formula) and that covered in a year of high school geometry. You can use this outline as you would any textbook—by beginning with Chapter 1 and proceeding sequentially through all the chapters, first reading the expository material, studying the examples, and then confirming each step of the solved problems. (Work them yourself; then check your answers with the solutions given.) When you think that you have mastered a technique illustrated by several solved problems, begin to solve supplementary exercises of that type. If your solutions to the supplementary exercises are correct, proceed to the next topic. If your solutions are incorrect, reread the examples, study the solved problems, and rework the supplementary exercises until you have surmounted the difficulty.

We have assumed that most of you are studying trigonometry for practical reasons: Either you want (or *will* want) to apply trigonometric methods in your mathematics courses such as calculus and linear algebra; or you need to use trigonometric methods to solve problems in the physical and natural sciences, or in navigation and surveying. Therefore, the examples and problems in this outline emphasize applied trigonometric calculations, instead of the theoretical details of trigonometry.

We have also assumed that you own a calculator. With few exceptions, a hand-held calculator will enable you to perform trigonometric calculations that once required tables of logarithms and trigonometric functions. Calculators also permit you to eliminate the tedious methods of computation that arise from the properties of logarithms. Therefore, we provide neither complete trigonometric tables nor logarithm tables, and we do not discuss the properties of logarithms. A hand-held calculator is essential for solving many of the problems. Even solutions that don't require the use of a calculator should be verified with a calculator. The calculator has become an essential mathematic tool, and you need to know how to use one efficiently.

The outline begins with a chapter on calculators, moves to a review of points, distances, and circles in the plane, and then continues with an introduction to angles and angle measurement, followed by a review of relations and functions (Chapters 1–4). An introduction to the trigonometry of right triangles leads to a discussion of trigonometric functions, trigonometric identities and equations, and inverse trigonometric functions (Chapters 5–13). In the concluding chapters, we explore solutions of oblique triangles, polar coordinates, vectors, and complex numbers (Chapters 14–17).

We gratefully acknowledge the advice of our colleagues and reviewers. Their suggestions and careful attention to detail were helpful in shaping the various drafts of this book.

Sacramento, California

JIM GEHRMANN

TOM LESTER

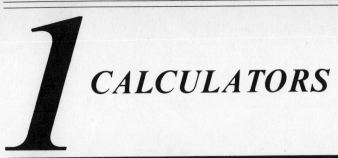

1 CALCULATORS

THIS CHAPTER IS ABOUT

- ☑ **Calculators and Trigonometry**
- ☑ **Desirable Features of a Calculator**
- ☑ **Function Keys**
- ☑ **Standard Algebraic and Reverse Polish Notation**
- ☑ **Arithmetic Operations on Hand-Held Calculators**

1-1. Calculators and Trigonometry

Computers and calculators have redefined the way in which numeric calculations in mathematics, engineering, and the sciences are performed. The speed, accuracy, and memory functions of the current generation of small computers and hand-held calculators make quick work of computations that are often tedious and very difficult to solve by hand.

In the past, trigonometry textbooks were forced to devote many pages to interpolation, the use of logarithms in calculations, techniques for finding trigonometric functions of given angles, and angles corresponding to given values of the trigonometric functions. With a hand-held calculator, logarithms are not needed to perform calculations, interpolation is replaced by use of a function key, and tables of trigonometric functions are held in the calculator's memory.

In this book, calculations that could be performed more simply on a calculator are presented in the manner of the most popular trigonometry textbooks. Logarithms or interpolation are not discussed: we'll assume that you'll use a hand-held calculator to find values of and perform computations with the trigonometric functions.

If you're using this book in preparation for a calculus course or as a supplement in a trigonometry course, don't bypass the "traditional" solution in favor of the direct calculator approach. You can use the calculator approach to verify your answers. If, however, you're using this book as a guide to solving numeric trigonometry problems, then use the appropriate calculator shortcuts.

1-2. Desirable Features of a Calculator

Table 1-1 relates features of hand-held calculators to the computational requirements of trigonometry. Features are rated according to desirability for trigonometric calculations in the right-hand column of the table.

TABLE 1-1. Calculator features.

Feature	Keys to look for	Desirability
1. Stores intermediate results	STO RCL M+ M− x → M	High
2. Provides one-stroke constants	π e	High
3. Computes powers, roots, reciprocals	x^2 $\sqrt{x}$ y^x $1/x$	Essential

TABLE 1-1 (*Continued*).   Calculator features.

Feature	Keys to look for	Desirability
4. Computes logarithms	$\boxed{\log}\ \boxed{\ln x}\ \boxed{e^x}$	Moderate
5. Accepts numbers in scientific notation	$\boxed{\text{EE}}\ \boxed{\text{EXP}}$	Moderate
6. Evaluates trigonometric functions and their inverses	$\boxed{\sin}\ \boxed{\cos}\ \boxed{\tan}\ \boxed{\text{INV}}$ $\boxed{\sin^{-1}}\ \boxed{\cos^{-1}}\ \boxed{\tan^{-1}}\ \boxed{\text{ARC}}$	Essential
7. Allows use of radians and degrees	$\boxed{\text{DRG}}\ \boxed{\text{DEG}}$ Rad $\leftrightarrow$ Deg	Essential

Calculators with memory keys, the ability to find powers and roots and to find values of the trigonometric functions and their inverses in both radians and degrees are available today for less than $20.00. If you do not already have a hand-held calculator with the essential features in Table 1-1, we recommend that you purchase or borrow one to continue your study of trigonometry.

1-3. Function Keys

The function keys execute such operations as squaring a number, taking the square root of a number, and changing the sign of a number. Table 1-2 lists some of the function keys that you will find on calculators. Use this table when you study the function keys on your calculator or when you buy a calculator, and later when you are solving trigonometric function problems.

Note: Your calculator may not have all of the function keys listed in Table 1-2.

1-4. Standard Algebraic and Reverse Polish Notation

Two standard notational systems are used in performing calculations on hand-held calculators: **Standard Algebraic Notation (SAN)** and **Reverse Polish Notation (RPN)**. A calculator with an ENTER key uses the RPN system; a calculator with an equals key ($=$) uses the SAN system.

On SAN calculators you press keys in the same order as you write and perform the operations with pencil and paper. For example, on a SAN calculator $3 + 4$ is evaluated by pressing $\boxed{3}$, then $\boxed{+}$, then $\boxed{4}$, and finally $\boxed{=}$. The calculator display shows the answer, 7. This sequence of keystrokes is denoted by $\boxed{3}\ \boxed{+}\ \boxed{4}\ \boxed{=}$.

On RPN calculators you enter the two numbers and then specify the operation. For example, $3 + 4$ is evaluated on a RPN machine by pressing $\boxed{3}$, then $\boxed{\text{ENTER}}$, then $\boxed{4}$, and finally $\boxed{+}$. This sequence is denoted by $\boxed{3}\ \boxed{\text{ENTER}}$ $\boxed{4}\ \boxed{+}$. The answer, 7, appears on the calculator display.

Note: Numbers that require multiple keystrokes or decimal points are enclosed within a single box: 3.47 is written as $\boxed{3.47}$, not as $\boxed{3}\ \boxed{\cdot}\ \boxed{4}\ \boxed{7}$.

The difference between the two systems is the order in which the operation symbols are entered: in SAN, the desired operation symbol is entered after entering the first number; in RPN, the operation symbol is entered after entering the second number. Even though keystrokes in RPN follow an unusual pattern, complicated operations are easier to perform with RPN.

1-5. Arithmetic Operations on Hand-Held Calculators

If you already have a calculator with the essential features shown in Table 1-1, imitate the steps of each of the following examples for the calculator type that you have. If you don't have a calculator with the essential features, try both RPN

TABLE 1-2. Function keys.

Key	Function
x^2	Computes the square of number in display
$\sqrt{x}$	Computes the square root of number in display
x^y	Computes the yth power of number in display
$x^{1/y}$	Computes the yth root of number in display
$1/x$	Computes the reciprocal of number in display
e^x	Raises e to power of number in display
→DEG	Converts number in display from radians to degrees
→RAD	Converts number in display from degrees to radians
+/− or CHS	Changes sign of number in display
AC or CLX	Clears the calculator for a new computation
ENTER	Used in entering numbers
DEG	Engages degree mode for trigonometric calculations
RAD	Engages radian mode for trigonometric calculations
sin	Computes sine of number in display
cos	Computes cosine of number in display
tan	Computes tangent of number in display
$\sin^{-1}$, INV sin, or arc sin	Computes inverse sine of number in display
$\cos^{-1}$, INV cos, or arc cos	Computes inverse cosine of number in display
$\tan^{-1}$, INV tan, or arc tan	Computes inverse tangent of number in display
f , g , or INV	Engages alternate functions on keys
M+ or SUM	Adds number in display to memory
log	Computes $\log_{10}$ of number in display
ln x	Computes natural log of number in display
M−	Subtracts number in display from memory
MR or RCL	Recalls number in memory
M in	Places number in display in memory
x → M or EXC	Exchanges numbers in memory and display

---

and SAN calculators on the following examples and problems. Consider the purchase of the model that you find easiest to use, within the restrictions of your budget.

A. The fundamental arithmetic operations

The following examples illustrate the keystrokes required in both the SAN and RPN systems for addition, subtraction, multiplication, division, and roots. (We'll use the standard symbols +, −, ×, and ÷; your calculator may have different symbols.) Refer to Table 1-2 and your instruction manual for other function keys. Before beginning a new computation, press the key that clears the display.

EXAMPLE 1-1: Evaluate the following expressions on your calculator: **(a)** $2 - 3$; **(b)** $2/3$; **(c)** $2 \cdot 3$; **(d)** $\sqrt{54}$.

Solution:

(a)

SAN Notation		RPN Notation	
Keystroke(s)	Display	Keystroke(s)	Display
2	2.	2	2.
−	2.	ENTER	2.0000
3	3.	3	3.
=	−1.	−	−1.0000

(b)

SAN Notation		RPN Notation	
Keystroke(s)	Display	Keystroke(s)	Display
2	2.	2	2.
÷	2.	ENTER	2.0000
3	3.	3	3.
=	0.666 666 7	÷	0.6667

(c)

SAN Notation		RPN Notation	
Keystroke(s)	Display	Keystroke(s)	Display
2	2.	2	2.
×	2.	ENTER	2.0000
3	3.	3	3.
=	6.	×	6.0000

(d)

SAN Notation		RPN Notation	
Keystroke(s)	Display	Keystroke(s)	Display
54	54.	54	54.
$\sqrt{x}$	7.348 469 2	f $\sqrt{x}$	7.3485

Note: Numbers with more than four places to the right or left of the decimal are printed with a space between each group of three in this book. Commas for spacing are not used.

In these simple calculations, SAN is probably easier for you because it corresponds to the way in which you do pencil-and-paper computations. Also, the number of keystrokes is the same for both notational systems; however, in the evaluation of more complex expressions, RPN is superior because it requires fewer keystrokes. RPN might be harder to master, but once mastered, it permits more efficient computation.

B. Complex algebraic expressions

You can see the difference in number of keystrokes required by the two notational systems when you evaluate more complex expressions.

EXAMPLE 1-2: Evaluate the following expressions: **(a)** $2(3 + 4)$; **(b)** $\sqrt{3^2 + 4^2}$; **(c)** $(3 + 4)/(5 - 9)$.

Solution:

(a)

SAN Notation		RPN Notation	
Keystroke(s)	Display	Keystroke(s)	Display
3	3.	3	3.
+	3.	ENTER	3.0000
4	4.	4	4.
=	7.	+	7.0000
×	7.	2	2.
2	2.	×	14.0000
=	14.		

(b)

SAN Notation		RPN Notation	
Keystroke(s)	Display	Keystroke(s)	Display
3	3.	3	3.
x^2	9.	g x^2	9.0000
+	9.	4	4.
4	4.	g x^2	16.0000
x^2	16.	+	25.0000
=	25.	f $\sqrt{x}$	5.0000
$\sqrt{x}$	5.		

(c)

SAN Notation		RPN Notation	
Keystroke(s)	Display	Keystroke(s)	Display
5	5.	3	3.
−	5.	ENTER	3.0000
9	9.	4	4.
=	−4.	+	7.0000
M+	−4.	5	5.
CLEAR	0.	ENTER	5.0000
3	3.	9	9.
+	3.	−	−4.0000
4	4.	÷	−1.7500
=	7.		
÷	7.		
MR	−4.		
=	−1.75		

Note that in parts **(a)** and **(b)** of this example, the RPN calculator required 1 less keystroke, while in part **(c)** the RPN calculator required 4 fewer keystrokes than a SAN calculator.

1-6. Examples

In the following examples both SAN and RPN solutions are shown.

EXAMPLE 1-3: Use your calculator to evaluate $5 - (2.465/3)$.

Solution:

SAN Notation		RPN Notation	
Keystroke(s)	Display	Keystroke(s)	Display
2.465	2.465	2.465	2.465
÷	2.465	ENTER	2.4650
3	3.	3	3.
=	0.821 667	÷	0.8217
+/−	−0.821 667	CHS	−0.8217
+	−0.821 667	5	5.
5	5.	+	4.1783
=	4.178 333		

EXAMPLE 1-4: Use your calculator to evaluate $3\left(\frac{2}{5}\right)$.

Solution:

SAN Notation		RPN Notation	
Keystroke(s)	Display	Keystroke(s)	Display
2	2.	2	2.0000
÷	2.	ENTER	2.0000
5	5.	5	5.
=	0.4	÷	0.4000
×	0.4	3	3.
3	3.	×	1.2000
=	1.2		

EXAMPLE 1-5: Use your calculator to evaluate $\frac{3}{7} + \frac{8}{11}$.

Solution:

SAN Notation		RPN Notation	
Keystroke(s)	Display	Keystroke(s)	Display
3	3.	3	3.
÷	3.	ENTER	3.0000
7	7.	7	7.
=	0.428 571 4	÷	0.4286
M+	0.428 571 4	8	8.
CLEAR	0.	ENTER	8.0000
8	8.	11	11.
÷	8.	÷	0.7273
11	11.	+	1.1558

SAN Notation		RPN Notation	
Keystroke(s)	Display	Keystroke(s)	Display
=	0.727 272 7		
+	0.727 272 7		
MR	0.428 571 4		
=	1.155 844 2		

EXAMPLE 1-6: Use your calculator to evaluate $(2.35 - 4)/3.67$.

Solution:

SAN Notation		RPN Notation	
Keystroke(s)	Display	Keystroke(s)	Display
2.35	2.35	2.35	2.35
−	2.35	ENTER	2.3500
4	4.	4	4.
=	−1.65	−	−1.6500
÷	−1.65	3.67	3.67
3.67	3.67	÷	−0.4496
=	−0.449 591 3		

EXAMPLE 1-7: Use your calculator to evaluate $\sqrt{12^2 + 5^2}$.

Solution:

SAN Notation		RPN Notation	
Keystroke(s)	Display	Keystroke(s)	Display
12	12.	12	12.
x^2	144.	g x^2	144.0000
+	144.	5	5.
5	5.	g x^2	25.0000
x^2	25.	+	169.0000
=	169.	f $\sqrt{x}$	13.0000
$\sqrt{x}$	13.		

EXAMPLE 1-8: Use your calculator to evaluate $1/(\frac{1}{2} + \frac{1}{3})$.

Solution:

SAN Notation		RPN Notation	
Keystroke(s)	Display	Keystroke(s)	Display
2	2.	2	2.
1/x	0.5	g 1/x	0.5000
+	0.5	3	3.
3	3.	g 1/x	0.3333
1/x	0.333 333 3	+	0.8333
=	0.833 333 3	g 1/x	1.2000
1/x	1.2		

EXAMPLE 1-9: Use your calculator to evaluate $3^{1/4}/7^5$.

Solution:

SAN Notation		RPN Notation	
Keystroke(s)	Display	Keystroke(s)	Display
7	7.	3	3.
x^y	7.	ENTER	3.0000
5	5.	4	4.
=	16 807.	g 1/x	0.2500
M+	16 807.	f y^x	1.3161
3	3.	7	7.
x^y	3.	ENTER	7.0000
4	4.	5	5.
1/x	0.25	f y^x	16 807.0000
=	1.316 074	÷	0.0001
÷	1.316 074		
MR	16 807.		
=	0.000 078 3		

The RPN calculator has rounded to the fourth digit to the right of the decimal. The calculator can be made to display more than four digits to the right of the decimal place (see your instruction manual).

SUMMARY

1. Calculators have simplified trigonometric computations.
2. The availability of a calculator that can square numbers, find square roots, and find values of the trigonometric and inverse trigonometric functions will prove invaluable to you in your study of trigonometry.
3. To perform a computation in Standard Algebraic Notation, the first number is entered, then the operation symbol, then the next number, and finally the equal sign. For example, you add 3 and 4 by using the keystroke sequence 3 + 4 = .
4. To perform a computation in Reverse Polish Notation, the numbers are entered followed by the operation symbol. For example, you add 3 and 4 by using the keystroke sequence 3 ENTER 4 + .

RAISE YOUR GRADES

Can you ...?

☑ use your calculator to add, subtract, multiply, and divide numbers
☑ use your calculator to find squares and square roots of numbers
☑ use your calculator to evaluate complex expressions involving sums, differences, products, quotients, roots, and powers of numbers

SOLVED PROBLEMS

PROBLEM 1-1 Use your calculator to evaluate $\frac{3}{4}\left(\frac{2}{9}\right)$.

Solution:

SAN Notation		RPN Notation	
Keystroke(s)	Display	Keystroke(s)	Display
3	3.	3	3.
÷	3.	ENTER	3.0000
4	4.	4	4.
×	0.75	÷	0.7500
2	2.	2	2.
÷	1.5	×	1.5000
9	9.	9	9.
=	0.166 666 7	÷	0.1667

PROBLEM 1-2 Use your calculator to evaluate $3^2 + \frac{2}{9}$.

Solution:

SAN Notation		RPN Notation	
Keystroke(s)	Display	Keystroke(s)	Display
2	2.	2	2.
÷	2.	ENTER	2.0000
9	9.	9	9.
=	0.222 222 2	÷	0.2222
+	0.222 222 2	3	3.
3	3.	f x^2	9.0000
x^2	9.	+	9.2222
=	9.222 222 2		

PROBLEM 1-3 Use your calculator to evaluate $2^5/16^3$.

Solution:

SAN Notation		RPN Notation	
Keystroke(s)	Display	Keystroke(s)	Display
16	16.	2	2.
x^y	16.	ENTER	2.0000
3	3.	5	5.
=	4096.	f y^x	32.0000
M+	4096.	16	16.
2	2.	ENTER	16.0000
x^y	2.	3	3.

	SAN Notation			RPN Notation	
	Keystroke(s)	Display		Keystroke(s)	Display
	5	5.		f y^x	4096.0000
	=	32.		÷	0.0078
	÷	32.			
	MR	4096.			
	=	0.007 812 5			

PROBLEM 1-4 A right triangle has a hypotenuse 20 feet long and one side 15 feet long. Use your calculator to find the length of the other side.

Solution: Recall the Pythagorean Theorem, $h^2 = a^2 + b^2$, where h is the length of the hypotenuse and a and b are the lengths of the other sides of the right triangle. Substitute 20 for h, 15 for a, and rearrange:

$$h^2 - a^2 = b^2 \qquad 20^2 - 15^2 = b^2 \qquad b = \sqrt{20^2 - 15^2}$$

Now use your calculator to solve for b:

	SAN Notation			RPN Notation	
	Keystroke(s)	Display		Keystroke(s)	Display
	20	20.		20	20.
	x^2	400.		g x^2	400.0000
	−	400.		15	15.
	15	15.		g x^2	225.0000
	x^2	225.		−	175.0000
	=	175.		f $\sqrt{x}$	13.2288
	$\sqrt{x}$	13.228 757			

Side b is 13.2288 feet long.

PROBLEM 1-5 One side of a right triangle is 20 cm long and the other side is 7 cm long. Use your calculator to find the length of the hypotenuse.

Solution: Use the Pythagorean Theorem (see Problem 1-4):

$$h^2 = a^2 + b^2 \qquad h^2 = 20^2 + 7^2 \qquad h = \sqrt{20^2 + 7^2}$$

Now use your calculator to solve for h:

	SAN Notation			RPN Notation	
	Keystroke(s)	Display		Keystroke(s)	Display
	20	20.		20	20.
	x^2	400.		g x^2	400.0000
	+	400.		7	7.
	7	7.		g x^2	49.0000
	x^2	49.		+	449.0000
	=	449.		f $\sqrt{x}$	21.1896
	$\sqrt{x}$	21.189 62			

The hypotenuse is 21.1896 cm long.

PROBLEM 1-6 Simple interest is given by the formula Interest = Principal × Rate × Time. Use your calculator to find the amount of interest that you would pay if you borrow $2000 for three months at a rate of 12% annually.

Solution: First, since interest is expressed in terms of a year, you must convert 3 months to $\frac{1}{4}$ or 0.25 year. Then calculate 2000(0.25)(0.12):

SAN Notation		RPN Notation	
Keystroke(s)	Display	Keystroke(s)	Display
2000	2000.	2000	2000.
×	2000.	ENTER	2000.0000
.25	0.25	.25	0.25
×	500.	×	500.0000
.12	0.12	.12	0.12
=	60.	×	60.0000

You would pay $60.00 in interest on the loan.

PROBLEM 1-7 Use your calculator to solve the equation $3/500 = 2.63/x$ for x.

Solution:

$$\frac{3}{500} = \frac{2.63}{x} \qquad 3x = 2.63(500) \qquad x = \frac{2.63(500)}{3}$$

SAN Notation		RPN Notation	
Keystroke(s)	Display	Keystroke(s)	Display
2.63	2.63	2.63	2.63
×	2.63	ENTER	2.6300
500	500.	500	500.
÷	1315.	×	1315.0000
3	3.	3	3.
=	438.333 33	÷	438.3333

The value of x is 438.3333.

PROBLEM 1-8 One of the angles of a triangle is 38.24° and another angle is 13.92°. Use your calculator to find the third angle.

Solution: Recall that the sum of the angles of a triangle is 180°. Let x be the measure of the third angle; then $x + 38.24° + 13.92° = 180°$, or $x = (180 - 38.24 - 13.92)°$.

SAN Notation		RPN Notation	
Keystroke(s)	Display	Keystroke(s)	Display
180	180.	180	180.
−	180.	ENTER	180.0000
38.24	38.24	38.24	38.24
−	141.76	−	141.7600
13.92	13.92	13.92	13.92
=	127.84	−	127.8400

The third angle measures 127.84°.

PROBLEM 1-9 Use your calculator to show that a triangle with sides 6, 8, and 10 inches long is a right triangle.

Solution: The Pythagorean Theorem states that if the sides a, b, and h, of a triangle satisfy the condition $h^2 = a^2 + b^2$, then the triangle is a right triangle. Is it true that $10^2 = 6^2 + 8^2$?

SAN Notation		RPN Notation	
Keystroke(s)	Display	Keystroke(s)	Display
6	6.	6	6.
x^2	36.	g x^2	36.0000
+	36.	8	8.
8	8.	g x^2	64.0000
x^2	64.	+	100.0000
=	100.		

Since $10^2 = 100$, the equation is satisfied and the triangle is a right triangle.

PROBLEM 1-10 A building casts a shadow 300 feet long and, at the same time, a flagpole 20 feet high casts a shadow 7.34 feet long. Find the height h of the building.

Solution: It can be shown by means of similar triangles that the ratio of the height of the building to the length of its shadow equals the ratio of the height of the flagpole to the length of its shadow:

$$\frac{h}{300} = \frac{20}{7.34}$$

$$h = \frac{300(20)}{7.34}$$

SAN Notation		RPN Notation	
Keystroke(s)	Display	Keystroke(s)	Display
300	300.	300	300.
×	300.	ENTER	300.0000
20	20.	20	20.
÷	6000.	×	6000.0000
7.34	7.34	7.34	7.34
=	817.438 69	÷	817.4387

The building is 817.4387 feet high.

PROBLEM 1-11 Use your calculator to evaluate $\dfrac{(3.2)^3 + 4 - \frac{32}{35}}{2 + \frac{17}{37}}$.

Solution:

SAN Notation		RPN Notation	
Keystroke(s)	Display	Keystroke(s)	Display
17	17.	3.2	3.2
÷	17.	ENTER	3.2000
37	37.	3	3.
+	0.459 459 5	f y^x	32.7680

SAN Notation		RPN Notation	
Keystroke(s)	Display	Keystroke(s)	Display
2	2.	4	4.
=	2.459 459 5	+	36.7680
M+	2.459 459 5	32	32.
CLEAR	0.	ENTER	32.0000
32	32.	35	35.
÷	32.	÷	0.9143
35	35.	−	35.8537
=	0.914 285 7	2	2.
+/−	−0.914 285 7	ENTER	2.0000
+	−0.914 285 7	17	17.
4	4.	ENTER	17.0000
+	3.085 714 3	37	37.
3.2	3.2	÷	0.4595
x^y	3.2	+	2.4595
3	3.	÷	14.5779
=	35.853 714		
÷	35.853 714		
MR	2.459 459 5		
=	14.577 884		

Note: With SAN, it's easier to evaluate the denominator first and store in memory; RPN allows you to work most problems exactly as written.

Supplementary Exercises

PROBLEM 1-12 Use your calculator to evaluate $(42.6^{1/5} \times 3)/(4^3 \times 7.391)$.

PROBLEM 1-13 Use your calculator to evaluate $(4.21^2 - 3.27)^{0.7}/(\frac{3}{7.7})^2$.

PROBLEM 1-14 Use your calculator to evaluate $(3^2 + 4^3 + 5^4 + 6^2)^{1/3}$.

PROBLEM 1-15 A right triangle has a hypotenuse 2.20 inches long and a side 0.984 inches long. Find the length of the other side.

PROBLEM 1-16 A triangle has one side of 10.32 cm, another of 7 cm, and a third of 14.35 cm; is it a right triangle?

PROBLEM 1-17 Simple interest is given by the formula Interest = Principle × Rate × Time. If you borrow $30 000 for 3 years at 12% annual interest, find the amount of interest that you would pay.

PROBLEM 1-18 Use your calculator to solve the equation $3^7/5.796^4 = 2.63/x^2$ for x.

PROBLEM 1-19   The sum of the angles of a quadrilateral (closed, four-sided figure) is 360°. Three of the angles are 38.2°, 113.2°, and 87.9°; find the fourth angle.

PROBLEM 1-20   A right triangle has sides 0.73 and 1.14 cm; find the length of the hypotenuse.

PROBLEM 1-21   A building 783 m high casts a shadow 93.2 m long; at the same time, a flagpole casts a shadow 7.68 m long. Find the height of the flagpole.

PROBLEM 1-22   Use your calculator to evaluate

$$\left[(3.2)^{1/3} + \frac{(5.17)(3.69)}{2.71} \right]^2 - 6.83 + \frac{2.17}{3.9^5}.$$

Answers to Supplementary Exercises

1-12:   0.013 431 4

1-13:   42.729 979

1-14:   9.020 529 3

1-15:   1.967 674 8

1-16:   No

1-17: $10 800.00

1-18:   1.165

1-19:   120.7°

1-20:   1.353 698 6 cm

1-21:   64.521 888 m

1-22:   65.647 093 4

2 DISTANCE AND CIRCLES

THIS CHAPTER IS ABOUT

☑ **The Cartesian Coordinate System**
☑ **The Distance Formula**
☑ **Circles Centered at the Origin**

2-1. The Cartesian Coordinate System

In trigonometry, you'll frequently find it necessary to describe the position of a point in a plane. You'll use the **Cartesian** or **rectangular** coordinate system to do this.

A. Coordinates

On a piece of paper, select any point and call it the **origin** (O). Draw a vertical line, the y **axis**, and a horizontal line, the x **axis**, through the origin (see Figure 2-1). Assign real number values to points on the x axis, starting with

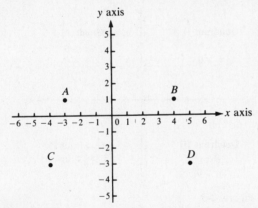

Figure 2-1

zero at the origin. Assign larger positive values as you move to the right of the origin and larger negative values as you move to the left of the origin. On the y axis, assign positive values above the origin and negative values below.

Let x be the location of a point with respect to the x axis and y the location with respect to the y axis. You can now describe a point in the plane with a pair of numbers, the **coordinates** of the point (x, y). Using this system, for any pair of numbers, (x, y), you can locate the corresponding point in the plane, and for any point in the plane, you can determine its coordinates.

You may refer to a point by giving its coordinates, for example, $(1, 3)$. You may also refer to a point by using a capital letter followed by the co-ordinates, for example, $A(1, 3)$, or by writing only the capital letter, A. All of these notations are used in this outline.

EXAMPLE 2-1: Find the coordinates of points A, B, C, and D of Figure 2-1.

Solution: The coordinates are $A(-3, 1)$, $B(4, 1)$, $C(-4, -3)$, and $D(5, -3)$. Note that you always list the x coordinate first.

EXAMPLE 2-2: Sketch the points with coordinates $(7, 3)$, $(0, 4)$, $(-3, -1)$, $(-2, 7)$, $(5, 0)$, and $(-2, -2)$ on a Cartesian coordinate plane.

Solution: The points are shown in Figure 2-2.

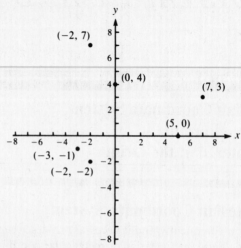

Figure 2-2

B. Quadrants

The coordinate axes divide the Cartesian plane into four quarters, or **quadrants**. The signs for points in each quadrant are shown in Figure 2-3.

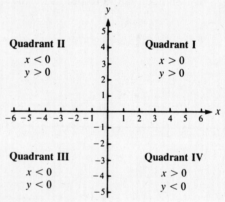

Figure 2-3
The Cartesian plane.

EXAMPLE 2-3: Find the quadrant in which each of the following points lies: $(7, -2)$, $(-8, -3)$, $(-3, 5)$, $(7, 4)$, and $(0, -7)$.

Solution: By comparing the signs of the coordinates with Figure 2-3, you can see that the first four points lie in the fourth, third, second, and first quadrants, respectively. Since the last point has a zero x coordinate, it lies on the negative y axis. The x and y axes form the boundaries of the four quadrants, but are not included in the quadrants.

EXAMPLE 2-4: Given $P(x, y)$, find the quadrants in which P may lie if: **(a)** $x < 0$; **(b)** $x > 0$ and $y > 0$; **(c)** $xy > 0$.

Solution:

(a) For $x < 0$, $P(x, y)$ may lie in either quadrant II or III, since they both have negative x coordinates.
(b) For $x > 0$ and $y > 0$, $P(x, y)$ is in quadrant I, since it has positive x and y coordinates.

(c) For $xy > 0$, $P(x, y)$ may lie in either quadrant I or III, since $xy > 0$ means $x > 0$ and $y > 0$, or $x < 0$ and $y < 0$.

2-2. The Distance Formula

If $A(x_1, y_1)$ and $B(x_2, y_2)$ are two points in a Cartesian plane, the **distance** between them, d_{AB}, is the length of the straight line segment joining the two points.

For points A and B in Figure 2-4, you can construct a right triangle with the hypotenuse joining points A and B. One leg is parallel to the x axis and passes through point A; the other leg is parallel to the y axis and passes through point B. The length of the horizontal leg is $x_2 - x_1$ and the length of the vertical leg is $y_2 - y_1$. From the Pythagorean Theorem,

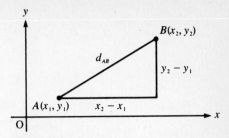

Figure 2-4
The distance formula.

DISTANCE FORMULA
$$d_{AB} = \sqrt{(x_2 - x_1)^2 + (y_2 - y_1)^2} \qquad (2\text{-}1)$$

Both points in Figure 2-4 lie in the first quadrant but the distance formula remains the same no matter where the two points lie. (You should verify this statement.)

EXAMPLE 2-5: Find the distance between (**a**) $(2, 3)$ and $(-2, 3)$; (**b**) $(-2, 3)$ and $(2, -3)$; (**c**) $(4, 3)$ and $(-4, -3)$.

Solution: Using Equation 2-1,

(a) $d = \sqrt{(2 - (-2))^2 + (3 - 3)^2} = \sqrt{4^2} = 4$

(b) $d = \sqrt{(-2 - 2)^2 + (3 - (-3))^2} = \sqrt{(-4)^2 + 6^2} = \sqrt{52} = 7.21$

(c) $d = \sqrt{(4 - (-4))^2 + (3 - (-3))^2} = \sqrt{8^2 + 6^2} = \sqrt{100} = 10$

EXAMPLE 2-6: Find the lengths of the sides of the triangle whose vertices are at $A(9, 4)$, $B(4, 8)$, and $C(0, 1)$.

Solution: Using Equation 2-1 for each side of the triangle,

$$d_{AB} = \sqrt{(9 - 4)^2 + (4 - 8)^2} = \sqrt{5^2 + (-4)^2} = \sqrt{41} = 6.40$$

$$d_{AC} = \sqrt{(9 - 0)^2 + (4 - 1)^2} = \sqrt{9^2 + 3^2} = \sqrt{90} = 9.49$$

$$d_{BC} = \sqrt{(4 - 0)^2 + (8 - 1)^2} = \sqrt{4^2 + 7^2} = \sqrt{65} = 8.06$$

EXAMPLE 2-7: Do the points $A(1, 4)$, $B(3, 8)$, and $C(4, 10)$ lie on a straight line?

Solution: From geometry, you recall that if the points lie on a straight line, then the distance from A to B plus the distance from B to C must equal the distance from A to C. Using Equation 2-1,

$$d_{AB} = \sqrt{(1 - 3)^2 + (4 - 8)^2} = \sqrt{(-2)^2 + (-4)^2} = \sqrt{20} = 4.47$$

$$d_{BC} = \sqrt{(3 - 4)^2 + (8 - 10)^2} = \sqrt{(-1)^2 + (-2)^2} = \sqrt{5} = 2.24$$

$$d_{AC} = \sqrt{(1 - 4)^2 + (4 - 10)^2} = \sqrt{(-3)^2 + (-6)^2} = \sqrt{45} = 6.71$$

Since $d_{AB} + d_{BC} = 6.71 = d_{AC}$, the points lie on a straight line.

EXAMPLE 2-8: Is the triangle with vertices $A(-1, 2)$, $B(2, -1)$, and $C(4, 3)$ an *isosceles triangle* (a triangle with two sides of equal length)?

Solution: From Equation 2-1,

$$d_{AB} = \sqrt{(-1 - 2)^2 + (2 - (-1))^2} = \sqrt{(-3)^2 + 3^2} = \sqrt{18}$$

$$d_{BC} = \sqrt{(2 - 4)^2 + (-1 - 3)^2} = \sqrt{(-2)^2 + (-4)^2} = \sqrt{20}$$

$$d_{AC} = \sqrt{(-1 - 4)^2 + (2 - 3)^2} = \sqrt{(-5)^2 + (-1)^2} = \sqrt{26}$$

Since no two sides have the same length, the triangle is not isosceles.

EXAMPLE 2-9: Find the coordinates of the midpoint of the line segment joining $A(a, 0)$ and $B(b, 0)$ where $b > a$.

Solution: The *midpoint* is the point P for which $d_{AP} = d_{PB}$. Choose coordinates $(x, 0)$ for P, since the y coordinate of both A and B is 0. Then, $d_{AP} = x - a$ and $d_{PB} = b - x$. Since $d_{AP} = d_{PB}$, $x - a = b - x$. Solving for x, you get $x = (a + b)/2$.

EXAMPLE 2-10: Find the point on the y axis that is equidistant from $A(9, 5)$ and $B(-4, -7)$.

Solution: Let the desired point be $P(0, y)$. Then,

$$d_{AP} = \sqrt{(9 - 0)^2 + (5 - y)^2} = \sqrt{9^2 + (5 - y)^2}$$
$$d_{PB} = \sqrt{(0 - (-4))^2 + (y - (-7))^2} = \sqrt{4^2 + (y + 7)^2}$$

Since $d_{AP} = d_{PB}$,

$$81 + (5 - y)^2 = 16 + (y + 7)^2$$
$$81 + 25 - 10y + y^2 = 16 + y^2 + 14y + 49$$

Simplifying,

$$106 - 10y = 65 + 14y$$
$$24y = 41$$
$$y = 1.71$$

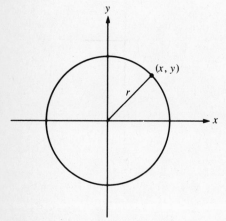

Figure 2-5
Circle centered at (0, 0).

2-3. Circles Centered at the Origin

A **circle** is the set of all points equidistant from a fixed point, the center of the circle. If (x, y) is a point on a circle with center at the origin and radius r, the distance between (x, y) and $(0, 0)$ is r (see Figure 2-5).

A. General equation

Applying the distance formula to (x, y) and $(0, 0)$ you get

$$\sqrt{(x - 0)^2 + (y - 0)^2} = r$$

Simplifying and squaring both sides of this equation leads to the equation of a circle with radius r centered at the origin:

CIRCLE CENTERED AT ORIGIN
$$x^2 + y^2 = r^2 \qquad (2-2)$$

EXAMPLE 2-11: Find the equation for a circle with center at the origin if:
(a) $r = 2$; (b) $r = 4$.

Solution: Substituting into Equation 2-2,

(a) $x^2 + y^2 = 4$ and   (b) $x^2 + y^2 = 16$

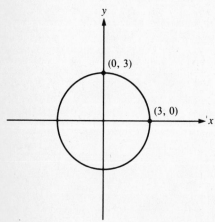

Figure 2-6

EXAMPLE 2-12: Find the equation of the circle shown in Figure 2-6.

Solution: Since the circle is centered at the origin, the radius is the distance between the origin and the point of intersection of the circle with the positive x axis (that is, $x = 3$). So the radius is 3 and the equation is $x^2 + y^2 = 9$.

EXAMPLE 2-13: The arch of a railroad tunnel forms a semicircle (see Figure 2-7). The height of the tunnel, from the midpoint of the floor to the top, is 15 feet; find the equation of the circle that defines the arch.

Solution: Consider the midpoint of the tunnel as the center of a circle and assign it coordinates $(0, 0)$. Then the point with coordinates $(0, 15)$ lies on the circle. This means that the radius is 15 and the equation is $x^2 + y^2 = 15^2$.

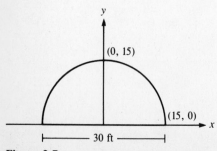

Figure 2-7

EXAMPLE 2-14: Write the equation of a circle with center at the origin and radius: (a) 1; (b) 2.5; (c) $2\sqrt{2}$.

Solution: Simply substitute the radius value into Equation 2-2:

(a) $x^2 + y^2 = 1^2$ or $x^2 + y^2 = 1$

(b) $x^2 + y^2 = 2.5^2$ or $x^2 + y^2 = 6.25$

(c) $x^2 + y^2 = (2\sqrt{2})^2$ or $x^2 + y^2 = 8$

EXAMPLE 2-15: Determine the radius of each of the following circles: (a) $x^2 + y^2 = 10$; (b) $x^2 + y^2 = 34$; (c) $x^2 + y^2 = 2.5$.

Solution: Comparing each equation with Equation 2-2,

(a) $r^2 = 10$ so $r = \sqrt{10} = 3.162$

(b) $r^2 = 34$ so $r = \sqrt{34} = 5.831$

(c) $r^2 = 2.5$ so $r = \sqrt{2.5} = 1.581$

B. Graphing

To graph circles centered at the origin, you must determine the radius. For example, suppose that you are asked to graph the circle whose equation is

$$x^2 + y^2 = 25$$

Since $r^2 = 25$, the radius of this circle is 5. Locate points $(5, 0)$, $(0, 5)$, $(-5, 0)$, and $(0, -5)$ on graph paper and sketch the circle passing through these points (see Figure 2-8).

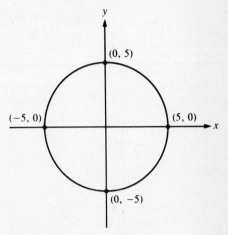

Figure 2-8
Graph of circle with $r = 5$.

SUMMARY

1. For a point in the Cartesian plane, the first, or x, coordinate determines the location of that point with respect to the horizontal, or x, axis, and the second, or y, coordinate determines the location of that point with respect to the vertical, or y, axis.
2. The origin of the Cartesian coordinate system is the point $(0, 0)$.
3. Points in quadrant I have positive x and y coordinates; points in quadrant II have negative x and positive y coordinates; points in quadrant III have negative x and y coordinates; and points in quadrant IV have positive x and negative y coordinates.
4. If (x_1, y_1) and (x_2, y_2) are the coordinates of two points, the distance between them is given by $\sqrt{(x_2 - x_1)^2 + (y_2 - y_1)^2}$.
5. A circle with radius r and center at the origin has the equation $x^2 + y^2 = r^2$.

RAISE YOUR GRADES

Can you...?

☑ place a point with given coordinates in the Cartesian plane
☑ determine the coordinates of a point from a sketch of its location in the Cartesian plane
☑ determine in which quadrant a point will lie, given the signs of the coordinates of the point
☑ find the distance between any two points
☑ determine the coordinate of a point, given the distance between the point and another point
☑ determine the equation of a circle centered at the origin, given a graph of the circle
☑ sketch the graph of a circle centered at the origin, given the equation of the circle

SOLVED PROBLEMS

PROBLEM 2-1 Sketch the following points on a Cartesian coordinate system: **(a)** $(2, 5)$; **(b)** $(-3, -3)$; **(c)** $(2, y)$, where y is any positive number; **(d)** $(x, -1)$, where x is any negative number.

Solution: See Figure 2-9.

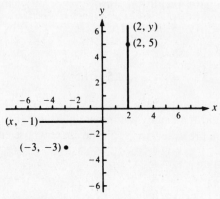

Figure 2-9 [See Section 2-1.]

PROBLEM 2-2 Name the quadrant(s) in which the following points can lie **(a)** $(t, 4t)$ for t positive; **(b)** $(t, -4t)$ for t positive; **(c)** (t, v) for t negative and v positive; **(d)** $(t, t - 2)$ for t negative.

Solution:

(a) Since t is positive, $4t$ is positive. The point lies in quadrant I.
(b) Since t is positive, $-4t$ is negative. The point lies in quadrant IV.
(c) Since the first coordinate is negative and the second coordinate is positive, the point lies in quadrant II.
(d) Since t is negative, $t - 2$ is negative. The point lies in quadrant III. [See Section 2-1.]

PROBLEM 2-3 Find the coordinates of the midpoint of the line segment joining $A(0, a)$ and $B(0, b)$, where $b > a$.

Solution: Let P be the midpoint with coordinates (x, y). Since A and B both have x coordinates equal to zero, $x = 0$. Then, $d_{AP} = y - a$ and $d_{PB} = b - y$. Since $d_{AP} = d_{PB}$, $y - a = b - y$. Solving for y, $y = (a + b)/2$. [See Section 2-2.]

PROBLEM 2-4 The midpoint of the line joining $(a, 0)$ and $(3, 0)$ is $(0, 0)$. Find a.

Solution: From Example 2-9, the x coordinate of the midpoint must be $(a + 3)/2$. The given x coordinate of the midpoint is 0. Thus $(a + 3)/2 = 0$ and $a = -3$. [See Section 2-2.]

PROBLEM 2-5 Find the distance between $(1, 3)$ and the midpoint of the segment joining $(0, 2)$ and $(0, 8)$.

Solution: The midpoint of the segment joining $(0, 2)$ and $(0, 8)$ has coordinates $(0, (2 + 8)/2) = (0, 5)$. The distance between $(1, 3)$ and $(0, 5)$ is $\sqrt{(1 - 0)^2 + (3 - 5)^2} = \sqrt{1^2 + (-2)^2} = \sqrt{5}$. [See Section 2-2.]

PROBLEM 2-6 Find the equation of the circle with center at (h, k) and radius r.

Solution: Let (x, y) be an arbitrary point on the circumference of the circle. The distance from (x, y) to (h, k) must be r, so $\sqrt{(x - h)^2 + (y - k)^2} = r$. Squaring both sides of this expression, you get $(x - h)^2 + (y - k)^2 = r^2$. This is the standard form of the equation of a circle with center at (h, k) and radius r. [See Section 2-3.]

PROBLEM 2-7 Find the equation of the circle that has as diameter the line segment joining $P(-4,0)$ and $Q(6,0)$.

Solution: The center of the circle is the midpoint of the line segment joining P and Q, and the radius is one-half the length of this line segment. The midpoint has coordinates $((-4+6)/2, 0) = (1, 0)$ and the length of the segment is $6 - (-4) = 10$. The radius is $\frac{10}{2} = 5$ and the equation is $(x - 1)^2 + (y - 0)^2 = 25$.   [See Section 2-3.]

PROBLEM 2-8 For each of the following, find the equation of the circle with center at $(0,0)$ and a radius of: **(a)** 7; **(b)** 5; **(c)** 6.4.

Solution: In each case the equation will have the form $x^2 + y^2 = r^2$; simply insert the correct value of r:

    (a) $x^2 + y^2 = 7^2 = 49$

    (b) $x^2 + y^2 = 5^2 = 25$

    (c) $x^2 + y^2 = 6.4^2 = 40.96$ [See Section 2-3.]

PROBLEM 2-9 Give the equation of the circle with center at the origin that crosses **(a)** the x axis at $(4, 0)$; **(b)** the y axis at $(0, -8)$; **(c)** the x axis at $(-2, 0)$; **(d)** the y axis at $(0, 8)$.

Solution: From the given condition, determine the radius of the circle and substitute this into the equation of a circle with center at the origin:

(a) The distance between $(0, 0)$ and $(4, 0)$ is 4; since $r = 4$, the equation is $x^2 + y^2 = 16$.
(b) The distance between $(0, 0)$ and $(0 - 8)$ is 8; since $r = 8$, the equation is $x^2 + y^2 = 64$.
(c) The distance between $(0, 0)$ and $(-2, 0)$ is 2; since $r = 2$, the equation is $x^2 + y^2 = 4$.
(d) The distance between $(0, 0)$ and $(0, 8)$ is 8; since $r = 8$, the equation is $x^2 + y^2 = 64$.

                [See Section 2-3.]

PROBLEM 2-10 In which quadrant(s) can point $P(x, y)$ lie if: **(a)** $xy < 0$; **(b)** $x < 0$ and $y > 0$; **(c)** $x/y > 0$; **(d)** $x > 0$ or $y > 0$?

Solution:

(a) If $xy < 0$, either $x < 0$ or $y < 0$ but not both; thus P is in quadrant II, where $x < 0$, or quadrant IV, where $y < 0$.
(b) If $x < 0$ and $y > 0$, P lies in quadrant II.
(c) If $x/y > 0$, either $x < 0$ and $y < 0$ or $x > 0$ and $y > 0$; thus P is in quadrant III or quadrant I.
(d) If $x > 0$ or $y > 0$, P lies in quadrant I, II, or IV.   [See Section 2-1.]

PROBLEM 2-11 The distance between $(x, 3)$ and $(5, 5)$ is 5. Find x.

Solution: From the distance formula, $5 = \sqrt{(x - 5)^2 + (3 - 5)^2}$. If you square both sides of this equation, $25 = (x - 5)^2 + (-2)^2$. Expanding the right side, $25 = x^2 - 10x + 25 + 4$. Then, $0 = x^2 - 10x + 4$. Applying the quadratic formula,

$$x = \frac{-b \pm \sqrt{b^2 - 4ac}}{2a}$$

where $ax^2 + bx + c = 0$,

$$x = \frac{10 \pm \sqrt{(-10)^2 - 4 \times 1 \times 4}}{2}$$

$$= \frac{10 \pm 2\sqrt{21}}{2} = 9.583 \quad \text{or} \quad 0.417$$ [See Section 2-2.]

PROBLEM 2-12 Find y, given that $(2, y)$ is equidistant from $(-1, 3)$ and $(4, 5)$.

Solution: The distance from $(2, y)$ to $(-1, 3)$ is $\sqrt{(2 - (-1))^2 + (y - 3)^2}$, and the distance from $(2, y)$ to $(4, 5)$ is $\sqrt{(2 - 4)^2 + (y - 5)^2}$. When you equate these two expressions and square both sides you get

$$(2+1)^2 + (y-3)^2 = (2-4)^2 + (y-5)^2$$

Computing the squares,

$$9 + y^2 - 6y + 9 = 4 + y^2 - 10y + 25$$

By combining like terms and subtracting y^2 from both sides of the equation, you can write $4y = 11$. Solving, you find $y = \frac{11}{4} = 2.75$.   [See Section 2-2.]

PROBLEM 2-13   Show that the triangle with vertices $A(0, -4)$, $B(0, 16)$, and $C(10\sqrt{3}, 6)$ is equilateral.

Solution: You must show that the lengths of the sides are equal. Using Equation 2-1,

$$d_{AB} = \sqrt{(0-0)^2 + (16-(-4))^2}$$
$$= \sqrt{20^2} = 20$$
$$d_{AC} = \sqrt{(10\sqrt{3}-0)^2 + (6-(-4))^2} = \sqrt{(10\sqrt{3})^2 + 10^2}$$
$$= \sqrt{400} = 20$$
$$d_{BC} = \sqrt{(10\sqrt{3}-0)^2 + (6-16)^2} = \sqrt{(10\sqrt{3})^2 + (-10)^2}$$
$$= \sqrt{400} = 20$$ [See Section 2-2.]

PROBLEM 2-14   Find the value of x such that $(x, 5)$ is 5 units to the right of $(-5, 5)$.

Solution: Since the y coordinates of the two points are equal, in order to be 5 units to the right of $(-5, 5)$, x must be 5 units greater than -5. This means that $x = -5 + 5 = 0$.   [See Section 2-2.]

PROBLEM 2-15   Find the length of the chord connecting the following pairs of points on the circle $x^2 + y^2 = 1$: **(a)** $(1, 0)$ and $(0, 1)$; **(b)** $(-\frac{1}{2}, \frac{\sqrt{3}}{2})$ and $(\frac{\sqrt{3}}{2}, \frac{1}{2})$.

Solution:

(a)   The length of the chord is the distance between $(1, 0)$ and $(0, 1)$:

$$d = \sqrt{(0-1)^2 + (1-0)^2} = \sqrt{(-1)^2 + 1^2} = \sqrt{2}$$

(b)   The length of the chord is the distance between $(-\frac{1}{2}, \frac{\sqrt{3}}{2})$ and $(\frac{\sqrt{3}}{2}, \frac{1}{2})$:

$$d = \sqrt{\left(\frac{\sqrt{3}}{2} - \left(-\frac{1}{2}\right)\right)^2 + \left(\frac{1}{2} - \frac{\sqrt{3}}{2}\right)^2}$$
$$= \sqrt{\left(\frac{\sqrt{3}+1}{2}\right)^2 + \left(\frac{1-\sqrt{3}}{2}\right)^2} = \sqrt{\frac{8}{4}} = \sqrt{2}$$ [See Section 2-2.]

PROBLEM 2-16   Find the y coordinates of the point(s) on the circle $x^2 + y^2 = 1$ for which $x = \frac{1}{2}$.

Solution: If you substitute $x = \frac{1}{2}$ into the equation for the circle, the equation becomes $(\frac{1}{2})^2 + y^2 = 1$, or $y^2 = \frac{3}{4}$. Then $y = \pm\frac{\sqrt{3}}{2}$.   [See Section 2-3.]

PROBLEM 2-17   Find the distance between the two points determined by the condition in Problem 2-16.

Solution: From the solution to Problem 2-16, the points are $(\frac{1}{2}, +\frac{\sqrt{3}}{2})$ and $(\frac{1}{2}, -\frac{\sqrt{3}}{2})$. The distance between these two points is

$$\sqrt{\left(\frac{1}{2} - \frac{1}{2}\right)^2 + \left(\frac{\sqrt{3}}{2} + \frac{\sqrt{3}}{2}\right)^2} = \sqrt{(\sqrt{3})^2} = \sqrt{3}$$ [See Section 2-2.]

PROBLEM 2-18   Find the x coordinates of the point(s) on the circle $x^2 + y^2 = 1$ for which $y = \frac{1}{4}$.

Solution: Substitute $y = \frac{1}{4}$ into the equation for the circle. That equation becomes $x^2 + (\frac{1}{4})^2 = 1$, or $x^2 = \frac{15}{16}$. Then $x = \pm\frac{\sqrt{15}}{4}$ [See Section 2-3.]

PROBLEM 2-19 Find the distance between the two points determined by the condition in Problem 2-18.

Solution: From the solution to Problem 2-18, the points are $(+\frac{\sqrt{15}}{4}, \frac{1}{4})$ and $(-\frac{\sqrt{15}}{4}, \frac{1}{4})$. The distance between these two points is

$$\sqrt{\left(\frac{\sqrt{15}}{4} + \frac{\sqrt{15}}{4}\right)^2 + \left(\frac{1}{4} - \frac{1}{4}\right)^2} = \frac{\sqrt{15}}{2} \qquad \text{[See Section 2-2.]}$$

PROBLEM 2-20 Find the point on the x axis that is equidistant from $(3, 4)$ and $(-2, -3)$.

Solution: Since the point lies on the $\dot{x}$ axis, its coordinates are $(x, 0)$. The distances from the two points are $\sqrt{(x-3)^2 + (0-4)^2}$ and $\sqrt{(x+2)^2 + (0+3)^2}$. Equating these two expressions and squaring, you find

$$(x-3)^2 + (0-4)^2 = (x+2)^2 + (0+3)^2$$

or

$$x^2 - 6x + 9 + 16 = x^2 + 4x + 4 + 9$$

Subtracting x^2 from both sides and combining terms, you get

$$-6x + 25 = 4x + 13$$
$$12 = 10x$$

So $x = \frac{6}{5} = 1.2$.   [See Section 2-2.]

Supplementary Exercises

PROBLEM 2-21 Using your calculator, find the radius of each of the following circles: (a) $x^2 + y^2 = 21\,815$; (b) $x^2 + y^2 = 432\,906$; (c) $x^2 + y^2 = 0.051\,98$.

PROBLEM 2-22 Tell whether the triangle with vertices $(1, 1)$, $(2, 2)$, and $(0, 2)$ is an equilateral triangle.

PROBLEM 2-23 Find the coordinates of the midpoint of the line segment joining $A(a, b)$ and $B(c, d)$. *Hint:* See Example 2-9 and Problem 2-3.

PROBLEM 2-24 Find the midpoint of the line segment joining (a) $(1, 0)$ and $(7, 0)$; (b) $(-2, 0)$ and $(-10, 0)$; (c) $(1, 1)$ and $(2, 2)$.

PROBLEM 2-25 The distance between $(1, 3)$ and the midpoint of the segment joining $(0, a)$ and $(0, 8)$ is 3. Find a.

PROBLEM 2-26 The midpoint of $(0, -4)$ and $(0, b)$ is $(0, 7)$. Find b.

PROBLEM 2-27 Write the equation of a circle with center at the origin and radius: (a) 13 feet; (b) 1.5 meters; (c) 0 inches; (d) -3 centimeters.

PROBLEM 2-28 Find the radius of each of the following circles: (a) $x^2 + y^2 = 10$; (b) $x^2 + y^2 = 0.25$; (c) $x^2 + y^2 = 500$.

PROBLEM 2-29 The line segment joining $P(0, 7)$ and $Q(0, -7)$ is a diameter of a circle; find the equation of the circle.

Answers to Supplementary Exercises

2-21: (a) 147.70; (b) 657.96; (c) 0.23

2-22: No; one side has length 2 and the other two sides have length $\sqrt{2}$.

2-23: $\big((a + c)/2, (b + d/2)\big)$

2-24: (a) (4, 0); (b) (−6, 0); (c) $(\frac{3}{2}, \frac{3}{2})$

2-25: $a = \pm 4\sqrt{2} - 2$

2-26: $b = 18$

2-27: (a) $x^2 + y^2 = 169$; (b) $x^2 + y^2 = 2.25$; (c) $x^2 + y^2 = 0$ (the only point on this degenerate circle is (0, 0)); (d) No equation

2-28: (a) $\sqrt{10}$; (b) 0.5; (c) $\sqrt{500}$, or $10\sqrt{5}$

2-29: $x^2 + y^2 = 49.$

3 ANGLES

THIS CHAPTER IS ABOUT

- ☑ **The Geometry of Angles**
- ☑ **Measuring Angles**
- ☑ **Arc Length**

3-1. The Geometry of Angles

Angles and triangles are central to the study of trigonometry. Before you can explore the trigonometric relationships, you'll need to review the geometry and measurement of angles.

A. Rotating a ray generates an angle.

A **ray** is a half-line with one fixed endpoint P. This point is the **vertex**, labeled P in Figure 3-1.

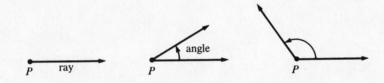

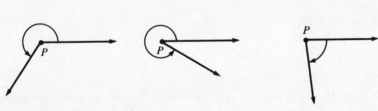

Figure 3-1
Rays form angles.

When you rotate the ray about P, you generate angles. The starting position of the ray is the **initial side** of the angle; the final position of the ray is the **terminal side** of the angle.

B. Intersecting lines form angles.

If you think of the point of intersection of two lines as the vertex and the half-lines as rays, any two intersecting lines form four angles (see Figure 3-2).

You designate one of the intersecting half-lines (rays) as the *initial side*; the other half-line is the *terminal side* (see Figure 3-3). Convenience will guide your choice in designating initial and terminal sides.

C. Two points on a circle define an angle.

Any two points on a circle define an angle as follows: Draw a line segment to each point from the center of the circle. Choose one line segment as the initial

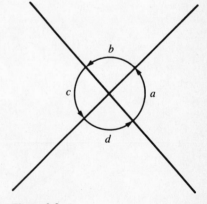

Figure 3-2
Intersecting lines form the four angles *a* through *d*.

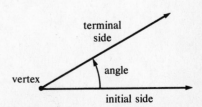

Figure 3-3
Initial and terminal sides of an angle.

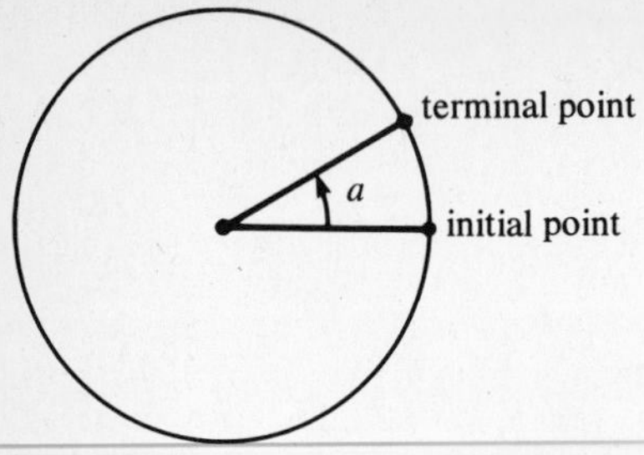

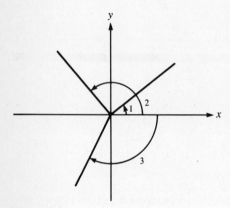

Figure 3-4
Initial and terminal points define an
angle.

side of the angle; the other line segment becomes the terminal side. The points
are the **initial point** and **terminal point**, respectively (see Figure 3-4).

D. Angles in rectangular coordinate systems

To draw an angle in standard position in a rectangular coordinate system,
place the vertex at the origin and the initial side along the positive *x* axis. Any
other ray with vertex at the origin can serve as the terminal side of the angle
(see Figure 3-5).

3-2. Measuring Angles

To measure an angle, you determine the rotation required to move a ray from the
initial side, or point, to the terminal side, or point. Consider a single ray attached
to a fixed vertex. Before you rotate the ray, the angle measure is zero. As you
rotate the ray in the counterclockwise direction, you generate angles with the
starting position of the ray as initial side and the current position of the ray as
terminal side (see Figure 3-6). The measure of the angle increases as you rotate the
ray. After rotating the ray through a complete circle, it again overlays the initial
side. The angular measure of one complete rotation is defined as 360 degrees.

Figure 3-5
Several angles in standard position.

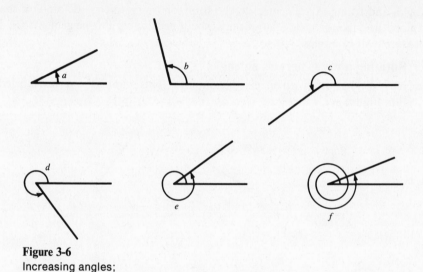

Figure 3-6
Increasing angles;
a < *b* < *c* < *d* < *e* < *f*.

A. Positive and negative angles

By definition, you generate **positive angles** through counterclockwise rotation of the ray; clockwise rotation generates **negative angles** (see Figure 3-7).

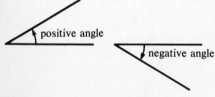

Figure 3-7
Positive and negative angles.

B. Degrees, minutes, and seconds

For angles in **standard position**, that is, vertex at the origin and initial side on
the positive *x* axis, a **degree** is defined as $\frac{1}{360}$ of a complete rotation. A
complete rotation of the terminal line contains 360 equal parts, or degrees (°).
If you imagine a point on the rotating ray tracing the circumference of a
circle, then that circle contains 360 degrees of arc. Each degree can be divided
into 60 **minutes** (′) and each minute can be divided into 60 **seconds** (″).

EXAMPLE 3-1: Express the following angles in degrees, minutes, and seconds:
(a) 52.4°; **(b)** 125.84°.

Solution:

(a) Since a degree contains 60′, 0.4° is 60(0.4) = 24′. Thus 52.4° is 52° 24′.

(b) 0.84° is 0.84(60) = 50.4′. However, a minute contains 60″, so 0.4′ is
60(0.4) = 24″. Thus 125.84° is 125° 50′ 24″.

EXAMPLE 3-2: Express the following angles in degrees: **(a)** $10'\,18''$; **(b)** $50°\,34'\,26''$.

Solution: $1° = 60'$ and $1' = 60''$, so $1° = 60' = (60 \times 60)'' = 3600''$. Dividing both sides of this expression by 3600, you determine that $1'' = \frac{1}{3600}°$.

(a) $\qquad 10'\,18'' = 10(60)'' + 18'' = 618'' = 618\left(\frac{1}{3600}\right)° \quad$ or $\quad 0.172°$.

(b) $\qquad 50°\,34'\,26'' = 50°[(34 \times 60)'' + 26''] \quad$ or $\quad 50°\,2066''$

$$2066'' = 2066\left(\frac{1}{3600}\right)° = 0.574°$$

$$50°\,34'\,26'' = 50.574°$$

C. Radians

Consider a point on a ray, with vertex at the origin, at distance r from the origin. As the ray is rotated the point traces a circle of radius r. Since the circumference of this circle is $2\pi r$, you can divide the circumference into 2π parts each of length r. The angle corresponding to each part is called a radian. A **radian** is defined as the angle subtended by an arc of a circle equal to the radius of that circle. The whole angle of a circle, that is, the angle that results when the ray rotates until the terminal side coincides with the initial side, is 2π radians. It is customary to express radian measures without the unit designation, that is, angle $\alpha = 2\pi$. (You will also encounter the abbreviation *rad* in some texts, including this one.) Figure 3-8 shows an angle of 1 radian for circles with various radii.

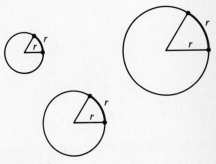

Figure 3-8
A one-radian angle for circles of various radii.

D. Converting between radians and degrees

You can relate radians and degrees by simple expressions. Since there are $360°$, or 2π radians, in a complete revolution of a ray,

$$360° = 2\pi \text{ radians} \tag{3-1}$$

Dividing both sides of Equation 3-1 by 360, you get

$$1 \text{ degree} = \frac{(2\pi)}{360} = \frac{\pi}{180} \text{ radians}$$

$$1 \text{ degree} = 0.0175 \text{ radians} \tag{3-2}$$

Dividing both sides of Equation 3-1 by 2π, you get

$$1 \text{ radian} = \frac{360}{2\pi} = \frac{180°}{\pi}$$

$$1 \text{ radian} = 57.30° \tag{3-3}$$

EXAMPLE 3-3: Convert to radian measure: **(a)** $60°$; **(b)** $135°$; **(c)** $330°$.

Solution: Multiplying by the number of radians in 1 degree, $\frac{1}{180}\pi°$, you get

$$\textbf{(a)} \quad 60° = 60\left(\frac{\pi}{180}\right) = \frac{\pi}{3}$$

$$\textbf{(b)} \quad 135° = 135\left(\frac{\pi}{180}\right) = \frac{3\pi}{4}$$

$$\textbf{(c)} \quad 330° = 330\left(\frac{\pi}{180}\right) = \frac{11\pi}{6}$$

EXAMPLE 3-4: Convert to degree measure: **(a)** $\frac{1}{4}\pi$; **(b)** $\frac{3}{2}\pi$; **(c)** $-\frac{1}{6}\pi$.

Solution: Multiplying by the number of degrees in 1 radian, $\frac{180}{\pi}$, you get

$$\textbf{(a)} \quad \frac{\pi}{4} = \frac{\pi}{4}\left(\frac{180}{\pi}\right) = 45°$$

$$\textbf{(b)} \quad \frac{3\pi}{2} = \frac{3\pi}{2}\left(\frac{180}{\pi}\right) = 270°$$

$$\textbf{(c)} \quad -\frac{\pi}{6} = -\frac{\pi}{6}\left(\frac{180}{\pi}\right) = -30°$$

EXAMPLE 3-5: Use your calculator to convert each of the degree-minute-second measures to radians, and radian measures to degree-minute-second measures: (a) $17°\,43'\,12''$; (b) $93°\,3'\,48''$; (c) $\frac{1}{2}$ rad; (d) 2.15 rad.

Solution: For parts (a) and (b), first convert minutes and seconds to fractions of a degree. Use the facts: $1' = \frac{1}{60}°$ and $1'' = \frac{1}{60}'$. Complete the conversion with $1° = \frac{1}{180}\pi$ radians.

(a)
$$12'' = \frac{12'}{60} = \frac{12°}{3600} = 0.0033°$$

$$43' = \frac{43°}{60} = 0.7167°$$

Thus,

$$17°\,43'\,12'' = 17° + 0.7167° + 0.0033° = 17.7200°$$

$$= 17.7200\left(\frac{\pi}{180}\right) \text{ rad} = 0.3093 \text{ rad}$$

(b)
$$48'' = \frac{48'}{60} = \frac{48°}{3600} = .0133°$$

$$3' = \frac{3°}{60} = .05°$$

Thus,

$$93°\,3'\,48'' = 93° + 0.05° + 0.0133° = 93.0633°$$

$$= 93.0633\left(\frac{\pi}{180}\right) \text{ rad} = 1.6243 \text{ rad}$$

In parts (c) and (d), first use 1 radian $= (180/\pi)°$ to convert radian measure to degrees; then convert decimal parts of a degree to minutes and seconds.

(c)
$$\frac{1}{2} \text{ rad} = \frac{1}{2}\left(\frac{180}{\pi}\right)° = 28.65°$$

Next change 0.65° to minutes:

$$0.65° = 0.65(60') = 39'$$

Therefore,

$$\frac{1}{2} \text{ rad} = 28°\,39'$$

(d)
$$2.15 \text{ rad} = 2.15\left(\frac{180}{\pi}\right)° = 123.19°$$

$$0.19° = 0.19(60') = 11.40'$$

Convert 0.4' to seconds:

$$0.4' = 0.4(60'') = 24''$$

Finally,

$$2.15 \text{ rad} = 123°\ 11'\ 24''$$

EXAMPLE 3-6: **(a)** How many radians does the minute hand of a clock sweep through during one-half hour? **(b)** How many degrees does the minute hand sweep through in 10 minutes?

Solution:

(a) During one-half hour, the minute hand traverses half of a circle, or π rad.
(b) In 10 minutes the minute hand advances through $\frac{10}{60}$, or $\frac{1}{6}$ of a circle. There are $360°$ in a full circle; therefore, the hand advances by $\frac{1}{6}$ of $360°$, or $60°$. Note that in 15 minutes, the minute hand moves through an angle of $90°$; in 5 feet, the minute hand moves through $30°$.

E. Acute, obtuse, quadrantal, and coterminal angles

Trigonometry has its share of special terminology. Let's look at four terms that are often used in discussing angles:

- An **acute angle** is a positive angle measuring between 0 rad and $\frac{1}{2}\pi$ rad, or $0°$ and $90°$.
- An **obtuse angle** is a positive angle measuring between $\frac{1}{2}\pi$ rad and π rad, or $90°$ and $180°$.
- A **quadrantal angle** measures 0 rad, $\frac{1}{2}\pi$ rad, π rad, or $\frac{3}{2}\pi$ rad, or $0°$, $90°$, $180°$, or $270°$.
- Two or more angles are **coterminal** if their terminal sides coincide when the angles are placed in standard position. For example, angles measuring $45°$ and $405°$ are coterminal; angles of $-45°$ and $315°$ are coterminal; angles of $\frac{1}{2}\pi$ rad and $\frac{5}{2}\pi$ rad are coterminal. Any two angles that differ by 2π radians or $360°$ are coterminal; thus if α is measured in radians, the angles $\alpha \pm 2n\pi$ with $n = 0, 1, 2, \ldots$ are coterminal, and if θ is measured in degrees, the angles $\theta \pm n(360°)$ with $n = 0, 1, 2, \ldots$ are coterminal.

3-3. Arc Length

Arc length is distance measured along the circumference of a circle. Given any two points on the circumference of a circle with the center at the origin, the arc length is the length of that portion of the circumference of the circle between the points (see Figure 3-9). You can measure arc length in a counterclockwise or clockwise direction. Note that two points define two arcs.

In Figure 3-10, you can calculate s, the length of the arc joining points (x_0, y_0) and (x_1, y_1) from the relationship

ARC LENGTH $\qquad\qquad s = r\alpha \qquad\qquad$ **(3-4)**

where r is the radius of the circle and α is the radian measure of the angle subtended by the two points. If you assume that the angle is positive, then the arc length is positive. The angle *must* be measured in radians for this formula. Note that r and s must be measured in the same units.

EXAMPLE 3-7: For a circle centered at $(0, 0)$ and with radius r, find the length of the arc joining the following points: **(a)** $(1, 0)$ and $(0, 1)$, $r = 1$; **(b)** $(1, 0)$ and $(\frac{\sqrt{2}}{2}, \frac{\sqrt{2}}{2})$, $r = 1$; **(c)** $(-2, 0)$ and $(0, 2)$, $r = 2$. Figure 3-11 illustrates these situations.

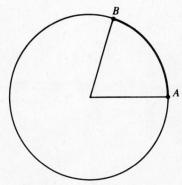

Figure 3-9
Arc joining two points of a circle.

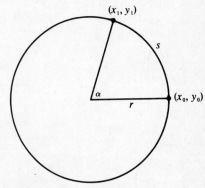

Figure 3-10
The arc length formula: $s = r\alpha$, α in radians.

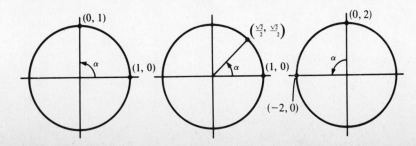

Figure 3-11

Solution:

(a) The angle is $\frac{1}{4}$ of a complete rotation, or $\frac{1}{2}\pi$ rad; thus, $s = 1(\frac{1}{2}\pi) = \frac{1}{2}\pi$.
(b) Since the x and y coordinates of the second point are equal and positive, the angle is $\frac{1}{8}$ of a rotation or $\frac{1}{4}\pi$ rad; thus, $s = 1(\frac{1}{4}\pi) = \frac{1}{4}\pi$.
(c) The angle is $\frac{1}{2}\pi$; thus, $s = 2(\frac{1}{2}\pi)$, or π.

Recall that **meridians of longitude** are half-circumferences of circles passing through the North and South Poles. Longitude $0°$, or the prime meridian, passes through the poles and Greenwich, England. Meridians west of Greenwich are designated west longitudes and meridians east of Greenwich are designated east longitudes. For example, Los Angeles, California, is approximately $120°$W longitude, while Baghdad is $45°$E longitude.

Parallels of latitude, or **parallels**, are circumferences of circles that lie in planes perpendicular to the line joining the North and South Poles and have centers on this line. All points on a parallel have the same latitude. The equator is latitude $0°$, the North Pole is latitude $90°$N, and the South Pole is latitude $90°$S. Parallels north of the equator are designated north latitudes, and parallels south of the equator are designated south latitudes.

EXAMPLE 3-8: Assume that the earth is a perfect sphere with radius of approximately 3960 miles. Suppose that two cities have the same longitude but are 1320 miles apart; find their difference in latitude.

Solution: Consider the great circle with circumference on the common longitude of the two cities. The radius of this circle is 3960 miles and the arc length is the distance between these two cities, or 1320 miles. You must find the angle subtended by this arc. Rearranging the equation $s = r\alpha$, you get $\alpha = s/r = \frac{1320}{3960} = \frac{1}{3}$ rad. Converting to degrees, $\frac{1}{3}(\frac{180}{\pi}) = \frac{60}{\pi} = 19.1°$. The cities differ in latitude by $19.1°$.

EXAMPLE 3-9: A whale at latitude $44°$N swims 400 miles due south. What is its new latitude? The earth is a sphere with radius 3960 miles.

Solution: From $s = r\alpha$, $400 = 3960\alpha$ and $\alpha = \frac{400}{3960}$ rad. To find the new latitude, convert the angle to degrees and subtract from $44°$. Thus $\alpha = \frac{400}{3960}(\frac{180}{\pi}) = 5.79°$. Converting $0.79°$ to minutes and seconds, $0.79(60) = 47.4'$. Converting $0.4'$ to seconds, $0.4(60) = 24''$. So the change in latitude is $5°\,47'\,24''$. The new latitude is $44° - 5°\,47'\,24'' = 43°\,59'\,60'' - 5°\,47'\,24'' = 38°\,12'\,36''$.

EXAMPLE 3-10: A wheel 6 feet in diameter is turning at 60 revolutions per minute (see Figure 3-12). Find the distance traveled by a point on the rim in one second.

Solution: Since 60 revolutions per minute equals 1 revolution per second, the point travels 2π radians per second. In 1 second, a point on the wheel will travel a distance of $s = r\alpha = 3(2\pi) = 6\pi$ feet.

EXAMPLE 3-11: You may remember from geometry that the area of a sector of a circle is proportional to the angle subtended by its arc (see Figure 3-13). The area of a circle is πr^2. The sector defined by an angle of measure $\frac{1}{2}\pi$ is $\frac{1}{4}$ of a circle, so its area is $\frac{1}{4}\pi r^2$. Expressing the proportion between the area of the sector A and the subtended angle α you arrive at

$$\frac{A}{\pi r^2/4} = \frac{\alpha}{\pi/2}$$

Cross-multiplying, you get

$$\frac{A\pi}{2} = \frac{\alpha(\pi r^2)}{4}$$

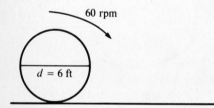

Figure 3-12

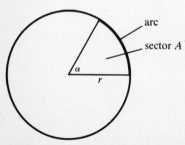

Figure 3-13

Multiplying by 2 and dividing by π, you get

$$A = \frac{\alpha r^2}{2} \qquad\qquad (3\text{-}5)$$

Using Equation 3-5, find the area of a sector with a central angle of $60°$ and radius of 4 inches.

Solution: A 60-degree angle is $60(\frac{1}{180}\pi) = \frac{1}{3}\pi$ rad. From Equation 3-5,

$$A = \frac{\pi}{3} \times 4^2 \times \frac{1}{2} = \frac{16\pi}{6} = \frac{8\pi}{3} = 8.38 \text{ square inches}$$

SUMMARY

1. Angles are generated by rotating a ray about a fixed point, by intersecting lines, and by drawing straight line segments from two points on the circumference of a circle to the center of the circle.
2. To draw an angle in standard position, place the initial side along the positive x axis with vertex at the origin.
3. Counterclockwise rotation of a ray produces positive angles and clockwise rotation produces negative angles.
4. A circle is divided into $360°$, or 2π, radians.
5. $\alpha(\text{radians}) = \frac{180}{\pi} \times \alpha(\text{degrees})$
6. $\alpha(\text{degrees}) = \frac{1}{180}\pi \times \alpha(\text{radians})$
7. $s = r\alpha \qquad$ Arc length = (radius)(angle in radians)
8. $A = \frac{1}{2}\alpha r^2 \qquad$ Area of a sector = $\frac{1}{2}$(angle in radians)(square of radius)

RAISE YOUR GRADES

Can you...?

☑ sketch an angle in standard position
☑ convert an angle from degrees, minutes, and seconds to degrees and fractions of a degree
☑ convert an angle from degrees and fractions of a degree to degrees, minutes, and seconds
☑ convert an angle from degrees to radians
☑ convert an angle from radians to degrees
☑ use $s = r\alpha$ to determine arc length, radius, or the angle subtended in radians, when given the other two quantities
☑ find the area of a sector of a circle using the formula $A = \frac{1}{2}\alpha r^2$

SOLVED PROBLEMS

PROBLEM 3-1 Find the degree measure of each of the following angles: **(a)** $\frac{3}{5}\pi$; **(b)** $\frac{9}{7}\pi$; **(c)** $\frac{14}{3}\pi$.

Solution: Use the relationship: 1 radian $= \frac{180°}{\pi}$:

$$\textbf{(a)} \quad \frac{3\pi}{5} = \left(\frac{3\pi}{5} \times \frac{180}{\pi}\right)^° = 108°$$

(b) $\dfrac{9\pi}{7} = \left(\dfrac{9\pi}{7} \times \dfrac{180}{\pi}\right)^{\circ} = 231.4^{\circ}$

(c) $\dfrac{14\pi}{3} = \left(\dfrac{14\pi}{3} \times \dfrac{180}{\pi}\right)^{\circ} = 840^{\circ}$ [See Section 3-2.]

PROBLEM 3-2 Find the radian measure of each of the following angles: **(a)** 115°; **(b)** 240°; **(c)** 310°; **(d)** −60°.

Solution: Use the relationship 1 degree = $\frac{1}{180}\pi$ rad:

(a) $115^{\circ} = 115\left(\dfrac{\pi}{180}\right)$ rad = 2.007 rad

(b) $240^{\circ} = 240\left(\dfrac{\pi}{180}\right)$ rad = 4.189 rad

(c) $310^{\circ} = 310\left(\dfrac{\pi}{180}\right)$ rad = 5.411 rad

(d) $-60^{\circ} = -60\left(\dfrac{\pi}{180}\right)$ rad = −1.047 rad   [See Section 3-2.]

PROBLEM 3-3 Find the arc length subtended by a central angle of 142° on a circle with radius 6 inches.

Solution: Use the relationship $s = r\alpha$, where α is the radian measure of the central angle: $142^{\circ} = 142(\frac{1}{180}\pi)$ rad = 2.48 rad. Then $s = 6(2.48) = 14.88$ inches.   [See Section 3-3.]

PROBLEM 3-4 A central angle A measures 65° 20′ and subtends an arc 18 inches long. Find the radius of the circle.

Solution: Use $s = r\alpha$:

$$\alpha = 65^{\circ}\,20' = 65.33^{\circ} = 65.33\left(\dfrac{\pi}{180}\right) \text{rad} = 1.14 \text{ rad}$$

$$r = \frac{s}{\alpha} = \frac{18}{1.14} = 15.79 \text{ inches}$$ [See Section 3-3.]

PROBLEM 3-5 Make a sketch of the following angles. Label both initial and terminal sides and indicate the direction of rotation. **(a)** 60°; **(b)** 150°; **(c)** 210°.

Solutions: See Figure 3-14.

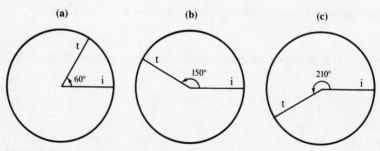

Figure 3-14

[See Section 3-2.]

PROBLEM 3-6 A car with wheels 30 inches in diameter moves 8 feet. For any point on the wheel, find the angle through which the point rotates, expressed in degrees and minutes.

Solution: The distance moved by any point on a rotating wheel gives the arc length. Imagine a string wrapped around the wheel, unwrapping as the wheel turns, as shown in Figure 3-15. Use $s = r\alpha = \frac{1}{2}d\alpha$, where d is the diameter of the wheel. Converting 8 feet to $8(12) = 96$ inches and substituting into the formula,

$$96 = \frac{1}{2}30\alpha$$

$$\alpha = \frac{96}{15} = 6.4 \text{ rad}$$

(Since the car moves forward, the tire rotates clockwise; thus the negative angle must be taken.) Changing to degrees,

$$-6.4 \text{ rad} = -6.4\left(\frac{180}{\pi}\right)$$

$$= -366.7° = -366° \, 42'$$

Figure 3-15

[See Section 3-3.]

PROBLEM 3-7 A racing car travels in a circular course around the judges' stand. A distance of 1 mile on the track subtends an angle of 120° at the judges' stand. What is the diameter of the track? See Figure 3-16.

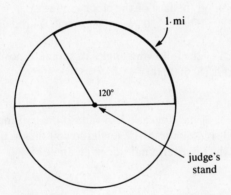

Figure 3-16

Solution: Again, use the fundamental relation, $s = r\alpha$ or $r = s/\alpha$. The diameter $d = 2r = 2s/\alpha$. First change 120° to radians:

$$120° = 120\left(\frac{\pi}{180}\right) = \frac{2\pi}{3} \text{ rad}$$

Then,

$$d = 2\left(\frac{1}{2\pi/3}\right) = \frac{3}{\pi} = 0.95 \text{ miles}$$

[See Section 3-3.]

PROBLEM 3-8 Find the area watered by a sprinkler system with a range of 200 feet and a turning angle of 135°.

Solution: The area covered is the sector of the circle shown in Figure 3-17. The area of the sector is given by $A = \frac{1}{2}\alpha r^2$, where α is in radians. So $135° = 135(\frac{1}{180}\pi) = 2.356$ rad. Therefore, the area covered is $\frac{1}{2}(2.356)(200^2) = 47\,120$ square feet.

[See Example 3-11.]

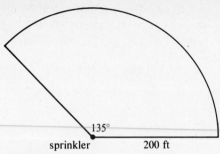

135°

sprinkler   200 ft

Figure 3-17

PROBLEM 3-9   The pendulum of a clock is 20 inches long and swings through an arc of 20° each second. How far does the tip of the pendulum move in 1 second?

Solution: First convert 20° to radians.

$$20° = 20\left(\frac{\pi}{180}\right) = \frac{\pi}{9} = 0.349 \text{ rad}$$

Then,

$$s = r\alpha = 20(0.349) = 6.98 \text{ inches} \qquad \text{[See Section 3-3.]}$$

PROBLEM 3-10   How many radians are contained in the angle between the hands of a clock at 4 P.M.?

Solution: Each hour division on the face of the clock contains $\frac{360}{12} = 30°$. The angle between the hands at 4 P.M. is $4(30) = 120°$. This is $120(\frac{1}{180}\pi) = \frac{2}{3}\pi = 2.094$ rad.   [See Section 3-3.]

PROBLEM 3-11   For each of the following angles in standard position, in which quadrant will the terminal side lie? **(a)** 3 rad; **(b)** 660°; **(c)** −330°.

Solution:

(a) 3 rad lies between $\frac{1}{2}\pi = 1.57$ rad and $\pi = 3.14$ rad; so the terminal side is in quadrant II.
(b) 660° is 360° + 300°. 300° is a quadrant-IV angle; so 660° lies in quadrant IV.
(c) An angle of −330° has the same terminal side as an angle of 360° − 330° = 30°; so −330° lies in quadrant I.   [See Section 3-2.]

PROBLEM 3-12   A wheel of radius 1 foot travels at π radians per minute. How many linear feet does the wheel travel in 2 minutes?

Solution: In two minutes the wheel will revolve through 2π radians, or one complete revolution of the wheel. The length of the circumference of the wheel is $s = r\alpha = 1(2\pi) = 2\pi = 6.28$ feet. The linear distance traveled in 2 minutes is 6.28 feet.   [See Section 3-3.]

PROBLEM 3-13   A wheel of radius 1 foot travels 40 linear feet in 3 minutes; what is the speed of the wheel in revolutions per minute?

Solution: From $s = r\alpha$, $40 = 1\alpha$, so the wheel revolves through an angle of 40 rad in 3 minutes. In 1 minute it must revolve through $\frac{40}{3}$ rad. There are 2π rad in a complete revolution, so divide $\frac{40}{3}$ rad by 2π rad to get the number of revolutions per minute. Thus, the speed is $(\frac{40}{3})/2\pi = 2.12$ revolutions per minute.   [See Section 3-3.]

PROBLEM 3-14   A wheel traveling at 10 revolutions per minute covers 33 linear feet in 1 minute; what is the radius of the wheel?

Solution: Again, the key formula is $s = r\alpha$. In 1 minute, the wheel travels through $10(2\pi) = 20\pi = 62.83$ rad. For $s = 33$, $r = \frac{33}{62.83} = 0.525$ feet.   [See Section 3-3.]

PROBLEM 3-15   Find the linear velocity of the earth about the sun in miles per hour assuming that the earth is 93 000 000 miles from the sun and makes one complete orbit in 365.25 days.

Solution: The earth is traveling on the circumference of a circle whose radius is 93 000 000 miles. In one day the earth moves through $\frac{1}{365.25}$ of this circumference and, since a day consists of 24 hours, in one hour the earth moves through $\frac{1}{(365.25 \times 24)} = \frac{1}{8766}$ of a complete circle. Since a complete circle contains 2π rad, in one hour the earth moves through $\frac{1}{8766}(2\pi) = 0.0007$ rad. Using $s = r\alpha$, $s = 93\,000\,000(.0007) = 65\,100$ miles. The linear velocity of the earth is 65 100 miles per hour.

[See Section 3-3.]

PROBLEM 3-16   What is the angle in degrees between the hour hand and minute hand of a clock when the time is 11:00 P.M.?

Solution: Since there are twelve hours on the face of a clock, each hour is equivalent to an angle of $\frac{360}{12} = 30°$. At eleven, the hands are 30° apart.   [See Section 3-2.]

PROBLEM 3-17   Two diameters are drawn on a circle with radius of 1 unit, as shown in Figure 3-18. Find α to the nearest second given that the arc is 4 units long.

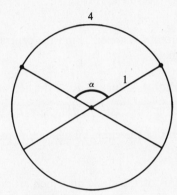

Figure 3-18

Solution: Using $s = r\alpha$, $4 = 1\alpha$; 4 rad $= 4(\frac{180}{\pi})° = 229.18° = 229°\,10'\,48''$.   [See Section 3-3.]

PROBLEM 3-18   Find the difference (a) in longitude between Los Angeles (120°W) and Denver (105°W); (b) in longitude between Los Angeles and New Delhi (75°E); (c) in latitude between Manila (15°N) and New Orleans (30°N); (d) in latitude between New Orleans and Perth, Australia (30°S).

Solution: In calculating differences in longitude, you subtract the smaller longitude from the larger longitude if they are both East or both West and add the longitudes if one is East and the other is West. In calculating differences in latitudes, you subtract the smaller latitude from the larger latitude if they are both North or South and add the latitudes if one is North and the other is South. (a) $120 - 105 = 15°$; (b) $120 + 75 = 195°$; (c) $30 - 15 = 15°$; (d) $30 + 30 = 60°$.

[See Section 3-2.]

Supplementary Exercises

PROBLEM 3-19   For each of the following angles in standard position, name the quadrant in which the terminal side will lie (a) $-130°$; (b) $312°$; (c) $\frac{3}{4}\pi$ rad; (d) 15 rad.

PROBLEM 3-20   Make a diagram showing each of the following angles in standard position: (a) $\frac{1}{4}\pi$ rad; (b) 230°; (c) 400°; (d) $-\frac{6}{8}\pi$ rad; (e) $-330°$.

PROBLEM 3-21   Express in radians: (a) 120°; (b) $-225°$; (c) 22° 30'; (d) 33° 45'; (e) $(\frac{50}{\pi})°$.

PROBLEM 3-22 Express the following radian measures in degrees: (a) $\frac{3}{4}\pi$; (b) $-\frac{11}{6}\pi$; (c) $\frac{15}{6}\pi$; (d) 7.

PROBLEM 3-23 Find the radian measure of each of the following angles; express your answer as a multiple of π: (a) 20°; (b) 270°; (c) $-330°$.

PROBLEM 3-24 Use your calculator to convert each of the following degree, minute, and second measures to radian measure and radian measure to degree, minute, and second measure: (a) 32° 31′ 14″; (b) 235° 12′ 50″; (c) 2 rad; (d) $\frac{8}{11}\pi$ rad.

PROBLEM 3-25 A string is wrapped around a cylinder 4 inches in diameter. The string completes $3\frac{1}{2}$ circumferences. How long is the string?

PROBLEM 3-26 A Ferris wheel spins at such a rate that a seat on the wheel travels at 78π feet per minute. The wheel turns $\frac{1}{3}\pi$ radians in 10 seconds; find the distance from the seat to the hub of the wheel.

PROBLEM 3-27 A circle has radius 4 meters. Find the central angle that subtends an arc of: (a) $\frac{2}{3}\pi$ meters; (b) π meters; (c) 4 meters.

PROBLEM 3-28 A circle has radius 32 inches. Find the arc subtended by an angle of: (a) 60°; (b) 135°; (c) $\frac{5}{4}\pi$ rad.

PROBLEM 3-29 Find the radius of the circle on which an arc of 20 inches is subtended by an angle of: (a) $\frac{1}{3}\pi$ rad; (b) 225°; (c) 30°.

PROBLEM 3-30 Find the radius of the circle on which an angle of 150° subtends an arc of: (a) 10 inches; (b) 2 meters; (c) π inches.

PROBLEM 3-31 Change each of the following angles measured in degree-decimal form to degree-minute form; express your answer to the nearest minute: (a) 38.25; (b) 195.33; (c) -42.78; (d) -211.56; (e) 537.98; (f) -1873.24.

PROBLEM 3-32 Change each of the following angles measured in degree-minute form to degree-decimal form: (a) 25° 15′; (b) 180° 17′; (c) 87° 55′; (d) 269° 35′; (e) 1066° 85′; (f) 357° 29′.

PROBLEM 3-33 For each of the following rotations of the positive x axis, give a degree and radian measure for the angle generated: (a) $\frac{1}{3}$ of a rotation clockwise; (b) $\frac{3}{5}$ of a rotation counterclockwise; (c) 15.35 rotations clockwise; (d) $2 + \sqrt{3}$ rotations counterclockwise.

PROBLEM 3-34 Let α be an angle with terminal side passing through the point $(-4, 0)$ on a circle of radius 4 with center at the origin. Find the (a) degree measure of angle α between $-1080°$ and $-1440°$; (b) radian measure of angle α between $(25 + \sqrt{2})$ and $(25 + \sqrt{2}) + 2\pi$.

PROBLEM 3-35 A space shuttle is sent into a circular orbit 2000 miles above the surface of the earth. Assume that the earth is a perfect sphere with radius 3960 miles and that the space shuttle travels at 22 000 miles per hour; how many revolutions about the earth does it make in 1 day?

PROBLEM 3-36 Find the linear velocity in miles per hour of the moon around the earth. Assume that the moon is 293 000 miles from the earth, completes one orbit every 28 days, and that its orbit is circular.

PROBLEM 3-37 A bicycle has wheels of 26 inch radius and the wheels are turning at the rate of 80 revolutions per minute. Find the speed in miles per hour.

PROBLEM 3-38 Two pullies with diameters 3 inches and 15 inches are connected by a belt, as shown in Figure 3-19. Find the velocity, in revolutions per minute, of the larger pulley when the smaller pulley is revolving at 2400 revolutions per minute.

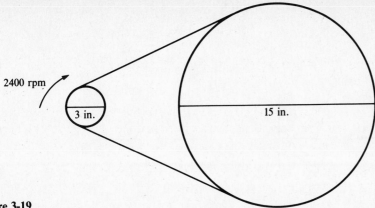

2400 rpm

3 in.

15 in.

Figure 3-19

PROBLEM 3-39 (a) For a circle with a radius 8 inches, find the arc length subtended by a central angle of 60°. (b) For a circle with a radius 10 inches, find the radian measure of the central angle subtended by an arc length of 3 inches.

PROBLEM 3-40 For a circle with a radius of $\frac{1}{2}$ foot, find the degree measure of the central angle corresponding to an arc length of 2 feet.

PROBLEM 3-41 The minute hand on a clock is 4 inches long. How far does the tip of the hand move between 7:35 and 8:00?

PROBLEM 3-42 A pendulum 10 feet in length swings through an arc of 38°. How long is the arc described by the midpoint of the pendulum?

PROBLEM 3-43 For a circle with a radius of 6 inches, the area of a sector is 13 square inches. Find the central angle, in radians, of the sector.

PROBLEM 3-44 A bicycle travels along a curve on a highway at 15 miles per hour. The curve is an arc of a circle with a radius of 1200 feet. Find the central angle that the bicycle subtends in 30 seconds.

PROBLEM 3-45 Give the latitude and longitude of a point 3000 miles east of the meridian through Greenwich, England, and 2400 miles north of the equator. Assume that the earth is a sphere of radius 3960 miles.

PROBLEM 3-46 Convert each of the following angle measures from radians to degrees: (a) $\frac{3}{4}\pi$; (b) $\frac{5}{6}\pi$; (c) $\frac{1}{3}\pi$; (d) $\frac{11}{12}\pi$.

PROBLEM 3-47 Convert each of the following angle measures to radians: (a) 200°; (b) 85°; (c) 210°; (d) 330°.

PROBLEM 3-48 Find the angle that subtends an arc of 12 inches on a circle with a diameter of 16 inches. Express your answer in both radians and degrees.

PROBLEM 3-49 For a circle with diameter of 8 units, find the length of an arc subtended by a central angle of $\frac{5}{3}\pi$.

PROBLEM 3-50 An angle of $\frac{7}{6}\pi$ rad subtends an arc of 8 units; find the radius of the circle.

PROBLEM 3-51 Find the difference: (a) in longitude between New York (74°W) and Moscow (37°E); (b) in longitude between New York (74°W) and Denver (105°W); (c) in latitude between Leningrad (60°N) and Durban, South Africa (30°S); (d) in latitude between Leningrad (60°N) and Turin, Italy (45°N).

PROBLEM 3-52 Find the radian measure of each of the following angles: (a) 195°; (b) 285°; (c) 110°; (d) 380°.

Answers to Supplementary Exercises

3-19: (a) III; (b) IV; (c) II; (d) II

3-20:

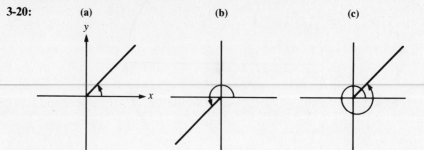

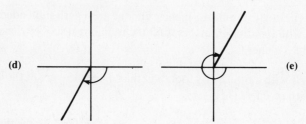

Figure 3-20

3-21: (a) $\frac{2}{3}\pi$; (b) $-\frac{5}{4}\pi$; (c) $\frac{1}{8}\pi$; (d) $\frac{3}{16}\pi$; (e) $\frac{5}{18}$

3-22: (a) 135°; (b) −330°; (c) 90° or 450°; (d) $\frac{1260}{\pi} \cong 401.1° \cong 41.1°$

3-23: (a) $\frac{1}{9}\pi$; (b) $\frac{3}{2}\pi$; (c) $-\frac{11}{6}\pi$

3-24: (a) 0.57 rad; (b) 4.11 rad; (c) 114° 35′ 30″; (d) 130° 54′ 33″

3-25: 14π in.

3-26: 39 ft

3-27: (a) $\frac{1}{6}\pi$ rad; (b) $\frac{1}{4}\pi$ rad; (c) 1 rad

3-28: (a) $\frac{32}{3}\pi$ in.; (b) 24π in.; (c) 40π in.

3-29: (a) $\frac{60}{\pi}$ in.; (b) $\frac{16}{\pi}$ in.; (c) $\frac{120}{\pi}$ in.

3-30: (a) $\frac{12}{\pi}$ in.; (b) $\frac{12}{5\pi}$ m; (c) $\frac{6}{5}$ in.

3-31: (a) 38° 15′; (b) 195° 20′; (c) −42° 47′; (d) −211° 34′; (e) 537° 59′; (f) −1873° 14′

3-32: (a) 25.25°; (b) 180.28°; (c) 87.92°; (d) 269.58°; (e) 1067.42°; (f) 357.48°

3.33: (a) −120° = $-\frac{2}{3}\pi$ rad; (b) 216° = 3.77 rad; (c) −5526° = −96.45 rad; (d) 1343.54 degrees = 23.45 rad

3.34: (a) $-(360 \times 3) - 180 = -1260°$; (b) $(2\pi \times 4) + \pi = 28.27$ rad

3-35: 14.10 revolutions of the earth

3-36: 2739.54 mi/h

3-37: 12.38 mi/h

3-38: 480 rpm

3-39: (a) $\frac{8}{3}\pi$; (b) $\frac{3}{10}$

3-40: 229° 11′

3-41: 10.5 in.

3-42: 3.32 ft

3-43: 0.72 rad

3-44: 0.55 rad = 31.51°

3-45: latitude = 34° 43′ 29″ N; longitude = 43° 24′ 21″ E

3-46: (a) 135°; (b) 150°; (c) 60°; (d) 165°

3-47: (a) 3.49 rad; (b) 1.48 rad; (c) 3.67 rad; (d) 5.76 rad

3-48: 1.5 rad = 85.94°

3-49: 20.94 units

3-50: 2.18 units

3-51: (a) 111°; (b) 31°; (c) 90°; (d) 15°

3-52: (a) 3.40 rad; (b) 4.97 rad; (c) 1.92 rad; (d) 6.63 rad

RELATIONS AND FUNCTIONS

☑ **Ordered Pairs**
☑ **Relations**
☑ **Functions**

4-1. Ordered Pairs

If x and y are any two real numbers, then (x, y) is an **ordered pair** of real numbers, where x is the **first component**, or **first coordinate**, and y is the **second component**, or **second coordinate**, of (x, y). Two ordered pairs, (x_1, y_1) and (x_2, y_2), are equal if and only if $x_1 = x_2$ and $y_1 = y_2$.

EXAMPLE 4-1: Tell which of the following ordered pairs are equal: **(a)** $(1, 2)$ and $(2, 1)$; **(b)** $(\sqrt{4}, 3)$ and $(2, \frac{9}{3})$; **(c)** $(8, \frac{5}{2})$ and $(3, 2.5)$.

Solution:

(a) Since $1 \neq 2$, $(1, 2) \neq (2, 1)$.

(b) Since $\sqrt{4} = 2$ and $3 = \frac{9}{3}$, $(\sqrt{4}, 3) = \left(2, \frac{9}{3}\right)$.

(c) Since $8 \neq 3$, $\left(8, \frac{5}{2}\right) \neq (3, 2.5)$.

A **set** is a collection of objects or numbers called **elements**. We'll denote sets with capital letters and elements with lowercase letters or numbers, contained in braces. If every element of set R is also an element of set S, then R is a **subset** of S. The **empty** set contains no elements.

If A and B are any two sets, the **Cartesian product** of A and B, written $A \times B$, is the set of all ordered pairs with first components from A and second components from B.

EXAMPLE 4-2: $A = \{1, 2, 3\}$ and $B = \{x, y\}$; find $A \times B$ and $B \times A$.

Solution: From the definition of Cartesian product,

$$A \times B = \{(1, x), (1, y), (2, x), (2, y), (3, x), (3, y)\}$$
$$B \times A = \{(x, 1), (y, 1), (x, 2), (y, 2), (x, 3), (y, 3)\}$$

4-2. Relations

A. A relation is a set of ordered pairs.

Any set of ordered pairs defines a **relation**. Any subset of a Cartesian product is also a *relation*. We'll denote relations with capital letters.

EXAMPLE 4-3: Each of the following statements defines a relation. The vertical bar is read "such that" and the text following the bar states the condition(s) the elements of each set must satisfy.

$R = \{(2,3),(2,5)\}$

$S = \{(x,y)\,|\,x \text{ is real and } y = 3\}$

$T = \{(x,y)\,|\,x \text{ is a positive integer and } y = x + 1\}$

$A = \{(\alpha_1,\alpha_2)\,|\,\alpha_1 \text{ is any letter of the alphabet between a and r and } \alpha_2 \text{ is the second letter of the alphabet after } \alpha_1\}$

$R_1 = \{(x,y)\,|\,x + y = 8\}$

$B = \{(x,y)\,|\,y > 3x - 2\}$

B. The domain is the set of first components.

In any relation, the **domain** is the set of all first components. If the domain is not stated, take it to be the set of all first components for which the relationship is defined. For example, if $S = \{(x,y)\,|\,y = 1/x\}$, the domain is the set of all real numbers except zero, since $1/0$ is undefined.

EXAMPLE 4-4: Find the domain of each relation in Example 4-3.

Solution: The domain of R is $\{2\}$; of S, $\{x\,|\,x \text{ is real}\}$; of T, $\{1,2,3,\ldots\}$; and of A, $\{a, b, c,\ldots, r\}$. Both $x + y = 8$ and $y > 3x - 2$ are defined for any real x, so the domain of R_1 and of B is $\{x\,|\,x \text{ any real number}\}$.

C. The range is the set of second components.

In any relation, the **range** is the set of all second components. If the range is not stated, take it to be the set of all second components for which the relationship is defined. For example, if $S = \{(x,y)\,|\,x = 1/y\}$, the range is the set of all real numbers except zero, since $1/0$ is undefined.

EXAMPLE 4-5: Find the range of each relation in Example 4-3.

Solution: The range of R is $\{3,5\}$; of S, $\{3\}$; of T, $\{2,3,4,\ldots\}$; and of A, $\{c, d, e,\ldots, s, t\}$. Both $x + y = 8$ and $y > 3x - 2$ are defined for any real y, so the domain of R_1 and of B is $\{y\,|\,y \text{ any real number}\}$.

For any relation, if $R = \{(x,y)\,|\,x \text{ and } y \text{ are real numbers}\}$, you may write the domain of R as $\{x\,|\,(x,y) \text{ is in } R\}$ and the range of R as $\{y\,|\,(x,y) \text{ is in } R\}$. The **graph** of relation R is the set of all points, (x,y), in the Cartesian plane such that (x,y) is in R.

EXAMPLE 4-6: Given the relation $\{(2,3),(-1,6),(8,2)\}$, find the domain, the range, and graph the relation.

Solution: The domain is $(2,-1,8\}$, the range is $\{3,6,2\}$. The graph of this relation is the set of three points, $(2,3), (-1,6)$, and $(8,2)$, as shown in Figure 4-1.

EXAMPLE 4-7: Is $\{4,(2,5),(\sqrt{2},7)\}$ a relation?

Solution: The set is not a relation since 4 is not an ordered pair.

EXAMPLE 4-8: Find the domain and range and graph the relation $R = \{(x,y)\,|\,x^2 + y^2 \leqslant 4\}$.

Solution: For any x in the domain of R, $x^2 \leqslant x^2 + y^2 \leqslant 4$. Thus $\sqrt{x^2} \leqslant \sqrt{4}$, or $|x| \leqslant 2$, so $-2 \leqslant x \leqslant 2$. The domain of R is $\{x\,|\,-2 \leqslant x \leqslant 2\}$ which corresponds to the closed interval $[-2,2]$. (We'll use brackets to denote closed intervals.) Similarly, the range of R is $\{y\,|\,-2 \leqslant y \leqslant 2\}$ which corresponds to the closed interval $[-2,2]$. If (a,b) is in R, it must lie on the circle $x^2 + y^2 = r^2$, where r

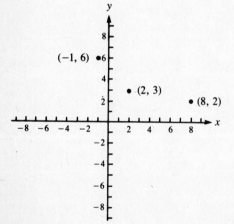

Figure 4-1

is a number in $[0, 2]$. Thus any point (a, b) in R lies on some circle of radius $0 \leqslant r \leqslant 2$. Conversely, any point (x, y) on $x^2 + y^2 = r^2$, for $0 \leqslant r \leqslant 2$, is also in the relation R. Therefore, the graph of R is the set $\{(x, y) \,|\, x^2 + y^2 = r^2, 0 \leqslant r \leqslant 2\}$, or, in words, the set of all points inside and on the circle $x^2 + y^2 = 4$ (see Figure 4-2).

4-3. Functions

A **function**, F, is any relation where the first component of each ordered pair in F is associated with a *unique* second component. This means that a function cannot contain the two ordered pairs (x_1, y_1) and (x_1, y_2) unless $y_1 = y_2$. We'll denote functions by either capital or small letters such as f, g, H, or K.

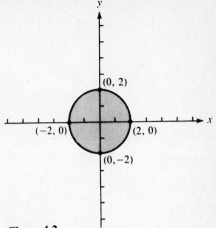

Figure 4-2

EXAMPLE 4-9: Is $R = \{(1, -2), (2, 4), (3, 5)\}$ a function?

Solution: Since each first component is paired with a unique second component, that is, 1 with second component -2, 2 with 4, and 3 with 5, R is a function.

EXAMPLE 4-10: Is $S = \{(3, 8), (4, 5), (3, 6)\}$ a function?

Solution: Since first component 3 is paired with second component 8 and second component 6, S is NOT a function.

EXAMPLE 4-11: Is $T = \{(1, -2), (2, 2), (3, -2), (4, 6)\}$ a function?

Solution: Because each first component is paired with a unique second component, T is a function. Note that first components 1 and 3 are associated with the *same* second component. The definition of a function allows two or more distinct first components to be paired with identical second components.

EXAMPLE 4-12: Is $R = \{(x, y) \,|\, y = x^3\}$ a function?

Solution: If (x_1, y_1) and (x_1, y_2) are in R, then $y_1 = (x_1)^3$ and $y_2 = (x_1)^3$. This shows that $y_1 = y_2$, so each first component is paired with a unique second component. Hence R is a function.

EXAMPLE 4-13: Is $R = \{(x, y) \,|\, x^2 + y^2 = 8\}$ a function?

Solution: You can see that the ordered pairs $(0, 2\sqrt{2})$ and $(0, -2\sqrt{2})$ are in R. Since $2\sqrt{2} \neq -2\sqrt{2}$, R is not a function.

A. The domain and range of a function are the sets of first and second components, respectively.

As with relations, the **domain** of a function, F, is the set of all first components of the ordered pairs in F and the **range** of a function is the set of second components of ordered pairs in F. The **graph** of a function is the set of all ordered pairs defined by the function.

If F is a function and (x, y) is in F, then you can write $y = F(x)$, where y is the **value** of F at x or the **image** of x under F. You can think of F as a machine in which x is the input and y is the output:

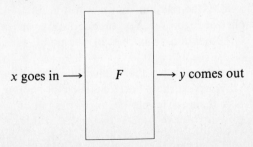

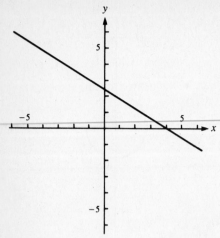

Figure 4-3

In this machine interpretation of a function, the domain of F is the set of all input number (x's) that produce a real number output (y); the range of F is the set of all possible output values (y's) that are produced by entering elements from the domain into the machine.

EXAMPLE 4-14: Find the domains and ranges of the following functions: **(a)** $\{(1,2),\ (2,3),\ (3,3),\ (4,9)\}$; **(b)** $\{(x, y)\,|\,2x + 3y = 8\}$; **(c)** $\{(x, y)\,|\,x + y = 1$ and $x - y = 3\}$.

Solution:

(a) The domain is $\{1, 2, 3, 4\}$ and the range is $\{2, 3, 9\}$.
(b) The domain is $\{x\,|\,x$ any real number$\}$ and the range is $\{y\,|\,y$ any real number$\}$. The graph of the function is shown in Figure 4-3.
(c) Solving the equations simultaneously,

$$\begin{array}{r} x + y = 1 \\ x - y = 3 \\ \hline 2x\quad\ = 4 \end{array}$$

so, $x = 2$ and $y = -1$. Therefore, the domain is $\{2\}$ and the range is $\{-1\}$.

B. One-to-one functions have unique first components.

A function, F, is **one-to-one** if $F(x_1) = F(x_2)$ only when $x_1 = x_2$. In words, we are demanding that every element in the domain be assigned to a *different* element of the range. Stated another way: Each second component must be paired with a unique first component.

EXAMPLE 4-15: Is $F(x) = x^3$ a one-to-one function?

Solution: Since $(x_1)^3 = (x_2)^3$ implies that $x_1 = x_2$, $F(x)$ is a one-to-one function. The domain and range are both equal to the set of all real numbers.

EXAMPLE 4-16: Is $F(x) = x^2$ a one-to-one function?

Solution: The function is not one-to-one because $F(-2) = 4 = F(2)$, while $-2 \neq 2$. The domain is the set of all real numbers and the range is the set of all nonnegative real numbers.

EXAMPLE 4-17: Is $F(x) = 1/x$ a one-to-one function?

Solution: The function is one-to-one since $F(x_1) = 1/x_1 = F(x_2) = 1/x_2$, which implies that $x_1 = x_2$. The domain of F is the set of all real numbers except zero and the range is the set of all real numbers except zero.

C. Vertical and horizontal line tests

You can often determine whether a relation is a function from the graph of the relation. If you can draw a vertical line that intersects the graph of the relation at *more* than one point, the relation is not a function; otherwise, it is a function. This is the **vertical line test**. Consider the relation graphed in Figure 4-2. The y axis, a vertical line, intersects the graph at all points on the interval $[-2, 2]$; therefore, the relation shown is not a function. Now consider the relation graphed in Figure 4-1. You cannot draw a vertical line that intersects the graph of the relation at more than one point. This relation must be a function.

You can also determine whether a function is one-to-one from the graph of the function. If you can draw a horizontal line that intersects the graph of the function at *more* than one point, the function is not one-to-one; otherwise, it is one-to-one. This is the **horizontal line test**. Consider the function graphed in Figure 4-3. Any horizontal line will intersect the graph at only one point, so the function is one-to-one. Now consider the function whose graph is shown in Figure 4-7. The x axis, a horizontal line, intersects the graph at three points, so this is not the graph of a one-to-one function.

SUMMARY

1. Two ordered pairs of numbers are equal if and only if the two first elements are equal to each other and the two second elements are equal to each other.
2. The Cartesian product of A and B is the set of all ordered pairs with first components from A and second components from B.
3. A relation with set A as domain and set B as range is any subset of the Cartesian product $A \times B$.
4. The domain of a relation is the set of all first components of ordered pairs in the relation.
5. The range of a relation is the set of all second components of ordered pairs in the relation.
6. A function is a relation in which each first component is associated with a unique second component.
7. The domain of a function is the set of all first components and the range is the set of all second components of ordered pairs of the function.
8. A function is one-to-one if for each second component there corresponds a unique first component.
9. Use the vertical line test to determine whether a relation is a function.
10. Use the horizontal line test to determine whether a function is one-to-one.

RAISE YOUR GRADES

Can you ...?

☑ find the Cartesian product of two sets
☑ find all relations defined on two sets
☑ find the domain and range of a relation
☑ determine whether a relation is a function
☑ find the domain and range of a function
☑ determine whether a function is one-to-one

SOLVED PROBLEMS

PROBLEM 4-1   Let $A = \{a, b, c\}$ and $B = \{1, 2\}$. **(a)** List the elements of $A \times B$. **(b)** Write three relations defined on $A \times B$. **(c)** What can you say about domains and ranges of relations defined on $A \times B$?

Solution:

(a) $A \times B = \{(a, 1), (a, 2), (b, 1), (b, 2), (c, 1), (c, 2)\}$
(b) $R_1 = \{(a, 1), (a, 2)\}$; $R_2 = \{(b, 1), (b, 2)\}$; $R_3 = \{(c, 1), (c, 2)\}$
You may define other relations; any subset of $A \times B$ is a relation.
(c) Since any subset of $A \times B$ is a relation, the domain of any relation is a subset of A and the range of any relation is a subset of B.

[See Section 4-2.]

PROBLEM 4-2 Determine the domain and range of the following relations: **(a)** $\{(5, 6), (-3, 2), (6, 3)\}$; **(b)** $\{(\pi, \pi), (0, 0), (0, \pi), (\pi, 0)\}$; **(c)** $\{(a, b), (b, c), (c, d), (d, e)\}$; **(d)** $\{(s, t) \mid s$ and t are positive integers and $t = s + 1\}$.

Solution: Recall that the domain is the set of first elements of all the ordered pairs and the range is the set of second elements of the ordered pairs.

(a) Domain $= \{5, -3, 6\}$ and range $= \{6, 2, 3\}$.
(b) Domain $= \{0, \pi\}$ and range $= \{0, \pi\}$.
(c) Domain $= \{a, b, c, d\}$ and range $= \{b, c, d, e\}$.
(d) Domain $= \{1, 2, 3, \ldots\}$ and range $= \{2, 3, 4, \ldots\}$.   [See Section 4-2.]

PROBLEM 4-3 List three ordered pairs from each of the following relations: **(a)** $\{(x, y) \mid y = x^2 + 2\}$; **(b)** $\{(x, y) \mid y + x = -1\}$; **(c)** $\{(x, y) \mid y^2 + x^2 = -1\}$; **(d)** $\{(x, y) \mid y^2 + x^2 = 1\}$.

Solution: You will often find it convenient to substitute integer values for x into the expression relating x and y. If you replace x by 0, 1, and 2, you get the following ordered pairs:

(a) $(0, 2), (1, 3)$, and $(2, 6)$;
(b) $(0, -1), (1, -2)$, and $(2, -3)$;
(c) Since the sum of two positive numbers cannot equal -1, no ordered pairs exist;
(d) $(0, 1), (1, 0)$, and $(-1, 0)$. Note: when $x = 2$ there is no corresponding y value.   [See Section 4-2.]

PROBLEM 4-4 Graph each of the following relations and give their domain and range: **(a)** $R = \{(x, y) \mid x^2 + y^2 \leqslant 9\}$; **(b)** $R = \{(x, y) \mid x - y + 2 \geqslant 0\}$; **(c)** $R = \{(x, y) \mid y \geqslant x^2\}$.

Solution: See Figures 4-4, 4-5, and 4-6 for graphs of the relations.

(a) The domain of this relation is $\{x \mid -3 \leqslant x \leqslant 3\}$ and the range is $\{y \mid -3 \leqslant y \leqslant 3\}$.
(b) The domain is $\{x \mid x$ any real number$\}$ and the range is $\{y \mid y$ any real number$\}$.
(c) The domain is $\{x \mid x$ any real number$\}$ and the range is $\{y \mid y \geqslant 0\}$.   [See Section 4-2.]

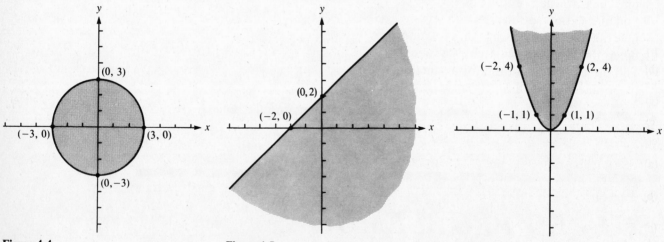

Figure 4-4   Figure 4-5   Figure 4-6

PROBLEM 4-5 Determine the domain and range of each relation in Figures 4-7 through 4-10.

Solution: Using the definitions of domain and range,

(a) Domain $= \{$All real numbers$\}$ and range $= \{$All real numbers$\}$
(b) Domain $= \{-1\}$ and range $= \{$All real numbers$\}$
(c) Domain $= \{$All real numbers$\}$ and range $= \{+3\}$
(d) Domain $= \{$All real numbers$\}$; range $= \{$All real numbers $\geqslant -1\}$ [See Section 4.2.]

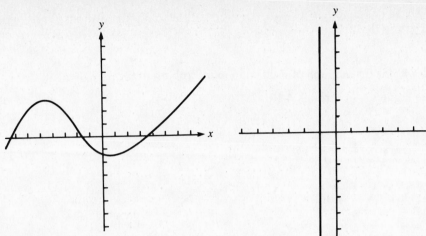

Figure 4-7

Figure 4-8

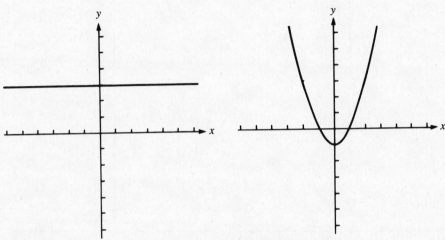

Figure 4-9

Figure 4-10

PROBLEM 4-6 Tell which of the following relations are functions: (a) $\{(\pi, \pi), (0, 0), (0, \pi), (\pi, 0)\}$; (b) $\{(-1, 1), (-1, 2), (-1, 3), (-1, 6)\}$; (c) $\{(1, -1), (2, -1), (3, -1), (6, -1)\}$; (d) $\{(x, y) \mid y = x^2 + 2 \text{ and } x \text{ is any real number}\}$; (e) $\{(x, y) \mid y + x = -1 \text{ and } x \text{ is any real number}\}$; (f) $\{(x, y) \mid y^2 + x^2 = 1\}$.

Solution: Recall that if each first component of an ordered pair corresponds to a unique second component, the relation is a function.

(a) Since first component π corresponds to second components π and 0, this relation is not a function.
(b) Since second components 1, 2, 3, and 6 are paired with first component -1, this relation is not a function.
(c) This relation is a function.
(d) Since to each x value there corresponds a unique y value, this relation is a function.
(e) This relation is a function.
(f) Since $x = 0$ corresponds to y values $+1$ and -1, this relation is not a function.

[See Section 4-3.]

PROBLEM 4-7 How many functions have domain in $A = \{a, b\}$ and range in $B = \{o, p\}$?

Solution: Consider the number of relations with first component in A and second component in B: $A \times B = \{(a, o), (b, o), (a, p), (b, p)\}$. Next, find the number of relations on $A \times B$ in which each first component of the ordered pairs is associated with a unique second component. The relations satisfying this condition are: $\{(a, o)\}, \{(a, p)\}, \{(b, o)\}, \{(b, p)\}, \{(a, o), (b, p)\}, \{(a, p), (b, o)\},$ $\{(a, p), (b, p)\}, \{(a, o), (b, o)\}$. You can see that each of these eight relations is a function.

[See Section 4-3.]

PROBLEM 4-8 Sketch a graph of each relation in Problem 4-3.

Solution: See Figures 4-11 through 4-14. [See Section 4-2.]

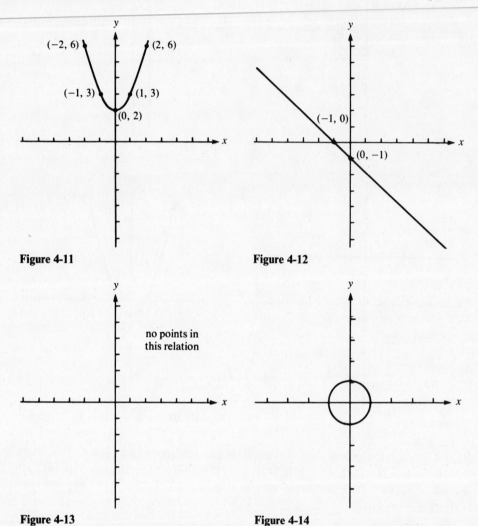

Figure 4-11 Figure 4-12

Figure 4-13 Figure 4-14

PROBLEM 4-9 Determine whether any of the graphs in Figures 4-15 through 4-18 are graphs of functions.

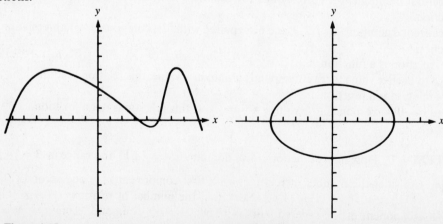

Figure 4-15 Figure 4-16

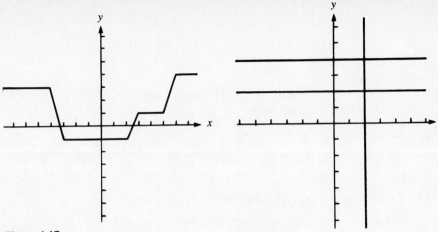

Figure 4-17   **Figure 4-18**

Solution: For the graphs of the relations shown in Figures 4-15 and 4-17, any vertical line will intersect the graph at only one point. So 4-15 and 4-17 are graphs of functions. Since you can draw vertical lines that intersect the graphs of Figures 4-16 and 4-18 at more than one point, 4-16 and 4-18 are not graphs of functions.   [See Section 4-3.]

PROBLEM 4-10 Graph each of the following functions and determine the domain and range:
(a) $f(x) = c$, where c is a constant; **(b)** $f(x) = x$; **(c)** $f(x) = |x|$.

Solution: See Figures 4-19 through 4-21 for the graphs. The domains and ranges are:
(a) domain = {all real numbers} and range = {c}; **(b)** domain = {all real numbers} and range = {all real numbers}; **(c)** domain = {all real numbers} and range = {all nonnegative real numbers}.   [See Section 4-3.]

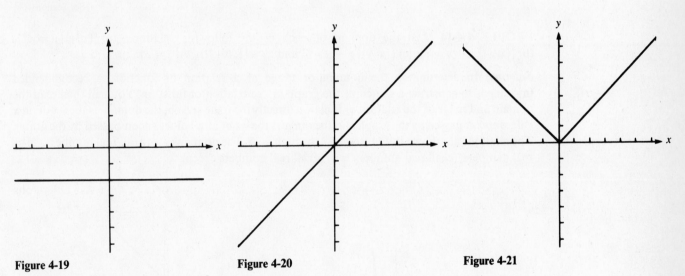

Figure 4-19   **Figure 4-20**   **Figure 4-21**

PROBLEM 4-11 You can say a function, f, is an **even function** if $f(-x) = f(x)$ for all x in the domain of f, and an **odd function** if $f(-x) = -f(x)$ for all x in the domain of f. Determine whether the following functions are even, odd, or neither even nor odd: **(a)** $f(x) = x^2 - 3x^4$; **(b)** $f(x) = x^3/(1 + x^2)$; **(c)** $f(x) = |x| + x$.

Solution:

(a) If you replace x by $-x$ in $f(x)$, then

$$f(-x) = (-x)^2 - 3(-x)^4 = x^2 - 3x^4 = f(x),$$

so this function is even.

(b) Replace x with $-x$:

$$f(-x) = \frac{(-x)^3}{1 + (-x)^2} = \frac{-x^3}{1 + x^2} = -f(x)$$

so this function is odd.

(c) Compute $f(-x) = |-x| + (-x) = |x| - x$. This is not equal to $f(x)$, nor is it equal to $f(-x)$, so this function is neither even nor odd.

PROBLEM 4-12 Tell which of the following ordered pairs are equal: **(a)** $(7, 2)$ and $(2, 7)$; **(b)** $(5, \frac{7}{2})$ and $(\frac{10}{2}, 5 + 1.5)$; **(c)** $(\frac{8}{2}, 5 - 3)$ and $(\frac{14}{2}, \frac{28}{7})$; **(d)** $(\sqrt{9}, \sqrt{49})$ and $(7, 3)$; **(e)** $(\sqrt{9}, -3)$ and $(3, -3)$.

Solution: Recall that $(x_1, y_1) = (x_2, y_2)$ if and only if $x_1 = x_2$ and $y_1 = y_2$.

(a) Let $x_1 = 7$ and $x_2 = 2$. Because $2 \neq 7$, these two ordered pairs are not equal.
(b) Let $x_1 = 5$ and $x_2 = \frac{10}{2} = 5$, while $y_1 = \frac{7}{2}$ and $y_2 = 6.5$. Because $\frac{7}{2} \neq 6.5$, $y_1 \neq y_2$, and the ordered pairs are not equal.
(c) Let $x_1 = \frac{8}{2} = 4$ and $x_2 = \frac{14}{2} = 7$. Because $4 \neq 7$, $x_1 \neq x_2$, and the ordered pairs are not equal.
(d) Let $x_1 = \sqrt{9} = 3$ and $x_2 = 7$. Because $3 \neq 7$, $x_1 \neq x_2$, and the ordered pairs are not equal.
(e) Let $x_1 = \sqrt{9} = 3$ and $x_2 = 3$, while $y_1 = -3$ and $y_2 = -3$. Because $x_1 = x_2$ and $y_1 = y_2$, the ordered pairs are equal. [See Section 4-1.]

PROBLEM 4-13 List two ordered pairs from each of the following relations: **(a)** $\{(x, y) \mid y = x^3\}$; **(b)** $\{(x, y) \mid y - x = 3\}$; **(c)** $\{(x, y) \mid y - x = 3 \text{ and } x > 1\}$; **(d)** $\{(x, y) \mid \frac{y}{x} = 3\}$.

Solution: You can find ordered pairs by selecting any value for x, substituting this value into the equation relating y and x, and solving for y. Then (x, y) is an ordered pair.

(a) In $y = x^3$, replace x by 0; $y = 0^3 = 0$. Then in $y = x^3$ replace x by 1; $y = 1^3 = 1$. You have found ordered pairs $(0, 0)$ and $(1, 1)$; **(b)** $(0, 3)$ and $(2, 5)$; **(c)** $(2, 5)$ and $(4, 7)$; **(d)** $(1, 3)$ and $(2, 6)$. [See Section 4-2.]

PROBLEM 4-14 List the domain of each of the following relations: **(a)** $\{(x, y) \mid y = x^3\}$; **(b)** $\{(x, y) \mid y - x = 3\}$; **(c)** $\{(x, y) \mid y - x = 3 \text{ and } x > 1\}$; **(d)** $\{(x, y) \mid \frac{y}{x} = 3\}$.

Solution: In determining the domain or range of a relation or function, a recommended first step is to construct a sketch of the graph of the relation or function. You can then read the domain and range of the relation or function directly from the sketch; the domain is the set of all x values encompassed by the graph and the range is the set of all y values encompassed by the graph. The relations are shown in Figures 4-22 through 4-25. The domains are **(a)** all real numbers; **(b)** all real numbers; **(c)** all real numbers > 1; **(d)** all real numbers except 0. [See Section 4-2.]

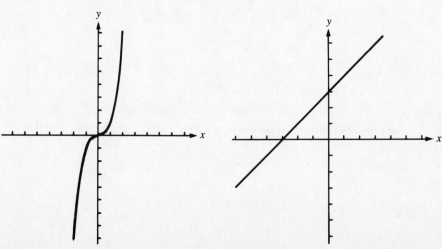

Figure 4-22   Figure 4-23

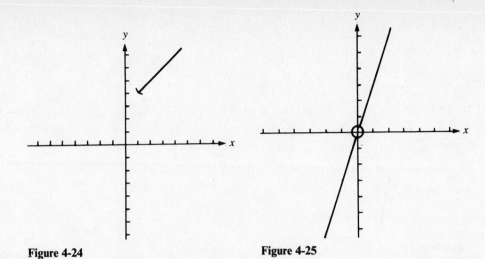

Figure 4-24   **Figure 4-25**

PROBLEM 4-15   List the range of each relation in Problem 4-14.

Solution: From Figures 4-22 through 4-25, you can find the ranges:
(a) all real numbers; (b) all real numbers; (c) all real numbers > 4; (d) all real numbers except 0.

[See Section 4-2.]

PROBLEM 4-16   Which of the following relations are functions? For relations (d) through (g), is *y* a function of *x*?

(a) $\{(3,1), (4,2), (2,4), (1,3), (3,3), (5,1)\}$
(b) $\{(3,1), (4,2), (2,4), (1,3), (5,3), (6,1)\}$
(c) $\{(1,1), (2,\frac{1}{2}), (3,\frac{1}{4}), (4,\frac{1}{8}), (5,\frac{1}{16}), \dots\}$
(d) $\{(x, y) \mid y - x = 3\}$

(e) $\{(x, y) \mid x - y = 3\}$
(f) $\{(x, y) \mid xy = 3\}$
(g) $\{(x, y) \mid x/y = 3\}$

Solution: A relation is a function if for each first component of an ordered pair there corresponds a unique second component.

(a) Since $(3,1)$ and $(3,3)$ show first component 3 paired with second components 1 and 3, the relation is not a function.
(b) Each first component corresponds to a unique second component, so the relation is a function.
(c) Each first component corresponds to a unique second component, so the relation is a function.

In problems where an algebraic expression connects *y* and *x*, you should first solve for *y* in terms of *x* and make a sketch of the graph. Imagine a series of lines parallel to the *y* axis through each *x* value: *y* is a function of *x* if, for each vertical line, there is no more than one point of intersection of the vertical line and the graph. See Figures 4-26 through 4-29 for parts (d) through (g). For relations

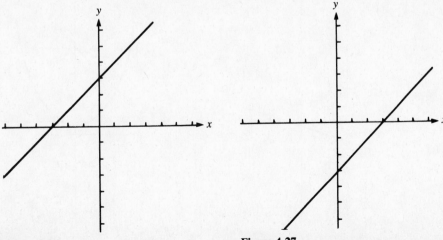

Figure 4-26   **Figure 4-27**

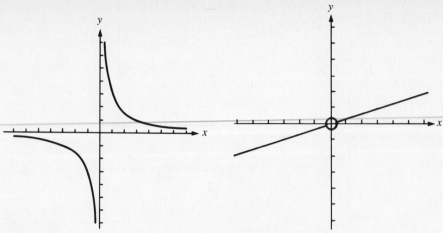

Figure 4-28   **Figure 4-29**

(**d**) through (**g**), you can see that any vertical line intersects the graph in a single point, so (**d**) through (**g**) are functions.

[See Section 4-3.]

PROBLEM 4-17   Tell which of the following functions are one-to-one. In (**c**) through (**e**), consider *y* a function of *x*.

(**a**) $\{(3, 1), (4, 2), (2, 4), (1, 3), (5, 3), (6, 1)\}$

(**b**) $\{(1, 1), (2, \frac{1}{2}), (3, \frac{1}{4}), (4, \frac{1}{8}), (5, \frac{1}{16}), \ldots\}$

(**c**) $\{(x, y)\,|\,2x + y = 4\}$

(**d**) $\{(x, y)\,|\,x - 3y = 1\}$

(**e**) $\{(x, y)\,|\,x^2 - 3y = 1\}$

Solution: A function is one-to-one if for each *y* value (second component of an ordered pair) there corresponds a unique *x* value (first component of an ordered pair). In problems like (**a**) and (**b**) in which the ordered pairs are shown, simply apply the definition just given. In problems like (**c**) through (**e**) first make a sketch of the graph of the function and then apply the horizontal line test.

(**a**) This function is not one-to-one because second component 3 corresponds to first components 1 and 5.

(**b**) This function is one-to-one since to each second component there corresponds a unique first component

(**c**) From the graph in Figure 4-30, you can see that any horizontal line intersects the graph in at most one point, so the function is one-to-one.

(**d**) From the graph in Figure 4-31, you can see that any horizontal line intersects the graph in at most one point, so the function is one-to-one.

(**e**) From the graph in Figure 4-32, you can see that any horizontal line above the *x* axis intersects the graph at two points, so the function is not one-to-one.

[See Section 4-3.]

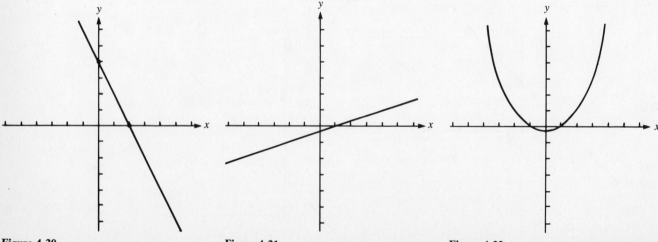

Figure 4-30   **Figure 4-31**   **Figure 4-32**

PROBLEM 4-18 Figure 4-33 contains diagrams that show pairings between elements in the domain (on the left) and elements in the range (on the right) for three relations. Which of these diagrams are diagrams of functions? Are any of the functions one-to-one?

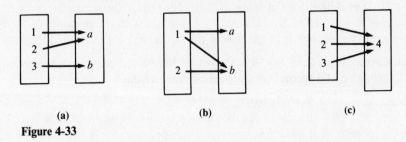

(a)   (b)   (c)

Figure 4-33

Solution:

(a) Since the arrow from each element in the domain goes to a single element in the range, this is a function. It is not one-to-one because *a* is paired with domain elements 1 and 2.

(b) This is not a function since 1 in the domain is paired with both *a* and *b*.

(c) This is a function since each element in the range is paired with a unique element in the domain. It is not one-to-one because 4 in the range is paired with more than one domain value.

[See Section 4-3.]

PROBLEM 4-19 Use the vertical line test to determine which of the relations in Figure 4-34 are functions of the form $y = f(x)$.

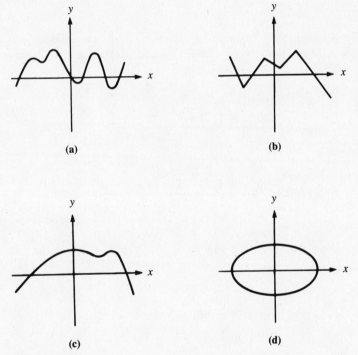

Figure 4-34

Solution: Since any vertical line drawn on the graphs of **(a)**, **(b)**, or **(c)** intersects the graph in at most one point, each of these relations are functions. Sketch **(d)** is not the graph of a function.

[See Section 4-3.]

PROBLEM 4-20 Given $f(x) = 2x + 1$, and $g(x) = x + 7$, where x is any real number, find the following functions: **(a)** $g(f(x))$; **(b)** $f(g(x))$; **(c)** $f(f(x))$; **(d)** $g(g(x))$.

Solution:

    (a) $g(f(x)) = g(2x + 1) = (2x + 1) + 7 = 2x + 8$

    (b) $f(g(x)) = f(x + 7) = 2(x + 7) + 1 = 2x + 15$

(c) $f(f(x)) = f(2x + 1) = 2(2x + 1) + 1 = 4x + 3$

(d) $g(g(x)) = g(x + 7) = (x + 7) + 7 = x + 14$ [See Section 4-3.]

PROBLEM 4-21   Let B be the set of integers 1 through 6 that appear on the faces of a die. Let $R = \{(x, y) | (x, y) \text{ is a member of } B \times B \text{ and } x + y = 7\}$. (a) List all elements of R. (b) Write a sentence relating the elements of R to the toss of the die. (c) In part (a), is y a function of x?

Solution: The Cartesian product $B \times B$ is the set of 36 ordered pairs with first component an integer from 1 through 6 and second component an integer from 1 through 6.

(a) $R = \{(1, 6), (2, 5), (3, 4), (4, 3), (5, 2), (6, 1)\}$.

(b) R is a representation for all rolls of the dice in which the sum on the dice is 7.

(c) Each first component of R has a unique second component, so y is a function of x.

[See Section 4-3.]

PROBLEM 4-22   Let B be the set of six numbers on the faces of a die and let R be all rolls of the die in which the sum is 2, 3, or 12. (a) List all elements of R; (b) Write the elements of part (a) in set notation; (c) If x denotes a first component of an ordered pair and y denotes a second component of an ordered pair in part (a), is y a function of x?

Solution:

(a) $\{(1, 1), (1, 2), (2, 1), (6, 6)\}$;

(b) $R = \{(x, y) | (x, y) \text{ is a member of } B \times B \text{ and } x + y = 2 \text{ or } x + y = 3 \text{ or } x + y = 12\}$;

(c) Since first component 1 has second components 1 and 2, y is not a function of x.

[See Section 4-3.]

PROBLEM 4-23   A function $f(x)$ and a function $g(x)$ are **inverses** of each other if $f(g(x)) = g(f(x)) = x$. Show that $f(x) = (2x + 1)/3$ and $g(x) = (3x - 1)/2$ are inverses of each other.

Solution:

$$f(g(x)) = \frac{2((3x - 1)/2) + 1}{3} = \frac{3x - 1 + 1}{3} = \frac{3x}{3} = x$$

and

$$g(f(x)) = \frac{3((2x + 1)/3) - 1}{2} = \frac{2x + 1 - 1}{2} = \frac{2x}{2} = x$$

[See Section 4-3 and Problem 4-20.]

PROBLEM 4-24   Find $f(0)$, $f(x + h)$, and $f(x + h) - f(x)$ for each of the following functions: (a) $f(x) = 2x - 3$; (b) $f(x) = x^2 + 1$.

Solution:

(a)
$$f(0) = 2(0) - 3 = -3$$
$$f(x + h) = 2(x + h) - 3 = 2x + 2h - 3$$
$$f(x + h) - f(x) = [2(x + h) - 3] - [2x - 3]$$
$$= (2x + 2h - 3) - (2x - 3) = 2h$$

(b)
$$f(0) = 0^2 + 1 = 1$$
$$f(x + h) = (x + h)^2 + 1 = x^2 + 2xh + h^2 + 1$$
$$f(x + h) - f(x) = x^2 + 2xh + h^2 + 1 - (x^2 + 1)$$
$$= 2xh + h^2$$

[See Section 4-3 and Problem 4-20.]

Supplementary Exercises

PROBLEM 4-25   Let $F = \{(-1, 3), (1, 5), (3, 7), (5, 8), (7, 11)\}$. Find (a) the domain of F; (b) the range of F; (c) $F(-1)$; (d) $F(1) + F(7)$; (e) the x value for which $F(x) = 7$.

PROBLEM 4-26   Let $G = \{(x, 2x)\}$ where x is any real number. Find (a) $G(-3)$; (b) $G(\sqrt{2}/2)$; (c) the x value for which $G(x) = -15$.

PROBLEM 4-27   Given that $H(t) = 1/(t + 1)$, where t is any real number except -1, find (a) $H(3)$; (b) $H(\frac{1}{2})$; (c) the t value for which $H(t) = 4$.

PROBLEM 4-28   Consider $\{(x, y) \mid x^2 + y^2 = 4; x \geq 0; y \leq 0\}$. Find the value (a) of y for which $x = 1$; (b) of y for which $x = -2$; (c) of x for which $y = -1$.

PROBLEM 4-29   What is the domain of $f(x) = 3\sqrt{x^2 - 4}$?

PROBLEM 4-30   What is the range of $f(x) = 3\sqrt{x^2 - 4}$?

PROBLEM 4-31   Given that $f(x) = x^2 - 1$, find (a) $f(2)$; (b) $f(x - 1)$; (c) $f(2x)$; (d) $f(x^2)$.

PROBLEM 4-32   Relation $y^2 = x$ defines two functions: $y = f(x) = +\sqrt{x}$ and $y = g(x) = -\sqrt{x}$. Find (a) the domain of f; (b) the range of f; (c) the domain of g; (d) the range of g.

PROBLEM 4-33   Let $g(u) = u - (1/u)$. Show that $g(u) + g(1/u) = 0$.

PROBLEM 4-34   Find an equation expressing the volume of a cube as a function of the length of an edge.

PROBLEM 4-35   Find an equation expressing the length of an edge of a cube as a function of the volume.

PROBLEM 4-36   Is the distance from a point in the Cartesian plane to the origin a function of the point? Explain.

PROBLEM 4-37   Given that f and g are functions and the range of f is contained in the domain of g, what is the domain of $g(f(x))$?

PROBLEM 4-38   What is the domain of $f(x) = x^2 + 7$?

PROBLEM 4-39   What is the range of $f(x) = x^2 + 7$?

PROBLEM 4-40   For the relation $A = \{(1, 1), (1, 2), (1, 3), (1, 4), (2, 2), (2, 3)\}$, find the domain and range of A.

PROBLEM 4-41   Let $A = \{$kilroy, was, here$\}$ and $B = \{$monkey, devour, bananas$\}$. (a) Write a relation R_1 with domain A and range B. (b) Write a relation R_2 with domain B and range A.

PROBLEM 4-42   Find the range of the absolute value function $f(x) = |x|$ for x any real number.

PROBLEM 4-43   For real numbers x and y, tell which of the following are functions: (a) $A = \{(x, y) \mid x \geq 0$ and $x^2 + y^2 = 25\}$; (b) $B = \{(x, y) \mid y \geq 0$ and $x^2 + y^2 = 25\}$; (c) $C = \{(x, y) \mid x \geq 0, y \geq 0,$ and $x^2 + y^2 = 25\}$; (d) $D = \{(x, y) \mid y < 0$ and $9x^2 + 4y^2 = 36\}$.

PROBLEM 4-44   Let $A = \{o, p\}$ and $B = \{q, r\}$. (a) Find the Cartesian product $B \times A$. (b) Write a relation defined on $B \times A$.

PROBLEM 4-45   Graph each relation and give its domain and range: (a) $R = \{(x, y) \mid x^2 + y^2 \leq 4$ and $x - y + 2 \geq 0$ and $y \geq x^2\}$; (b) $R = \{(x, y) \mid x \leq -y^2$ and $2x - y \geq 2\}$.

PROBLEM 4-46 Explain by means of examples why a relation may fail to be a function.

PROBLEM 4-47 Explain why the relation $R = \{(x, y)\,|\,x^2 + y^2 = r^2\}$ is not a function. *Hint:* Consider the graph.

PROBLEM 4-48 Given that $A = \{a, b, c\}$ and $B = \{1, 2\}$, write two functions with domain in A and range in B.

PROBLEM 4-49 For the following, graph each function and give its domain and range: (a) $f(x) =$ greatest integer less than or equal to x; (b) $f(x) = +\sqrt{x}$.

PROBLEM 4-50 Tell which functions are even, which are odd, and which are neither even nor odd: (a) $f(x) =$ greatest integer less than or equal to x; (b) $f(x) = (x + 1)^2$; (c) $f(x) = 2x - 3$.

PROBLEM 4-51 In which of the following is y a function of x: (a) $y = x^2 - x$; (b) $x = y^2 - y$; (c) $x + 2y = 3$; (d) $xy + 2 = 3$?

PROBLEM 4-52 For $R = \{(x, y)\,|\,x$ and y are any of $1, 2, 3, 4, 5,$ or $6,$ and $x + y$ is even$\}$ (a) list all elements of R; (b) write a sentence describing R in terms of rolling a pair of dice; (c) in R, can y be considered a function of x?

Answers to Supplementary Exercises

4-25: (a) $\{-1, 1, 3, 5, 7\}$; (b) $\{3, 5, 7, 8, 11\}$; (c) $F(-1) = 3$; (d) $F(1) + F(7) = 16$; (e) $x = 3$

4-26: (a) -6; (b) $\sqrt{2}$; (c) $x = -\frac{15}{2}$

4-27: (a) $\frac{1}{4}$; (b) $\frac{2}{3}$; (c) $t = -\frac{3}{4}$

4-28: (a) $-\sqrt{3}$; (b) Can't be found since x is specified as nonnegative; (c) $\sqrt{3}$

4-29: All real numbers ≥ 2 or ≤ -2

4-30: All real numbers ≥ 0

4-31: (a) 3; (b) $x^2 - 2x$; (c) $4x^2 - 1$; (d) $x^4 - 1$

4-32: (a) All real numbers ≥ 0; (b) All real numbers ≥ 0; (c) All real numbers ≥ 0; (d) All real numbers ≤ 0

4-33: $g(1/u) = (1/u) - \bigl(1/(1/u)\bigr) = (1/u) - u$ so $g(u) + g(1/u) = u - (1/u) + (1/u) - u = 0$

4-34: $V = f(e) = e^3$ where e is the length of an edge

4-35: $e = f(V) = V^{1/3}$

4-36: Yes; for each point in the Cartesian plane there exists one and only one distance to the origin.

4-37: The domain of f

4-38: All real numbers

4-39: All real numbers ≥ 7

4-40: Domain $= \{1, 2\}$ and range $= \{1, 2, 3, 4\}$

4-41: (a) $\{(\text{kilroy, monkey}), (\text{was, devour}), (\text{here, bananas})\}$; (b) $\{(\text{bananas, kilroy}), (\text{devour, here}), (\text{monkey, was})\}$; many other answers are possible for this problem.

4-42: The range is the set of all nonnegative real numbers.

4-43: (a) Not a function; (b) a function; (c) a function; (d) a function

4-44: (a) $B \times A = \{(q, o), (q, p), (r, o), (r, p)\}$; (b) $\{(q, o), (q, p), (r, o)\}$; many other answers are possible.

4-45: (a) Domain is all real numbers from -1 to $\sqrt{2}$ and the range is all real numbers from 0 to 2; (b) Domain and range are the empty set. The graphs are shown in Figures 4-35 and 4-36.

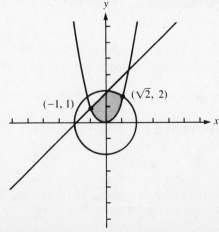

Figure 4-35

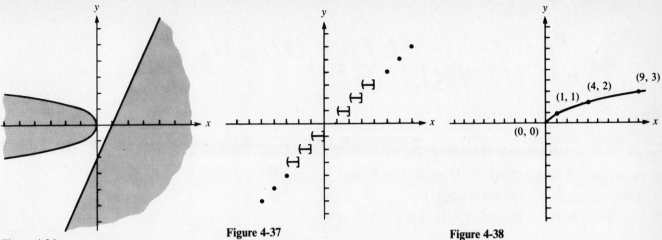

Figure 4-36

Figure 4-37

Figure 4-38

4-46: A relation will fail to be a function whenever two ordered pairs with the same second component have the same first component. Some examples are $\{(1, 1), (1, 2), (1, 3), (1, 4)\}$, $\{(x, y) \mid x < y$ with x and y any real numbers$\}$, and $\{(x, y) \mid x = |y|$ with y any real number$\}$.

4-47: The graph is a circle centered at the origin and with radius r. So any vertical line drawn within the circumference of the circle will intersect the circle at two points. This means that at the x value through which the vertical line is drawn there are two corresponding y values. This violates the definition of a function.

4-48: Two functions are $F_1 = \{(a, 1), (b, 1), (c, 1)\}$ and $F_2 = \{(a, 1), (b, 2), (c, 2)\}$. You may define other functions.

4-49: **(a)** Domain = all real numbers and range = $\{\ldots, -3, -2, -1, 0, 1, 2, 3, \ldots\}$;
(b) Domain = all real numbers $\geqslant 0$ and range = all real numbers $\geqslant 0$. See Figures 4-37 and 4-38 for graphs.

4-50: **(a)** Neither; **(b)** Neither; **(c)** Neither

4-51: **(a)** Yes; **(b)** No; **(c)** Yes; **(d)** Yes

4-52: **(a)** $\{(1, 1), (1, 3), (1, 5), (2, 2), (2, 4), (2, 6), (3, 1),$ $(3, 3), (3, 5), (4, 2), (4, 4), (4, 6), (5, 1), (5, 3), (5, 5), (6, 2),$ $(6, 4), (6, 6)\}$; **(b)** R is the set of faces when a pair of dice are rolled and the sum on the faces is an even number; **(c)** Since first component 1 is paired with second components 1, 3, and 5, you can't consider y a function of x.

5 TRIGONOMETRY OF THE RIGHT TRIANGLE

5-1. The Six Trigonometric Ratios for Right Triangles

Before you can investigate the trigonometric functions introduced in Chapter 6, you'll need some new terminology. Let's begin by looking at a specific kind of triangle.

A **right triangle** contains one 90° angle and two acute (less than 90°) angles. Look at the right triangle in Figure 5-1, where α is one of the acute angles. By definition, y is the length of the **side opposite α**, that is, the side that is neither the initial nor terminal side of α, x is the length of the **side adjacent to α**, that is, the initial side of α. The **hypotenuse** is the terminal side of α and its length is designated as r.

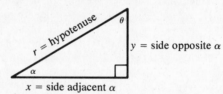

Figure 5-1
Right-triangle notation.

From the geometry of similar triangles, you know that the **ratios** of the lengths of the sides, x/r, y/r, y/x, and their reciprocals, r/x, r/y, x/y, are dependent on α only, not on the lengths of the sides. Each of these ratios has been given a specific name, defined in the next section.

A. The trigonometric ratios defined

The six trigonometric ratios are the **sine**, **cosine**, **tangent**, **cosecant**, **secant**, and **cotangent** of an angle α. They are defined as follows:

$$\sin \alpha = \frac{\text{side opposite } \alpha}{\text{hypotenuse}} = \frac{y}{r} \qquad \csc \alpha = \frac{\text{hypotenuse}}{\text{side opposite } \alpha} = \frac{r}{y}$$

$$\cos \alpha = \frac{\text{side adjacent } \alpha}{\text{hypotenuse}} = \frac{x}{r} \qquad \sec \alpha = \frac{\text{hypotenuse}}{\text{side adjacent } \alpha} = \frac{r}{x}$$

$$\tan \alpha = \frac{\text{side opposite } \alpha}{\text{side adjacent } \alpha} = \frac{y}{x} \qquad \cot \alpha = \frac{\text{side adjacent } \alpha}{\text{side opposite } \alpha} = \frac{x}{y}$$

These definitions apply only to the acute angles of a right triangle. You'll learn the formal definitions of the trigonometric functions, which apply to angles in general, in Chapter 6. Note that the cosecant, secant, and cotangent are the reciprocals of sine, cosine, and tangent, respectively.

EXAMPLE 5-1: Use the right triangle in Figure 5-2 to find the six trigonometric ratios of angle α and angle θ.

Solution: For α:

$$\sin \alpha = \frac{\text{side opposite } \alpha}{\text{hypotenuse}} = \frac{3}{5} \qquad \csc \alpha = \frac{\text{hypotenuse}}{\text{side opposite } \alpha} = \frac{5}{3}$$

$$\cos \alpha = \frac{\text{side adjacent } \alpha}{\text{hypotenuse}} = \frac{4}{5} \qquad \sec \alpha = \frac{\text{hypotenuse}}{\text{side adjacent } \alpha} = \frac{5}{4}$$

$$\tan \alpha = \frac{\text{side opposite } \alpha}{\text{side adjacent } \alpha} = \frac{3}{4} \qquad \cot \alpha = \frac{\text{side adjacent } \alpha}{\text{side opposite } \alpha} = \frac{4}{3}$$

For θ:

$$\sin \theta = \frac{\text{side opposite } \theta}{\text{hypotenuse}} = \frac{4}{5} \qquad \csc \theta = \frac{\text{hypotenuse}}{\text{side opposite } \theta} = \frac{5}{4}$$

$$\cos \theta = \frac{\text{side adjacent } \theta}{\text{hypotenuse}} = \frac{3}{5} \qquad \sec \theta = \frac{\text{hypotenuse}}{\text{side adjacent } \theta} = \frac{5}{3}$$

$$\tan \theta = \frac{\text{side opposite } \theta}{\text{side adjacent } \theta} = \frac{4}{3} \qquad \cot \theta = \frac{\text{side adjacent } \theta}{\text{side opposite } \theta} = \frac{3}{4}$$

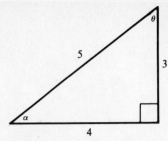

Figure 5-2

B. Numeric range of the trigonometric ratios

Because the hypotenuse of any right triangle is never shorter than either of the other two sides, values of the sine and cosine of an acute angle can never be greater than $+1$ nor less than -1. Similarly, the values of cosecant and secant, since they are reciprocals of sine and cosine, can never take values between -1 and $+1$. Because the sides of a right triangle, excluding the hypotenuse, can be in any ratio to each other, the values of tangent and cotangent can vary from arbitrarily large negative values to arbitrarily large positive values.

C. Evaluating trigonometric ratios

Given one trigonometric ratio, you can use the Pythagorean relationship to determine all the others.

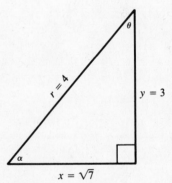

Figure 5-3

EXAMPLE 5-2: For a right triangle, $\sin \alpha = \frac{3}{4}$, find the other trigonometric ratios of α and construct a diagram.

Solution: Figure 5-3 illustrates the situation when $\sin \alpha = \frac{3}{4}$. Using the Pythagorean Theorem $(x^2 = r^2 - y^2)$, $x = \sqrt{4^2 - 3^2} = \sqrt{7}$. From the definitions of Section 5-1A, $\cos \alpha = \frac{\sqrt{7}}{4}$, $\tan \alpha = \frac{3\sqrt{7}}{7}$, $\cot \alpha = \frac{\sqrt{7}}{3}$, $\sec \alpha = \frac{4\sqrt{7}}{7}$, and $\csc \alpha = \frac{4}{3}$.

EXAMPLE 5-3: For a right triangle, $\sin \alpha = \frac{1}{2}$. Find $\csc \alpha$, $\sec \alpha$, and $\cot \alpha$.

Solution: Construct a right triangle with angle α, side opposite angle α of length 1, and the hypotenuse of length 2 (see Figure 5-4). Using the Pythagorean Theorem to find x, $x^2 + 1^2 = 2^2$. Simplifying, $x^2 + 1 = 4$, so $x^2 = 3$ and $x = \sqrt{3}$. Now $\csc \alpha = 1/\sin \alpha = \frac{2}{1} = 2$; $\sec \alpha = \text{hypotenuse/side adjacent} = \frac{2}{\sqrt{3}} = \frac{2\sqrt{3}}{3}$, $\cot \alpha = \text{side adjacent/side opposite} = \frac{\sqrt{3}}{1} = \sqrt{3}$.

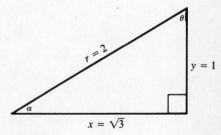

Figure 5-4

D. Complementary angles

In the right triangle of Figure 5-5, A and B are the acute angles. Since the sum of the angles of a triangle is π radians or $180°$, $A + B + 90° = 180°$ (in radians, $A + B + \frac{1}{2}\pi = \pi$, $A + B = \frac{1}{2}\pi$).

The two acute angles in a right triangle with a total measure of $90°$ are **complementary angles**. The sine of an angle equals the cosine of its complementary angle. You can prove that this relationship holds for all of the **co-ratios** of complementary angles (sine and *co*sine, tangent and *co*tangent, secant and *co*secant) by using the definitions of Section 5-1A. If $A + B = \frac{1}{2}\pi$, then:

$$\sin A = \frac{y}{r} = \cos B \qquad \csc A = \frac{r}{y} = \sec B$$

$$\cos A = \frac{x}{r} = \sin B \qquad \sec A = \frac{r}{x} = \csc B$$

$$\tan A = \frac{y}{x} = \cot B \qquad \cot A = \frac{y}{x} = \tan B$$

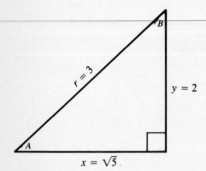

Figure 5-5

EXAMPLE 5-4: Given that $\sin A = \frac{2}{3}$ for the right triangle in Figure 5-5, find all six trigonometric ratios of A and B.

Solution: From the Pythagorean Theorem, $x^2 = r^2 - y^2 = 9 - 4 = 5$, so $x = \sqrt{5}$. Therefore,

$$\sin A = \frac{2}{3} = \cos B \qquad \csc A = \frac{3}{2} = \sec B$$

$$\cos A = \frac{\sqrt{5}}{3} = \sin B \qquad \sec A = \frac{3\sqrt{5}}{5} = \csc B$$

$$\tan A = \frac{2\sqrt{5}}{5} = \cot B \qquad \cot A = \frac{\sqrt{5}}{2} = \tan B$$

EXAMPLE 5-5: Given the following ratios, find their respective co-ratios: **(a)** $\sin 23°$; **(b)** $\cos 45°$; **(c)** $\sec \frac{1}{6}\pi$; **(d)** $\cot 40°$; **(e)** $\tan 63°$; **(f)** $\csc \frac{1}{3}\pi$.

Solution: Using ratio α = co-ratio $(90° - \alpha)$ and ratio α = co-ratio $(\frac{1}{2}\pi - \alpha)$, you get

(a) $\sin 23° = \cos(90 - 23)° = \cos 67°$

(b) $\cos 45° = \sin(90 - 45)° = \sin 45°$

(c) $\sec \dfrac{\pi}{6} = \csc\left(\dfrac{\pi}{2} - \dfrac{\pi}{6}\right) = \csc \dfrac{\pi}{3}$

(d) $\cot 40° = \tan(90 - 40)° = \tan 50°$

(e) $\tan 63° = \cot(90 - 63)° = \cot 27°$

(f) $\csc \dfrac{\pi}{3} = \sec\left(\dfrac{\pi}{2} - \dfrac{\pi}{3}\right) = \sec \dfrac{\pi}{6}$

EXAMPLE 5-6: Given that A is an acute angle with $\tan(A + 42°) = \cot A$, find A.

Solution: If $\tan(A + 42)° = \cot A$, then $(A + 42)°$ and A must be complementary angles. In other words, $A + 42° + A = 90°$, or $2A = (90 - 42)° = 48°$. Therefore, $A = 24°$.

E. Values for selected quadrantal and acute angles

Consider a series of right triangles where angle α gets smaller and smaller. When α is very close to zero, the length of the side adjacent to α is very close to the length of the hypotenuse. The side opposite α is close to zero. If we could have a triangle where $\alpha = 0$, it seems reasonable to assume that y would equal 0 and x would equal r. Formal proof of this statement will not be given here. When $\alpha = 0$, you may assume that $x = r$ and $y = 0$ so that when you evaluate the six ratios, you get

$$\sin 0 = \frac{y}{r} = \frac{0}{r} = 0 \qquad \csc 0 = \frac{r}{y} = \frac{r}{0}, \text{ which is undefined}$$

$$\cos 0 = \frac{x}{r} = \frac{r}{r} = 1 \qquad \sec 0 = \frac{r}{x} = \frac{r}{r} = 1$$

$$\tan 0 = \frac{y}{x} = \frac{0}{r} = 0 \qquad \cot 0 = \frac{x}{y} = \frac{r}{0}, \text{ which is undefined}$$

Similarly, if we could have a triangle where $\alpha = \frac{1}{2}\pi$ radians, it's reasonable to assume that y would equal r and x would equal 0. Substituting $\alpha = \frac{1}{2}\pi$, $x = 0$, and $y = r$ into the definitions of the six ratios gives

$$\sin\frac{\pi}{2} = \frac{y}{r} = \frac{r}{r} = 1 \qquad \csc\frac{\pi}{2} = \frac{r}{y} = \frac{r}{r} = 1$$

$$\cos\frac{\pi}{2} = \frac{x}{r} = \frac{0}{r} = 0 \qquad \sec\frac{\pi}{2} = \frac{r}{x} = \frac{r}{0}, \text{ which is undefined}$$

$$\tan\frac{\pi}{2} = \frac{y}{x} = \frac{r}{0}, \text{ which is undefined} \qquad \cot\frac{\pi}{2} = \frac{x}{y} = \frac{0}{r} = 0$$

You could have derived the ratios for $\frac{1}{2}\pi$ by using the co-ratio relationships of Section 5-1D. For example, $\sin\frac{1}{2}\pi = \cos(\frac{1}{2}\pi - \frac{1}{2}\pi) = \cos 0 = 1$. Prove to yourself that the other entries in the last display can be determined in this manner.

You can use the axioms of geometry to determine the trigonometric ratios for angles of $\frac{1}{6}\pi$ and $\frac{1}{3}\pi$ radians (30° and 60°). Consider the equilateral triangle with sides of length 2, shown in Figure 5-6. Bisecting any angle produces two right triangles with acute angles of $\frac{1}{6}\pi$ and $\frac{1}{3}\pi$ radians (30 and 60 degrees). From the Pythagorean Theorem, the length of the bisector equals $\sqrt{2^2 - 1^2}$ or $\sqrt{3}$. Using either right triangle, you determine that the trigonometric ratios for $\frac{1}{6}\pi$ radians are

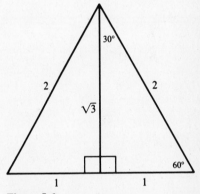

Figure 5-6
'30-60-90' triangle.

$$\sin\frac{\pi}{6} = \frac{1}{2} \qquad \csc\frac{\pi}{6} = \frac{2}{1}$$

$$\cos\frac{\pi}{6} = \frac{\sqrt{3}}{2} \qquad \sec\frac{\pi}{6} = \frac{2\sqrt{3}}{3}$$

$$\tan\frac{\pi}{6} = \frac{\sqrt{3}}{3} \qquad \cot\frac{\pi}{6} = \frac{\sqrt{3}}{1}$$

Since angles of $\frac{1}{6}\pi$ and $\frac{1}{3}\pi$ radians are complementary, you use the co-ratio relationships to determine that:

$$\sin\frac{\pi}{3} = \cos\frac{\pi}{6} = \frac{\sqrt{3}}{2} \qquad \csc\frac{\pi}{3} = \sec\frac{\pi}{6} = \frac{2\sqrt{3}}{3}$$

$$\cos\frac{\pi}{3} = \sin\frac{\pi}{6} = \frac{1}{2} \qquad \sec\frac{\pi}{3} = \csc\frac{\pi}{6} = \frac{2}{1}$$

$$\tan\frac{\pi}{3} = \cot\frac{\pi}{6} = \sqrt{3} \qquad \cot\frac{\pi}{3} = \tan\frac{\pi}{6} = \frac{\sqrt{3}}{3}$$

Consider the square with sides of length 1 in Figure 5-7. If you join opposite vertices of the square with a diagonal line, you produce two right triangles where each acute angle equals $\frac{1}{4}\pi$ radians (45°). The hypotenuse

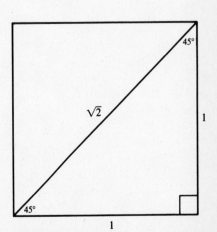

Figure 5-7
'45-45-90' triangle.

of these right triangles is equal to $\sqrt{1^2 + 1^2} = \sqrt{2}$. Since $\frac{1}{4}\pi + \frac{1}{4}\pi = \frac{1}{2}\pi$, the angles are complementary. Using the co-ratio relationships, you can determine that:

$$\sin\frac{\pi}{4} = \cos\frac{\pi}{4} = \frac{1}{\sqrt{2}} = \frac{\sqrt{2}}{2}$$

$$\tan\frac{\pi}{4} = \cot\frac{\pi}{4} = \frac{1}{1} = 1$$

$$\csc\frac{\pi}{4} = \sec\frac{\pi}{4} = \frac{\sqrt{2}}{1} = \sqrt{2}$$

The results of this section are summarized in Table 5-1.

TABLE 5-1: Values of the trigonometric ratios for selected quadrantal and acute angles.

α (radians)	α (degrees)	$\sin\alpha$	$\cos\alpha$	$\tan\alpha$	$\csc\alpha$	$\sec\alpha$	$\cot\alpha$
0	0	0	1	0	undefined	1	undefined
$\frac{\pi}{6}$	30	$\frac{1}{2}$	$\frac{\sqrt{3}}{2}$	$\frac{\sqrt{3}}{3}$	2	$\frac{2\sqrt{3}}{3}$	$\sqrt{3}$
$\frac{\pi}{4}$	45	$\frac{\sqrt{2}}{2}$	$\frac{\sqrt{2}}{2}$	1	$\sqrt{2}$	$\sqrt{2}$	1
$\frac{\pi}{3}$	60	$\frac{\sqrt{3}}{2}$	$\frac{1}{2}$	$\sqrt{3}$	$\frac{2\sqrt{3}}{3}$	2	$\frac{\sqrt{3}}{3}$
$\frac{\pi}{2}$	90	1	0	undefined	1	undefined	0

F. Using tables and calculators

Table 5-1 lists the trigonometric ratios for a few selected angles. You'll need to solve many problems that involve angles not in this table. You can consult a more extensive table (see your textbook for the procedures you'll need to use the trigonometric function tables) or you can use the $\boxed{\sin}$, $\boxed{\cos}$, or $\boxed{\tan}$ keys on your calculator.

Many tables and most calculators don't give values for $\csc\alpha$, $\sec\alpha$, or $\cot\alpha$—they can be found by finding $\sin\alpha$, $\cos\alpha$, or $\tan\alpha$ and taking the reciprocal of the result. Remember that the reciprocal of zero is undefined.

To find the angle α for which $\sin\alpha = m$, where m is a number between -1 and 1, use the $\boxed{\sin^{-1}}$ or $\boxed{\text{arc sin}}$ key on your calculator. The notation $\sin^{-1}$ (arc sin) is equivalent to **the angle whose sine is**. On some calculators, you can select either degrees or radians. To find α in degrees, put your calculator in 'degree mode,' enter m, the value of the trigonometric ratio, and $\boxed{\sin^{-1}}$, $\boxed{\text{arc sin}}$, or $\boxed{\text{INV}}$ $\boxed{\sin}$. Read the result on the display. To find α in radians, put your calculator in 'radian mode', enter m, press $\boxed{\sin^{-1}}$, $\boxed{\text{arc sin}}$, or $\boxed{\text{INV}}$ $\boxed{\sin}$, and read the result on the display. If your calculator does not have a radian mode, you'll need to convert radian measure to *decimal* degrees before you can begin. The $\boxed{\cos^{-1}}$ (arc cos) and $\boxed{\tan^{-1}}$ (arc tan) keys work in the same way as the $\boxed{\sin^{-1}}$ (arc sin) key. Methods for finding the values of the trigonometric ratios for general angles are discussed in Chapter 6, and $\sin^{-1}$, $\cos^{-1}$, and $\tan^{-1}$ are discussed in Chapter 12.

EXAMPLE 5-7: Find the acute angles, α (in degrees), for which: (a) $\sin\alpha = 0.237$; (b) $\tan\alpha = 0.237$; (c) $\tan\alpha = 2.9$; (d) $\cos\alpha = 0.987$; (e) $\csc\alpha = 21.1$.

Solution: First, fix your calculator in degree mode. Then:

(a) Enter 0.237, press $\boxed{\sin^{-1}}$ or $\boxed{\arcsin}$, read $\alpha = 13.71°$.
(b) Enter 0.237, press $\boxed{\tan^{-1}}$ or $\boxed{\arctan}$, read $\alpha = 13.33°$.
(c) Enter 2.9, press $\boxed{\tan^{-1}}$ or $\boxed{\arctan}$, read $\alpha = 70.97°$.
(d) Enter 0.987, press $\boxed{\cos^{-1}}$ or $\boxed{\arccos}$, read $\alpha = 9.25°$.
(e) Enter 21.1, press $\boxed{1/x}$ or $\boxed{x^{-1}}$, press $\boxed{\sin^{-1}}$ or $\boxed{\arcsin}$, read $\alpha = 2.72°$.

EXAMPLE 5-8: Find the acute angles, α (in radians), for which: (a) $\sin \alpha = 0.642$; (b) $\tan \alpha = 0.111$; (c) $\tan \alpha = 4.0$; (d) $\cos \alpha = 0.557$; (e) $\sec \alpha = 1.43$.

Solution: First, fix your calculator in radian mode. Then:

(a) Enter 0.642, press $\boxed{\sin^{-1}}$, read $\alpha = 0.697$ rad.
(b) Enter 0.111, press $\boxed{\tan^{-1}}$, read $\alpha = 0.111$ rad.
(c) Enter 4.0, press $\boxed{\tan^{-1}}$, read $\alpha = 1.326$ rad.
(d) Enter 0.557, press $\boxed{\cos^{-1}}$, read $\alpha = 0.980$ rad.
(e) Enter 1.43, press $\boxed{1/x}$ or $\boxed{x^{-1}}$, press $\boxed{\cos^{-1}}$, read $\alpha = 0.796$ rad.

5-2. How to Solve Right Triangles

You solve right triangles by determining the two acute angles and the lengths of the three sides of the triangle. You can do this given either the length of one side and the measure of one angle, or the lengths of two sides. A trigonometric ratio of an acute angle involves three quantities: the lengths of two sides and an angle. Thus, given two of these quantities, you can find the third.

A. Given the lengths of two sides

Given the lengths of two sides of a right triangle, you use the Pythagorean Theorem to determine the length of the remaining side. You can then find any trigonometric ratio for either of the two unknown angles (for ease of calculation, choose sine or cosine). Consult a table or use your calculator to find the angle. Once you know one of the acute angles, you can find the other angle because the sum of the acute angles is 90°, or $\frac{1}{2}\pi$ radians.

EXAMPLE 5-9: Solve the triangle shown in Figure 5-8.

Solution: From the Pythagorean Theorem, $x^2 + 5^2 = 13^2$, so $x = 12$. Note that $\sin \alpha = \frac{5}{13}$; thus, $\alpha = \sin^{-1}\left(\frac{5}{13}\right) = 22.6°$. The expression $\sin^{-1}\left(\frac{5}{13}\right)$ means that you are to calculate the quotient $\frac{5}{13}$ and then press $\boxed{\sin^{-1}}$ or $\boxed{\text{INV}}$ $\boxed{\sin}$ on your calculator. From $\alpha + \theta + 90° = 180°$, you find that $\theta = 67.4°$. As a check, $\sin \theta = \frac{12}{13}$, so $\theta = \sin^{-1}\left(\frac{12}{13}\right) = 67.4°$.

B. Given one acute angle and the length of one side

Since the sum of the three angles is 180°, if you know one acute angle, then you can readily find the third angle. Once all the angles are known, choose a trigonometric ratio of one of the angles that involves the known side. The ratio will also involve an unknown side. You can then find the remaining side from the Pythagorean Theorem.

EXAMPLE 5-10: Solve the triangle of Figure 5-9.

Solution: The unknown angle is $(90 - 19)° = 71°$. Label the side opposite the 19° angle as y; then $\sin 19° = y/6$. Rearranging, $y = 6 \sin 19° = 6(0.3256) = 1.9536$. From the Pythagorean Theorem, $6^2 = x^2 + 1.9536^2$, so $x = 5.6730$.

EXAMPLE 5-11: Solve the triangle in Figure 5-10.

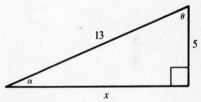

Figure 5-8

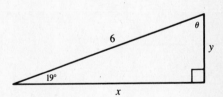

Figure 5-9

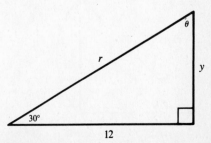

Figure 5-10

Solution: Since $\cos 30° = 12/r$, $r = 12/\cos 30° = 12/(\sqrt{3}/2) = 8\sqrt{3}$. To find y, $12^2 + y^2 = r^2$, so $12^2 + y^2 = (8\sqrt{3})^2$; $y = \sqrt{48} = 4\sqrt{3}$. Angle θ is $(90 - 30)° = 60°$.

EXAMPLE 5-12: For an acute angle, A, where $\sin A = \frac{1}{2}p$, draw two different right triangles that contain A and show that each yields the same values of the cosine and tangent ratios.

Solution: When $\sin A = \frac{1}{2}p$, you can see that the ratio of the side opposite A to the hypotenuse of the right triangle is $\frac{1}{2}p$. Any right triangle with this ratio is a solution. Two choices are shown in Figure 5-11.

For Figure 5-11(l):

$$b^2 = 4 - p^2$$

$$b = \sqrt{4 - p^2}$$

$$\cos A = \frac{b}{2} = \frac{\sqrt{4 - p^2}}{2}$$

$$\tan A = \frac{p}{b} = \frac{p}{\sqrt{4 - p^2}}$$

For Figure 5-11(r):

$$\frac{b^2}{4} = 1 - \frac{p^2}{4}$$

$$b = \sqrt{4 - p^2}$$

$$\cos A = \frac{b/2}{1} = \frac{\sqrt{4 - p^2}}{2}$$

$$\tan A = \frac{(p/2)}{(b/2)} = \frac{p}{b} = \frac{p}{\sqrt{4 - p^2}}$$

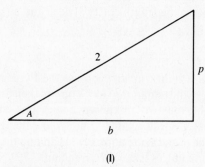

(l)   (r)

Figure 5-11
Triangles with a given sine ratio.

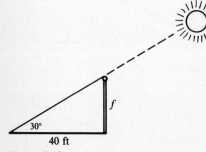

30°
40 ft
Figure 5-12

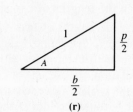

5-3. Applications of Right-Angle Trigonometry

The typical applications of right-triangle trigonometry fall into several general categories, including angles of depression and elevation, navigational bearings, and vectors. Vector problems are explained in Chapter 16. Let's examine some of the other applications.

EXAMPLE 5-13: When the angle between the ground and the sun is 30°, a flagpole casts a shadow of 40 feet long (see Figure 5-12). Find the height of the flagpole.

Solution: Let f be the height of the pole; then $\tan 30° = f/40$, or $f = 40 \tan 30° = 40(0.5773) = 23.092$ feet.

EXAMPLE 5-14: Find the length of a side of a square inscribed in a circle with diameter of 10 inches.

Solution: Figure 5-13 shows the situation. Since $\tan \alpha = x/x = 1$, $\alpha = 45°$. Then, $\cos 45° = x/10$ and $x = 10 \cos 45° = 10(0.707) = 7.07$ inches.

Figure 5-13

EXAMPLE 5-15: A vertical pole 20 feet high casts a shadow 12 feet in length. Find the **angle of elevation** of the sun (the angle between the horizon and the sun).

Solution: From Figure 5-14, you determine that $\tan \alpha = \frac{20}{12} = 1.6667$. With your calculator in degree mode, enter 1.6667 and press $\boxed{\text{INV}}$ $\boxed{\tan}$, $\boxed{\tan^{-1}}$, or $\boxed{\text{arc tan}}$; read $\alpha = 59.04°$. To work the problem without a calculator, consult a table of trigonometric functions. You should find that the angle whose tangent is closest to 1.6667 is 59°.

EXAMPLE 5-16: The top of a plateau is 100 feet above the floor of a valley. If the **angle of depression** (the angle measured from the horizontal down to the object) of a car moving toward the observer is 10°, how far is the car from the observer?

Solution: From the right triangle shown in Figure 5-15, $\cos 80° = 100/d$; $d = 100/\cos 80° = 575.9$ feet.

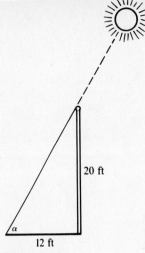

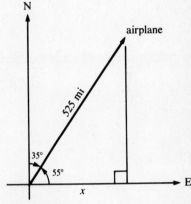

Figure 5-14

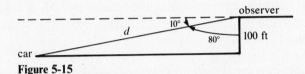

Figure 5-15

EXAMPLE 5-17: A **bearing** is the angle, measured in a clockwise direction, between due north and an object of interest. If an airplane is traveling at 525 miles per hour on a bearing of 35°, determine how far the airplane moves to the east in one hour.

Solution: The triangle shown in Figure 5-16 demonstrates the situation. In one hour, the airplane travels 525 miles at an angle 35° east of north. Solving for x, the easterly component, $\cos 55° = x/525$, so $x = 525 \cos 55° = 301.13$ miles.

Figure 5-16

SUMMARY

1. For a right triangle where y is the side opposite α, x is the side adjacent to α, and r is the hypotenuse,

$$\sin \alpha = \frac{y}{r} \qquad \csc \alpha = \frac{r}{y}$$

$$\cos \alpha = \frac{x}{r} \qquad \sec \alpha = \frac{r}{x}$$

$$\tan \alpha = \frac{y}{x} \qquad \cot \alpha = \frac{y}{x}$$

2. For an acute angle of a right triangle, the values of sine and cosine lie between -1 and $+1$, inclusive; the values of secant and cosecant are less than or equal to -1 or greater than or equal to $+1$; the values of tangent and cotangent are not restricted.
3. The trigonometric ratio of an acute angle α in a right triangle equals the co-ratio of the complementary angle.
4. The symbols $\sin^{-1}$ and arc sin mean "the angle whose sine is."
5. To solve a right triangle, find the measures of *all* angles and the lengths of *all* sides.
6. Construct a clearly labeled diagram as your first step in the solution of any problem.

7. Establish the relationship of the quantities in the diagram and convert these to equations containing trigonometric ratios.

8. Use tables or your calculator to complete the solution of the problem.

RAISE YOUR GRADES

Can you ... ?

☑ find all of the trigonometric ratios of an angle, given one of the ratios for that angle

☑ find any trigonometric ratio of the complement of an angle, given one of the ratios for that angle

☑ find all of the trigonometric ratios for angles of measure $0, \frac{1}{6}\pi, \frac{1}{4}\pi,$ $\frac{1}{3}\pi,$ or $\frac{1}{2}\pi$ radians, without using your calculator

☑ use your calculator to find any trigonometric ratio of a given acute angle

☑ use your calculator to determine the measure of an acute angle, given the value of its sine, cosine, tangent, cosecant, secant, or cotangent

☑ solve any right triangle

☑ apply your knowledge of the trigonometry of right triangles to solve problems involving bearings and angles of elevation and depression

SOLVED PROBLEMS

PROBLEM 5-1 Let A and B be acute angles such that $A + B = 90°$. If $\sin A = \frac{2}{5}$, find all six trigonometric ratios of B.

Solution: Since $\sin A = \frac{2}{5}$, and A and B are acute, draw the right triangle with side opposite angle A of length 2 and hypotenuse of length 5 as shown in Figure 5-17.

From the Pythagorean theorem, $b^2 = c^2 - a^2 = 25 - 4 = 21$ and $b = \sqrt{21}$. From the definitions of the trigonometric ratios for angle B, you determine that

$$\sin B = \frac{\sqrt{21}}{5} \qquad \csc B = \frac{5\sqrt{21}}{21}$$

$$\cos B = \frac{2}{5} \qquad \sec B = \frac{5}{2}$$

$$\tan B = \frac{\sqrt{21}}{2} \qquad \cot B = \frac{2\sqrt{21}}{21}$$

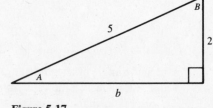

Figure 5-17

Note that the reciprocal relationships hold for the ratios in each column.   [See Section 5-1A.]

PROBLEM 5-2 Let A be one acute angle in a right triangle. Given $\tan A = \frac{3}{4}$, find all trigonometric ratios of A.

Solution: Draw the right triangle as shown in Figure 5-18.

From the Pythagorean Theorem, $r^2 = a^2 + b^2 = 9 + 16 = 25$ and $r = 5$. Therefore,

$$\sin A = \frac{3}{5} \qquad \csc A = \frac{5}{3}$$

$$\cos A = \frac{4}{5} \qquad \sec A = \frac{5}{4}$$

$$\tan A = \frac{3}{4} \qquad \cot A = \frac{4}{3} \qquad \text{[See Section 5-1A.]}$$

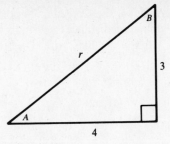

Figure 5-18

PROBLEM 5-3 The sides of the **isosceles triangle** (two sides have equal length) shown in Figure 5-19 measure 30, 45, and 45 units. Find the measures of the angles in degrees.

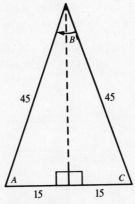

Figure 5-19

Solution: The dashed line in Figure 5-19 is the altitude, and it bisects the base; the two right triangles thus formed each have a hypotenuse of 45 units and base of $\frac{1}{2}(30) = 15$ units. Since $\cos A = \frac{15}{45} = \frac{1}{3} = \cos C$, $A = C = \cos^{-1}(\frac{1}{3}) = 70.5°$. Then, $B + 70.5 + 70.5 = 180$ or $B = 39°$.

[See Section 5-2.]

PROBLEM 5-4 Find the length of the chord subtended by a central angle of 110° in a circle with a radius of 24 units (see Figure 5-20).

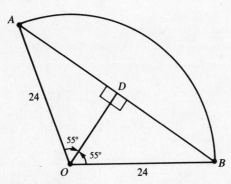

Figure 5-20

Solution: In triangle OAB, sides OA and OB are both radii of the circle and are 24 units in length. Altitude OD bisects the angle subtended by the chord, forming congruent right triangles OAD and OBD, each with $\theta = \frac{1}{2}(110) = 55°$. Then, $\sin 55° = AD/24$; $AD = 24 \sin 55° = 24(0.8192) = 19.6596$. The length of chord AB is $2(19.6596) = 39.3192$ units.   [See Section 5-2.]

PROBLEM 5-5 A butterfly sits 530 feet above ground level on the Washington Monument. If the angle of elevation from an observer to the butterfly is 20° 37′, how far is the observer from the base of the Monument?

Solution: First, sketch the positions of monument, observer, and butterfly, letting x represent the unknown distance (see Figure 5-21). Note that $\alpha = 20° \ 37' = 20.62°$ (see Chapter 3). By definition, $530/x = \tan 20.62°$ and $x = 530/\tan 20.62° = \frac{530}{0.3762} = 1408.8$ feet.   [See Section 5-2.]

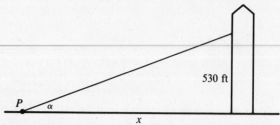

530 ft

P α

x

Figure 5-21

PROBLEM 5-6 If you walk 150 feet toward the base of a building, and the angle of elevation to the top of the building increases from $32° \ 20'$ to $45° \ 40'$, how tall is the building?

Solution: First, sketch the situation (see Figure 5-22). Let $A = 45° \ 40'$ and $B = 32° \ 20'$. You need to find an expression for the distance x to solve the problem. You can write two equations involving h, x, and the two angles of elevation:

$$(1) \ \frac{h}{x} = \tan 45° \ 40' \qquad (2) \ \frac{h}{150 + x} = \tan 32° \ 20'$$

Solving equation **(1)** for x, $x = h/\tan 45° \ 40'$.
Substitute this value of x into equation **(2)** and simplify:

$$h/[150 + (h/\tan 45° \ 40')] = \tan 32° \ 20'$$

$$h = 150 \tan 32° \ 20' + \frac{h \tan 32° \ 20'}{\tan 45° \ 40'}$$

$$= 150 \tan 32° \ 20' + h \frac{\tan 32° \ 20'}{\tan 45° \ 40'}$$

$$= 94.95 + 0.62h$$

$$0.38h = 94.95$$

$$h = 249.9 \text{ feet}$$ [See Section 5-3.]

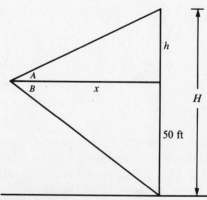

h

B A

|← 150 ft →|← x →|

Figure 5-22

PROBLEM 5-7 From a point 50 feet above level ground, the angles of elevation and depression of the top and bottom of a tower are $A = 25° \ 32'$ and $B = 36° \ 20'$, respectively. Find the height of the tower (see Figure 5-23).

h

A
B x

H

50 ft

Figure 5-23

Solution: The key distance is x. As in Problem 5-6, it isn't necessary to solve for x. First, $25° \ 32' = 25.53°$, and $36° \ 20' = 36.33°$. Notice that $\tan 25.53° = h/x$ and $\tan 36.33° = 50/x$, so

$h = x \tan 25.53°$ and $x = 50/\tan 36.33°$. Substituting the value of x into the first equation, $h = 50 \tan 25.53°/\tan 36.33° = \frac{23.88}{0.7354} = 32.47$ feet. Solving for the total height, $H = 32.47 + 50 = 82.47$ feet. [See Section 5-3.]

PROBLEM 5-8 A ski-jump ramp, 30 feet in length from where it leaves the water to the top, is inclined at an angle of 10°. If the tow rope from the boat to the skier is 60 feet long, how far is the boat from the base of the high end of the ramp when the skier is at the top of the ramp?

Solution: Figure 5-24 shows the situation. Since $\sin 10° = y/30$, $y = 30 \sin 10° = 30(0.174) = 5.22$ feet. You want to find x, the distance from the boat to the base of the high end of the ramp. Since $\sin \theta = y/60 = \frac{5.22}{60} = 0.0868$, $\theta = \sin^{-1} 0.0868 = 5.0°$. So, $x = y/\tan 5.0° = \frac{5.22}{0.0875} = 59.66$ feet. You can also find x by applying the Pythagorean Theorem to the triangle BAT. [See Section 5-2.]

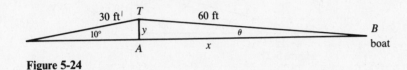

Figure 5-24

PROBLEM 5-9 An airplane flies directly above an observer at a speed of 400 miles/hour. One minute later, the observer must look up at an angle of 47 degrees to see the airplane. What is the altitude of the airplane?

Solution: Figure 5-25 illustrates the problem. Since the airplane is flying at 400 miles/hour, it travels $\frac{400}{60}$, or 6.67 miles in one minute. In the right triangle, $\theta = 43°$. Solving for y, the altitude of the plane, $\tan 43° = 6.67/y$ or $y = 6.67/\tan 43° = \frac{6.67}{0.9325} = 7.15$ miles. [See Section 5-3.]

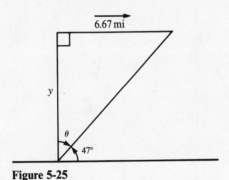

Figure 5-25

PROBLEM 5-10 To measure the width of a river without crossing, a surveyor stands on one bank and sights straight across to a tree on the opposite bank. The surveyor then paces off 100 yards along the bank and sights the same tree, measuring an angle of 50° from the bank to his line of sight to the tree (see Figure 5-26). How wide is the river?

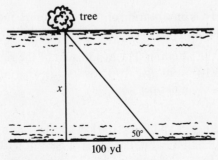

Figure 5-26

Solution: In the right triangle of Figure 5-26, x is the width of the river. Then $\tan 50° = x/100$ or $x = 100 \tan 50° = 100(1.1918) = 119.18$ yards. [See Section 5-2.]

PROBLEM 5-11   From the top of a cliff 150 feet above the level of the ocean, the angle of depression of a boat coming directly toward shore is 15°. After two minutes the angle of depression is 20°. What is the speed of the boat in miles per hour?

Solution: Figure 5-27 illustrates the relationships described in the problem. Initially, $\tan 75° = (x + y)/150$. At the second sighting, $\tan 70° = y/150$. Substituting $150 \tan 70°$ for y in the first equation, $x = 150(\tan 75° - \tan 70°) = 150(3.732 - 2.747) = 147.75$ feet in 2 minutes or $\frac{147.75}{2} = 73.875$ feet per minute. To convert this to miles per hour, multiply by the number of miles per foot $\left(\frac{1}{5280}\right)$, and multiply by the number of minutes per hour (60), to get speed in miles per hour: $73.875\left(\frac{1}{5280}\right)(60) = 0.84$ miles per hour.   [See Section 5-3.]

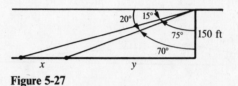

Figure 5-27

PROBLEM 5-12   A car travels on a bearing of 30° for 2 hours at 55 miles per hour. How far north and how far east does it travel?

Solution: Figure 5-28 demonstrates the situation. In 2 hours, the car will travel $2(55) = 110$ miles in the direction of travel. Since $\sin 30° = e/110$, $e = 110 \sin 30° = 110(\frac{1}{2}) = 55$ miles. Since $\cos 30° = n/110$, $n = 110 \cos 30° = 110(0.866) = 95.26$ miles.   [See Section 5-3.]

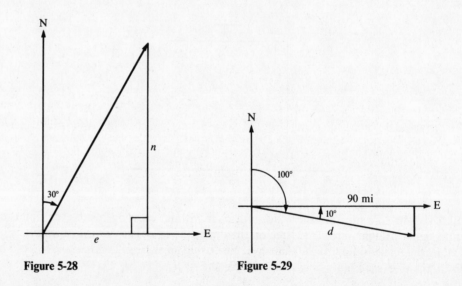

Figure 5-28   **Figure 5-29**

PROBLEM 5-13   If a car travels on a bearing of 100° for 3 hours and in that time moves 90 miles to the east of its initial position, what is the average speed of the car in miles per hour?

Solution: Figure 5-29 shows the situation. The hypotenuse of the right triangle, d, is the distance traveled in 3 hours. Since $\cos 10° = 90/d$, $d = 90/\cos 10° = \frac{90}{0.9848} = 91.39$ miles in 3 hours. The speed must then be $\frac{91.39}{3} = 30.46$ miles per hour.   [See Section 5-3.]

PROBLEM 5-14   Two buildings, A and B, are 100 feet apart. The angle of elevation from the top of building A to the top of building B is 20°. The angle of elevation from the base of building B to the top of building A is 50°. How tall is building B?

Solution: From the diagram in Figure 5-30, $\tan 50° = a/100$ and $\tan 20° = x/100$. Note that $a + x$ is the height of B; $a = 100 \tan 50°$, $x = 100 \tan 20°$, so $a + x = 100(\tan 50° + \tan 20°) = 100(1.192 + 0.364) = 100(1.556) = 155.6$ feet.   [See Section 5-3.]

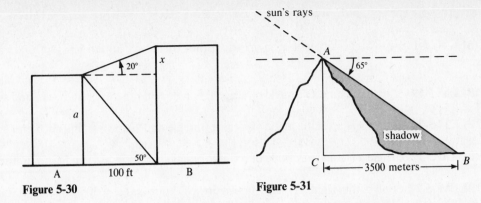

Figure 5-30

Figure 5-31

PROBLEM 5-15   Figure 5-31 shows a method for determining the height of a mountain on another planet. The angle of the sun's rays (angle BAC) can be calculated by astronomers and BC can be measured from the Earth. If angle BAC is $25°$ and BC is measured to be 3500 meters, how high is the mountain?

Solution: Since $\tan BAC = BC/AC$, $AC = BC/\tan BAC = 3500/\tan 25° = \frac{3500}{0.4663} = 7505.9$ meters.   [See Section 5-2.]

PROBLEM 5-16   Let A be an acute angle such that $\cos A = \sin 20°$. Find A.

Solution: Since $\cos A = \sin(90 - A)°$,   and $\cos A = \sin 20°$, $\sin(90 - A)° = \sin 20°$.   Then $(90 - A)° = 20°$, or $A = (90 - 20)° = 70°$.   [See Section 5-1F.]

PROBLEM 5-17   A fence 12 feet in height is 5 feet from a house. Find the length of the shortest ladder which will reach from the ground to the top of a window 20 feet off the ground, just touching the top of the fence.

Solution: Figure 5-32 shows the situation. Let l be the length of the shortest ladder satisfying the given conditions. You can formulate two equations involving the unknown quantities, α and x. Using the smaller right triangle, $\tan \alpha = 12/x$; from the larger right triangle, $\tan \alpha = 20/(x + 5)$. Therefore, $12/x = 20/(x + 5)$. Simplifying $12x + 60 = 20x$, $60 = 8x$, and $x = 7.5$ feet. From the Pythagorean Theorem, $l = \sqrt{(12.5)^2 + 20^2} = 23.58$ feet.   [See Section 5-3.]

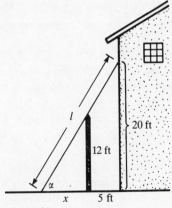

Figure 5-32

Supplementary Exercises

PROBLEM 5-18 Find the other trigonometric ratios of α if: **(a)** $\tan \alpha = \frac{3}{4}$; **(b)** $\cos \alpha = \frac{4}{5}$; **(c)** $\csc \alpha = \frac{3}{2}$.

PROBLEM 5-19 Angle α is an acute angle of a right triangle and $\tan \alpha = \frac{4}{3}$; find $\cos \alpha$ and $\sin \alpha$.

PROBLEM 5-20 Angle α is an acute angle of a right triangle and $\cos \alpha = \frac{2}{7}$; find $\cot \alpha$ and $\csc \alpha$.

PROBLEM 5-21 Angle α is an acute angle of a right triangle and $\csc \alpha = \frac{3}{2}$; find $\tan \alpha$ and $\sec \alpha$.

PROBLEM 5-22 Solve the right triangle in Figure 5-33 for: **(a)** $\alpha = 30°$ and $h = 8$ units; **(b)** $\theta = 45°$ and $b = 10$ units; **(c)** $\theta = 30°$ and $h = 5$ units.

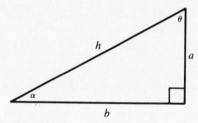

Figure 5-33

PROBLEM 5-23 Find the length of a side of a hexagon (six-sided polygon) inscribed in a circle with a radius of 5 inches.

PROBLEM 5-24 A taut kite string extends 300 feet from a child's hand to the kite. The kite string makes an angle of 30° with the horizon. What is the altitude of the kite (above the level of the child's hand)?

PROBLEM 5-25 Find the height of a building if the angle of elevation to the top of the building changes from 35° to 25° when the observer moves 50 feet farther away from the base of the building.

PROBLEM 5-26 A window washer is working on a building. An observer stands 500 feet from the base of the building. The angle of elevation from the observer to the worker is 30°. The worker climbs an additional 300 feet, and the observer moves 100 feet farther away. Find the new angle of elevation from the observer to the worker.

PROBLEM 5-27 A musician stands on the edge of an elevated stage. From a point $3\sqrt{3}$ meters from the base of the stage, the angle of elevation to the top of the musician's head is 60°, while the angle of elevation to his feet is 30°. Find both the height of the stage and the height of the musician.

PROBLEM 5-28 From the top of building A, the angle of elevation to the top of building B is 15°. From the base of building B, the angle of elevation to the top of building A is 65°. Building B is 140 feet tall. Find the distance between the two buildings.

PROBLEM 5-29 Find the perimeter and area of a seven-sided regular polygon: **(a)** inscribed in a circle with a radius of 6 units; **(b)** circumscribed about a circle with a radius of 6 units.

PROBLEM 5-30 Find the perimeter and area of an *n*-sided regular polygon: **(a)** inscribed in a circle with a radius of *r* units; **(b)** circumscribed about a circle with a radius of *r* units.

PROBLEM 5-31 The shadow cast by a policeman standing 28 feet from a street light is 14 feet. The light is 18 feet in height. Find the height of the policeman.

PROBLEM 5-32 Given an isosceles triangle having equal sides of length *l* and third side of length *b*, show that the area is $A = \frac{1}{4}b\sqrt{4l^2 - b^2}$ (see Figure 5-34).

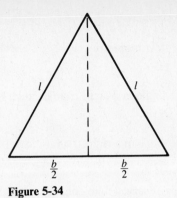

Figure 5-34

PROBLEM 5-33 Given a parallelogram having adjacent sides of lengths a and b with an included angle α, show that the area is $A = ab \sin \alpha$ (see Figure 5-35).

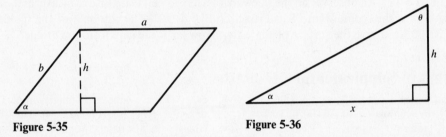

Figure 5-35   **Figure 5-36**

PROBLEM 5-34 Given the right triangle in Figure 5-36, show that the area is $A = \frac{1}{2}x^2 \tan \alpha$.

PROBLEM 5-35 Find the length of the base and the altitude of an isosceles triangle whose equal angles measure $68°$, and whose equal sides measure 42 units.

PROBLEM 5-36 The base of an isosceles triangle measures 15.8 units and the vertex angle (the angle opposite the base), is $67°$. Find the length of the two equal sides.

PROBLEM 5-37 From a point 150 feet from the base of a flagpole, the angle of elevation to the top of the flagpole is $29° 40'$. Find the height of the pole.

PROBLEM 5-38 The height of a building is 76 feet. Determine the angle of elevation of the sun when the shadow cast by the building measures 26 feet.

PROBLEM 5-39 A twenty-foot flagpole is located on the top edge of a building. From a point on the ground, the angle of elevation to the top of the building is $37° 35'$. The angle of elevation to the top of the flagpole is $5° 25'$ greater than that to the top of the building. Find the height of the building. The situation is shown in Figure 5-37.

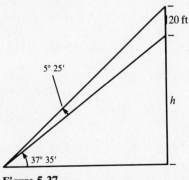

Figure 5-37

PROBLEM 5-40 Given that $A + B = 90°$ and $\cos B = \frac{3}{5}$, find all the trigonometric ratios of A.

PROBLEM 5-41   Let A be an acute angle in a right triangle such that $\sin A = \cos 65°$. Find A.

PROBLEM 5-42   Find all the trigonometric ratios of an acute angle B if the sine of the complementary angle, A, equals $\frac{7}{25}$.

PROBLEM 5-43 A is an acute angle in a right triangle such that $\sin(A + 30)° = \cos(A - 60)°$. Find A.

PROBLEM 5-44 A is an acute angle in a right triangle such that $\tan A = \frac{1}{5}$. Find the values for $\cos A$ and $\sin A$.

PROBLEM 5-45   An airplane is observed flying on a bearing of 50°. After one hour, the airplane has moved 450 miles to the east of its original point of observation. Find the speed of the airplane in miles per hour.

PROBLEM 5-46   An observer on the sidewalk of The Golden Gate Bridge measures the angle of depression of a ship coming into San Francisco Bay as 20°. The sidewalk of the Golden Gate Bridge is 300 feet above the surface of the water. How far is the ship from the bridge?

Answers to Supplementary Exercises

5-18: **(a)** $\sin \alpha = \frac{3}{5}$; $\cos \alpha = \frac{4}{5}$; $\csc \alpha = \frac{5}{3}$; $\sec \alpha = \frac{5}{4}$; $\cot \alpha = \frac{4}{3}$; **(b)** same answers as part (a); **(c)** $\sin \alpha = \frac{2}{3}$; $\cos \alpha = \frac{\sqrt{5}}{3}$; $\tan \alpha = \frac{2\sqrt{5}}{5}$; $\sec \alpha = \frac{3\sqrt{5}}{5}$; $\cot \alpha = \frac{\sqrt{5}}{2}$

5-19: $\sin \alpha = \frac{4}{5}$; $\cos \alpha = \frac{3}{5}$

5-20: $\cot \alpha = \frac{2\sqrt{5}}{15}$; $\csc \alpha = \frac{7\sqrt{5}}{15}$

5-21: $\tan \alpha = \frac{2\sqrt{5}}{5}$; $\sec \alpha = \frac{3\sqrt{5}}{5}$

5-22: **(a)** unknown acute angle = 60°; side $a = 4$ and $b = 4\sqrt{3}$ units; **(b)** unknown acute angle = 45°; unknown side = 10 units; hypotenuse = $10\sqrt{2}$ units; **(c)** unknown angle = 60°; side $a = 4.33$ and $b = 2.5$ units

5-23: 5 in.

5-24: 150 ft

5-25: 69.79 ft

5-26: 44.45°

5-27: 3 m and 6 m, respectively.

5-28: 58.03 ft

5-29: **(a)** perimeter = 36.45 units; area = 98.51 square units; **(b)** perimeter = 40.45 units; area = 121.36 square units

5-30: **(a)** perimeter = $2nr\sin(360/2n)$; area = $nr^2\sin(360/2n)\cos(360/2n)$; **(b)** perimeter = $2nr\tan(360/2n)$; area = $nr^2\tan(360/2n)$

5-31: 6 ft

5-32: From the Pythagorean Theorem, $h^2 = l^2 - (\frac{1}{2}b)^2$ (see Figure 5-34). The area of a triangle is $\frac{1}{2} \times$ base $\times$ height $= \frac{1}{2}b\sqrt{l^2 - (b/2)^2} = \frac{1}{4}b\sqrt{4l^2 - b^2}$

5-33: Line segment h is the height of the parallelogram (see Figure 5-35). Since $\sin \alpha = h/a$, $h = a\sin \alpha$. The area of a parallelogram is base $\times$ height $= ba\sin \alpha$.

5-34: Since $\tan \alpha = h/x$, $h = x\tan \alpha$ (see Figure 5-36). The area of a triangle is $\frac{1}{2} \times$ base $\times$ height $= \frac{1}{2}x \cdot x\tan \alpha = \frac{1}{2}x^2\tan \alpha$.

5-35: base = 31.5 units and altitude = 38.9 units

5-36: 14.3 units

5-37: 85.45 ft

5-38: 71° 7′

5-39: 94.5 ft

5-40: $\sin A = \frac{3}{5}$; $\cos A = \frac{4}{5}$; $\tan A = \frac{3}{4}$; $\csc A = \frac{5}{3}$; $\sec A = \frac{5}{4}$; $\cot A = \frac{4}{3}$

5-41: $A = 25°$

5-42: $\sin B = \frac{24}{25}$; $\cos B = \frac{7}{25}$; $\tan B = \frac{24}{7}$; $\csc B = \frac{25}{24}$; $\sec B = \frac{25}{7}$; $\cot B = \frac{7}{24}$

5-43: $A = 60°$

5-44: $\sin \alpha = \frac{\sqrt{26}}{26}$; $\cos \alpha = \frac{5\sqrt{26}}{26}$

5-45: 587.43 mi/h

5-46: 824.24 ft

6 GENERAL TRIGONOMETRY

THIS CHAPTER IS ABOUT

☑ **Trigonometric Ratios for General Angles**
☑ **The Six Trigonometric Functions**
☑ **Evaluating the Trigonometric Functions**

6-1. Trigonometric Ratios for General Angles

The definitions of the trigonometric ratios for a right triangle can be extended to any general angle in standard position. For angle α in standard position (see Chapter 3), if (x, y) is any point on the terminal side of α, excluding the origin, and $r = \sqrt{x^2 + y^2}$, then

$$\sin \alpha = \frac{y}{r} \qquad \csc \alpha = \frac{r}{y}$$

$$\cos \alpha = \frac{x}{r} \qquad \sec \alpha = \frac{r}{x} \qquad \textbf{(6-1)}$$

$$\tan \alpha = \frac{y}{x} \qquad \cot \alpha = \frac{x}{y}$$

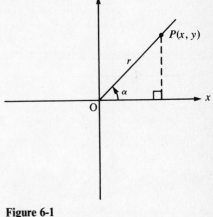

Figure 6-1
Notation for an angle in standard position.

If α is an angle of measure $0 \pm 2n\pi$, where n is any integer, the initial and terminal sides coincide, that is, they are both the positive x axis. The coordinates of any point on the terminal side of such an angle are $(x, 0)$. Then $\csc \alpha$ and $\cot \alpha$ are undefined. Similarly, if α is an angle of measure $\frac{1}{2}\pi \pm 2n\pi$ for n any integer, the terminal side will coincide with the positive y axis. The coordinates of any point on the terminal side of such an angle are $(0, y)$, so $\sec \alpha$ and $\tan \alpha$ are undefined.

You can form a right triangle from any angle in standard position. The line segment from the origin to (x, y) is the hypotenuse and the line segments from (x, y) to $(x, 0)$ and $(x, 0)$ to $(0, 0)$ are the sides. For this reason, you can consider Equations 6-1 as an extension of the trigonometric ratios for right triangles. These ratios are independent of the lengths of the sides of the triangle and depend on only the angle α. Note that the reciprocal relationships remain true for any angle where defined:

$$\csc \alpha = \frac{1}{\sin \alpha}$$

$$\sec \alpha = \frac{1}{\cos \alpha}$$

$$\cot \alpha = \frac{1}{\tan \alpha}$$

EXAMPLE 6-1: The point $P(-12, -5)$ is on the terminal side of α. Find $\sin \alpha$, $\cos \alpha$, and $\tan \alpha$.

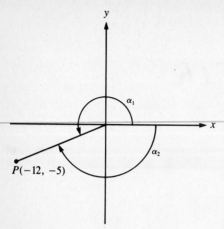

Figure 6-2

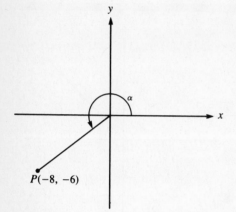

Figure 6-3

Solution: There are an infinite number of angles with P on their terminal sides. Two possible angles, positive angle α_1, and negative angle α_2, are shown in Figure 6-2. In Figure 6-2,

$$(x, y) = (-12, -5) \quad \text{and} \quad r = \sqrt{x^2 + y^2} = \sqrt{(-12)^2 + (-5)^2} = \sqrt{169} = 13.$$

Hence,

$$\sin \alpha = \frac{y}{r} = \frac{-5}{13}$$

$$\cos \alpha = \frac{x}{r} = \frac{-12}{13}$$

$$\tan \alpha = \frac{y}{x} = \frac{-5}{-12} = \frac{5}{12}$$

EXAMPLE 6-2: Find the values of the trigonometric functions given that $P(-8, -6)$ lies on the terminal side of α (see Figure 6-3).

Solution: Since $r = \sqrt{(-8)^2 + (-6)^2} = \sqrt{100} = 10$,

$$\sin \alpha = \frac{-6}{10} \qquad \csc \alpha = \frac{10}{-6}$$

$$\cos \alpha = \frac{-8}{10} \qquad \sec \alpha = \frac{10}{-8}$$

$$\tan \alpha = \frac{-6}{-8} \qquad \cot \alpha = \frac{-8}{-6}$$

6-2. The Six Trigonometric Functions

You will use the following definitions of the trigonometric functions in your subsequent work in trigonometry and in applications of trigonometry in the Calculus.

A. Defined for a general angle

The value of a **trigonometric function** for any real number x is equal to the corresponding trigonometric ratio of an angle of x radians. If the trigonometric ratio is undefined for a certain x, the trigonometric function is undefined for that number.

From Equations 6-1, you see that the domain of the sine and cosine functions is the set of all real numbers; the range is $[-1, 1]$. The domains and ranges of the other trigonometric functions are discussed in Chapter 8.

Note: The trigonometric functions ($\sin \alpha$, $\cos \alpha$, etc.) repeat their values for each increase or decrease of 2π in α, that is the trigonometric functions are the same for α and $\alpha \pm 2n\pi$, where n is any integer. This follows because α and $\alpha \pm 2n\pi$ have the same terminal side. Such behavior is called **periodicity** and will be treated in more detail in Chapter 8.

EXAMPLE 6-3: Find the sine, secant, and tangent of: **(a)** π; **(b)** 1.25; **(c)** 2; **(d)** −1.

Solution:

(a) Recall that angles given without the degree sign are in radians. An angle of π radians in standard position has its terminal side on the negative x axis. Choose one point on the terminal side; for convenience let's use $(-1, 0)$, so

$$r = \sqrt{x^2 + y^2} = \sqrt{(-1)^2 + 0^2} = 1$$

Then, from Equations 6-1,

$$\sin \pi = \frac{y}{r} = \frac{0}{1} = 0$$

$$\sec \pi = \frac{x}{r} = \frac{-1}{1} = -1$$

$$\tan \pi = \frac{y}{x} = \frac{0}{-1} = 0$$

(b) Place your calculator in radian mode. Enter $\boxed{1.25}$ $\boxed{\sin}$. You get sin 1.25 = 0.9490. Enter $\boxed{1.25}$ $\boxed{\cos}$ to get cos 1.25 = 0.3153. Then sec 1.25 = 1/cos 1.25 = $\frac{1}{0.3153}$ = 3.1714. Enter $\boxed{1.25}$ $\boxed{\tan}$ to get tan 1.25 = 3.0096. Now try solving part **(a)** again with this calculator method.

(c) With your calculator in radian mode, enter $\boxed{2.0}$ $\boxed{\sin}$. You get sin 2.0 = 0.9093. Enter $\boxed{2.0}$ $\boxed{\cos}$ to get cos 2.0 = −0.4161. Then sec 2.0 = 1/cos 2.0 = $\frac{1}{-0.4161}$ = −2.4030. Enter $\boxed{2.0}$ $\boxed{\tan}$ to get tan 2.0 = −2.1850.

(d) Keep your calculator in radian mode, enter $\boxed{1}$ $\boxed{+/-}$ $\boxed{\sin}$ to get sin(−1.0) = −0.8415. Enter $\boxed{1}$ $\boxed{+/-}$ $\boxed{\cos}$ to get cos(−1.0) = 0.5403. Then sec(−1.0) = 1/cos(−1.0) = $\frac{1}{0.5403}$ = 1.851. Enter $\boxed{1}$ $\boxed{+/-}$ $\boxed{\tan}$ to get tan(−1.0) = −1.5574.

B. Defined by the wrapping function

Let U be the **unit circle**, a circle of radius 1 centered at the origin of the Cartesian plane. Suppose that you start at point $P(1,0)$ on the unit circle and move along the circumference a distance t in the counterclockwise direction, for $t > 0$. Let (x, y) be the point at which you stop. For $t < 0$, move distance t in the clockwise direction; again call the ending point (x, y). In this manner, you can assign a unique point (x, y) on U to each real number t. This procedure is illustrated in Figure 6-4a.

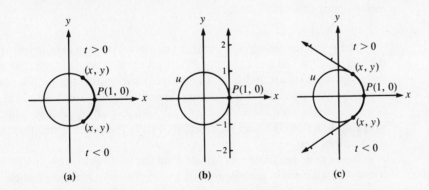

(a)   (b)   (c)

Figure 6-4
Distance t and related point (x, y) on the unit circle.

EXAMPLE 6-4: Find the point (x, y) on the unit circle U corresponding to:
(a) $t = \frac{1}{4}\pi$; **(b)** $t = \frac{1}{2}\pi$; **(c)** $t = -\pi$.

Solution:

(a) From the relationship $s = r\alpha$, with $r = 1$ (unit circle), $\frac{1}{4}\pi = 1\alpha$ or $\alpha = \frac{1}{4}\pi$. Recall that $\tan \frac{1}{4}\pi = 1$; from Equations 6-1, $\tan \alpha = y/x$, $1 = y/x$, and $x = y$. Since the unit circle has radius 1, $x^2 + y^2 = x^2 + x^2 = 1$, so $x = \frac{\sqrt{2}}{2}$. The point associated with $t = \frac{1}{4}\pi$ is $\left(\frac{\sqrt{2}}{2}, \frac{\sqrt{2}}{2}\right)$ [see Figure 6-5a].

(b) The point corresponding to $t = \frac{1}{2}\pi$ lies $\frac{1}{4}$ of the circumference away from $(1,0)$ in the counterclockwise direction; this point is $(0, 1)$. Using the method in part **(a)**, $\frac{1}{2}\pi = 1\alpha$ or $\alpha = \frac{1}{2}\pi$. Recall that $\sin \frac{1}{2}\pi = 1$ and $\sin \alpha = y/r$, so $y = 1$. Since $x^2 + y^2 = 1$, $x = 0$ (see Figure 6-5b).

(c) The point corresponding to $t = -\pi$ is $\frac{1}{2}$ of the circumference away from $(1,0)$ in the clockwise direction. It is $(-1,0)$, (see Figure 6-5c).

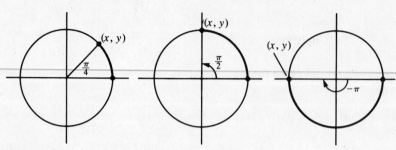

Figure 6-5

EXAMPLE 6-5: Find the point on the unit circle U corresponding to: (a) $t = \frac{3}{4}\pi$; (b) $t = \frac{3}{2}\pi$; (c) $t = -\frac{1}{2}\pi$.

Solution: Make a sketch of the unit circle to guide you through the solutions that follow.

(a) $\frac{3}{4}\pi = \frac{1}{2}\pi + \frac{1}{4}\pi$, so this point is $\frac{1}{4}\pi$ radians in the counterclockwise direction beyond the point corresponding to $t = \frac{1}{2}\pi$. This puts it halfway between the point corresponding to $t = \frac{1}{2}\pi$ and $t = \pi$; therefore, it must have the same coordinates (but different signs) as the point corresponding to $t = \frac{1}{4}\pi$. From the last example, the point associated with $t = \frac{1}{4}\pi$ is $\left(\frac{\sqrt{2}}{2}, \frac{\sqrt{2}}{2}\right)$. The point corresponding to $t = \frac{3}{4}\pi$ must have coordinates $\left(-\frac{\sqrt{2}}{2}, \frac{\sqrt{2}}{2}\right)$.

(b) Since the distance around the unit circle is 2π, $\frac{3}{2}\pi$ is three-quarters of the way around the circumference in the counterclockwise direction. Your sketch of the unit circle shows that the point corresponding to $t = \frac{3}{2}\pi$ is $(0, -1)$.

(c) This point is one-quarter of the way around the unit circle in the clockwise direction. It has coordinates $(0, -1)$. This example may also be solved by the technique of Example 6-4(a).

It'll help you to think of real number t as a point on the t axis. If you attach the origin of the t axis to the point $P(1,0)$ on the unit circle U, you can wrap the positive t axis around U in a counterclockwise direction and wrap the negative t axis around U in a clockwise direction (see Figures 6-4b and c). The correspondence established between real numbers and points is called the **wrapping function** (W), and we say that $W(t)$ is the point corresponding to real number t.

You can now designate every point P on the unit circle in two ways, using either the Cartesian coordinates (x, y) or the real number t such that $W(t) = P(x, y)$. You can also redefine the trigonometric functions as a result of the wrapping function:

$$\sin t = y \qquad \csc t = \frac{1}{y}$$

$$\cos t = x \qquad \sec t = \frac{1}{x} \qquad \qquad \textbf{(6-2)}$$

$$\tan t = \frac{y}{x} \qquad \cot t = \frac{x}{y}$$

Note: Cosecant and cotangent are not defined when $y = 0$; secant and tangent are not defined when $x = 0$.

Recall from Section 6-2A that the trigonometric functions repeat their values for each increase or decrease of 2π in α. Similarly,

$$W(t) = W(t + 2\pi n), \ n \text{ any integer}$$

so $W(t)$ is not a one-to-one function. Every point on the unit circle corresponds to infinitely many real numbers.

EXAMPLE 6-6: Let t be a real number such that $W(t) = \left(\frac{3}{5}, \frac{4}{5}\right)$. Find $\sin t$, $\cos t$, and $\tan t$.

Solution: From the definition of the wrapping function, $(x, y) = W(t) = \left(\frac{3}{5}, \frac{4}{5}\right)$. Hence,

$$\sin t = y = \frac{4}{5}$$

$$\cos t = x = \frac{3}{5}$$

$$\tan t = \frac{y}{x} = \frac{\frac{4}{5}}{\frac{3}{5}} = \frac{4}{3}$$

Since the wrapping function approach to trigonometric functions associates a point on the unit circle with each real number, the trigonometric functions are also called **circular functions**.

C. Signs of the trigonometric functions

Instead of committing to memory the sign of each trigonometric function in each quadrant, you'll find it easier to remember where the signs of the trigonometric functions are positive. Using the definitions in Equations 6-2, the positive signs of the trigonometric functions by quadrant are:

Quadrant II: $x < 0, y > 0$ Quadrant I: $x > 0, y > 0$

$\sin \alpha = \dfrac{y}{r} > 0$ $\sin \alpha = \dfrac{y}{r} > 0$

$\cos \alpha = \dfrac{x}{r} > 0$

$\tan \alpha = \dfrac{y}{x} > 0$

Quadrant III: $x < 0, y < 0$ Quadrant IV: $x > 0, y < 0$

$\tan \alpha = \dfrac{y}{x} > 0$ $\cos \alpha = \dfrac{x}{r} > 0$

You can determine the signs of cosecant, secant, and cotangent from the reciprocal relationships. For example, $\tan t > 0$ in quadrant III, so $\cot t$ is also positive there.

D. Values of the trigonometric functions for selected numbers

Table 6-1 lists values of the trigonometric functions for certain real numbers t that often occur in calculations. For convenience, these numbers are also designated as angles in degrees. Using the definitions of this chapter, you should be able to fill in all entries of this table and to determine the values for reciprocal functions: cosecant, secant, and cotangent.

TABLE 6-1: Values of the trigonometric functions for selected t.

t	Equivalent Degrees	$W(t)$	$\sin t$	$\cos t$	$\tan t$
0	0	$(1, 0)$	0	1	0
$\pi/6$	30	$(\sqrt{3}/2, 1/2)$	$1/2$	$\sqrt{3}/2$	$1/\sqrt{3}$
$\pi/4$	45	$(\sqrt{2}/2, \sqrt{2}/2)$	$\sqrt{2}/2$	$\sqrt{2}/2$	1
$\pi/3$	60	$(1/2, \sqrt{3}/2)$	$\sqrt{3}/2$	$1/2$	$\sqrt{3}$
$\pi/2$	90	$(0, 1)$	1	0	undef.
π	180	$(-1, 0)$	0	-1	0
$3\pi/2$	270	$(0, -1)$	-1	0	undef.
2π	360	$(1, 0)$	0	1	0

Because the values of the trigonometric functions given in Table 6-1 are used in many trigonometric calculations, you should know all of them; however, instead of memorizing the entire table, *memorize only the values of sine and cosine* for each entry in the table. You can determine the values of the other trigonometric functions from the values of sine and cosine.

You'll often want to find trigonometric functions of numbers greater than 2π, or 360°. The procedures for doing this are presented in Chapter 7. You'll need a more extensive table of trigonometric functions, or a calculator, to find trigonometric functions of numbers between 0 and 2π that are not listed in Table 6-1.

EXAMPLE 6-7: Find the sine and cosine of the following real numbers without using a calculator: **(a)** 0.5236; **(b)** 0.7845; **(c)** 7.3304; **(d)** 39.2699.

Solution: First reduce these numbers to multiples of π by division of each number by π; then use Table 6-1.

(a) Since $0.5236/\pi = 0.16667 = \frac{1}{6}$, $0.5236 = \frac{1}{6}\pi$. Then, $\sin 0.5236 = \sin\frac{1}{6}\pi = \frac{1}{2}$ and $\cos\frac{1}{6}\pi = 0.8660$.

(b) Since $0.7845/\pi = 0.2499 = \frac{1}{4}$, $0.7845 = \frac{1}{4}\pi$. Then, $\sin(\frac{1}{4}\pi) = 0.707$ and $\cos(\frac{1}{4}\pi) = \cos\frac{1}{4}\pi = 0.707$.

(c) Since $7.3304/\pi = 2.33333 = 2 + \frac{1}{3}$, $7.3304 = 2\pi + \frac{1}{3}\pi$. Thus, $\sin(2\pi + \frac{1}{3}\pi) = \sin\frac{1}{3}\pi = 0.8660$ and $\cos(2\pi + \frac{1}{3}\pi) = \cos\frac{1}{3}\pi = \frac{1}{2}$.

(d) Since $39.2699/\pi = 12.5 = 6(2) + \frac{1}{2}$, $39.2699 = 6(2\pi) + \frac{1}{2}\pi$. Then, $\sin(6 \times 2\pi + \frac{1}{2}\pi) = \sin\frac{1}{2}\pi = 1$ and $\cos(6 \times 2\pi + \frac{1}{2}\pi) = 0$.

You can also solve this example with a calculator. Place your calculator in radian mode, enter the number, and press the appropriate trigonometric function key.

6-3. Evaluating the Trigonometric Functions

You can use the relationships you've learned so far to find all of the trigonometric functions of α, given only one of them.

A. Using ratios

You learned techniques for finding all of the trigonometric ratios, given only one, in Section 5-1C. You can use the same methods for evaluating the trigonometric functions of α. You'll need one additional piece of information: the quadrant in which the terminal side of α lies.

1. Given sine, cosine, or their reciprocals: From Equations 6-1, you know y/r (if sine is given), x/r (if cosine is given), r/x (if secant is given), or r/y (if cosecant is given). Since the Pythagorean relationship can be rewritten as $(x/r)^2 + (y/r)^2 = 1$, you can solve for the unknown ratio. Complete the process by attaching the sign that corresponds to the quadrant defined by α. You can then use Equations 6-1 to find all of the trigonometric functions of α.

EXAMPLE 6-8: The terminal side of angle α lies in the first quadrant and $\sin \alpha = \frac{1}{2}$; find the values of the other trigonometric functions of α.

Solution: Since α is in the first quadrant, $y/r = \sin \alpha = \frac{1}{2}$. Then $(x/r)^2 + (\frac{1}{2})^2 = 1$ or $(x/r) = \frac{\sqrt{3}}{2}$. ($x/r$ is positive since the angle lies in the first quadrant). By Equations 6-1,

$$\cos \alpha = \frac{x}{r} = \frac{\sqrt{3}}{2}$$

$$\csc \alpha = \frac{r}{y} = \frac{1}{y/r} = \frac{1}{1/2} = 2$$

$$\sec \alpha = \frac{r}{x} = \frac{1}{x/r} = \frac{1}{\sqrt{3}/2} = \frac{2}{\sqrt{3}} = \frac{2\sqrt{3}}{3}$$

$$\cot \alpha = \frac{x}{y} = \frac{x/r}{y/r} = \frac{\sqrt{3}/2}{1/2} = \sqrt{3}$$

$$\tan \alpha = \frac{y}{x} = \frac{y/r}{x/r} = \frac{1/2}{\sqrt{3}/2} = \frac{\sqrt{3}}{3}$$

EXAMPLE 6-9: For $\cos \alpha = -\frac{2}{3}$ and the terminal side of angle α in the third quadrant, find $\csc \alpha$, $\sec \alpha$, and $\cot \alpha$.

Solution: $x/r = \cos \alpha = -2/3$. Since $(-2/3)^2 + (y/r)^2 = 1$, $y/r = -\sqrt{5}/3$. (y/r is negative since the terminal side of the angle lies in the third quadrant.) Using Equations 6-1,

$$\csc \alpha = \frac{r}{y} = \frac{1}{y/r} = \frac{1}{-\sqrt{5}/3} = \frac{3}{-\sqrt{5}} = \frac{-3\sqrt{5}}{5}$$

$$\sec \alpha = \frac{r}{x} = \frac{1}{r/x} = \frac{1}{-2/3} = -\frac{3}{2}$$

$$\cot \alpha = \frac{x}{y} = \frac{x/r}{y/r} = \frac{-2/3}{-\sqrt{5}/3} = \frac{2\sqrt{5}}{5}$$

EXAMPLE 6-10: For $\cos \alpha = -\frac{1}{3}$ and $\sin \alpha = \frac{\sqrt{8}}{3}$, find $\csc \alpha$, $\sec \alpha$, and $\cot \alpha$.

Solution: Since $x/r = -\frac{1}{3}$ and $y/r = \frac{\sqrt{8}}{3}$,

$$\csc \alpha = \frac{r}{y} = \frac{1}{y/r} = \frac{1}{\sqrt{8}/3} = \frac{3\sqrt{8}}{8}$$

$$\sec \alpha = \frac{r}{x} = \frac{1}{x/r} = \frac{1}{-1/3} = -3$$

$$\cot \alpha = \frac{x}{y} = \frac{x/r}{y/r} = \frac{-1/3}{\sqrt{8}/3} = \frac{-\sqrt{8}}{8}$$

You can also solve the last three examples by assuming that $r = 1$ and using Equations 6-2. Compute the solutions to this example using this approach.

EXAMPLE 6-11: If $\sin \alpha$ is positive and $\cot \alpha$ is negative, in which quadrant does α lie?

Solution: Since $\sin \alpha$ is positive in the first and second quadrants and $\cot \alpha$ is negative in the second and fourth quadrants, α must lie in the second quadrant.

2. **Given the value of tangent or cotangent:** From Equations 6-1, you know either y/x (if tangent is given) or x/y (if cotangent is given). If y/x is known, divide both sides of $x^2 + y^2 = r^2$ by x to get $1^2 + (y/x)^2 = (r/x)^2$. Then solve for r/x, attaching the sign according to the quadrant of the terminal line defined by α. If x/y is known, divide both sides of $x^2 + y^2 = r^2$ by y to get $(x/y)^2 + 1^2 = (r/y)^2$. Then solve for r/y, attaching the proper sign. You can then use Equations 6-1 to find all of the trigonometric functions of α.

EXAMPLE 6-12: Angle α is an angle in the fourth quadrant and $\tan \alpha = -\frac{4}{7}$; find $\sin \alpha$, $\cos \alpha$, and $\sec \alpha$.

Solution: Since $\tan \alpha = y/x = -\frac{4}{7}$, $1^2 + (y/x)^2 = (r/x)^2$ becomes $1^2 + \left(-\frac{4}{7}\right)^2 = (r/x)^2$. Thus, $r/x = \sqrt{1 + \frac{16}{49}} = \frac{\sqrt{65}}{7}$ (positive, since x is in quadrant IV). Then,

$$\sin \alpha = \frac{y}{r} = \frac{y/x}{r/x} = \frac{-4/7}{\sqrt{65}/7} = \frac{-4\sqrt{65}}{65}$$

$$\cos \alpha = \frac{x}{r} = \frac{1}{\sqrt{65}/7} = \frac{7\sqrt{65}}{65}$$

$$\sec \alpha = \frac{r}{x} = \frac{\sqrt{65}}{7}$$

EXAMPLE 6-13: Angle α is an angle whose terminal side lies in the second quadrant and $\cot \alpha = -\frac{5}{4}$; find $\sin \alpha$, $\cos \alpha$, and $\tan \alpha$.

Solution: You know that $\cot \alpha = x/y = -\frac{5}{4}$. Then divide $x^2 + y^2 = r^2$ by y to get $(x/y)^2 + 1^2 = (r/y)^2$. Substitute $x/y = -\frac{5}{4}$ into this equation to get $\left(-\frac{5}{4}\right)^2 + 1^2 = (r/y)^2$, or $r/y = \frac{\sqrt{41}}{4}$ (positive, since y is positive in quadrant II). Then,

$$\sin \alpha = \frac{y}{r} = \frac{4}{\sqrt{41}} = \frac{4\sqrt{41}}{41}$$

$$\cos \alpha = \frac{x}{r} = \frac{x/y}{r/y} = \frac{-5/4}{\sqrt{41}/4} = \frac{-5}{\sqrt{41}} = \frac{-5\sqrt{41}}{41}$$

$$\tan \alpha = \frac{y}{x} = \frac{1}{x/y} = \frac{1}{-5/4} = -\frac{4}{5}$$

B. **Using the wrapping function**

From Equations 6-2 and the equation for the unit circle, you can easily find all the trigonometric functions of t, given only one of them and the quadrant in which t lies.

1. **Given sine or cosecant:** From Equations 6-2, you know y. Solve the equation of the unit circle for x, then attach the proper sign. Use Equations 6-2 to find all of the trigonometric functions of t.

EXAMPLE 6-14: For $\sin t = \frac{\sqrt{8}}{3}$ and $\cos t < 0$, find the values of the other trigonometric functions of t.

Solution: You know that $\sin t > 0$ only in quadrants I and II and $\cos t < 0$ only in quadrants II and III, so t must lie in quadrant II. You also know that $y = \frac{\sqrt{8}}{3}$. Rewrite the equation for the unit circle to isolate x and substitute for the value of y:

$$x^2 + y^2 = 1$$
$$x = \pm\sqrt{1 - y^2}$$
$$= \pm\sqrt{1 - \left(\frac{\sqrt{8}}{3}\right)^2}$$
$$= \pm\sqrt{1 - \frac{8}{9}} = \pm\frac{1}{3}$$

Since $\cos t = x < 0$, $x = -\frac{1}{3}$. Therefore, $W(t) = \left(-\frac{1}{3}, \frac{\sqrt{8}}{3}\right)$. Now use Equations 6-2 to complete the solution:

$$\sin t = y = \frac{\sqrt{8}}{3} \qquad\qquad \csc t = \frac{1}{y} = \frac{1}{\sqrt{8}/3} = \frac{3\sqrt{8}}{8}$$

$$\cos t = x = -\frac{1}{3} \qquad\qquad \sec t = \frac{1}{x} = \frac{1}{-1/3} = -3$$

$$\tan t = \frac{y}{x} = \frac{\sqrt{8}/3}{-1/3} = -\sqrt{8} \qquad \cot t = \frac{x}{y} = \frac{-1/3}{\sqrt{8}/3} = \frac{-\sqrt{8}}{8}$$

2. Given cosine or secant: From Equations 6-2, you know x. Proceed as in 1.

EXAMPLE 6-15: For $\cos t = -\frac{2}{3}$ and t in quadrant III, find the values of the other trigonometric functions of t.

Solution: You know that $x = -\frac{2}{3}$ and that $y < 0$. Solving the equation of the unit circle for y,

$$y = \pm\sqrt{1 - x^2}$$
$$= \pm\sqrt{1 - \left(-\frac{2}{3}\right)^2}$$
$$= \pm\sqrt{1 - \frac{4}{9}} = \pm\frac{\sqrt{5}}{3}$$

Since $\sin t = y < 0$, $y = -\frac{\sqrt{5}}{3}$. Therefore, $W(t) = \left(-\frac{2}{3}, -\frac{\sqrt{5}}{3}\right)$. Now use Equations 6-2 to complete the solution:

$$\sin t = y = -\frac{\sqrt{5}}{3} \qquad\qquad \csc t = \frac{1}{y} = \frac{1}{-\sqrt{5}/3} = \frac{-3\sqrt{5}}{5}$$

$$\cos t = x = -\frac{2}{3} \qquad\qquad \sec t = 1/x = \frac{1}{-1/3} = -\frac{3}{2}$$

$$\tan t = \frac{y}{x} = \frac{-\sqrt{5}/3}{-2/3} = \frac{\sqrt{5}}{2} \qquad \cot t = \frac{x}{y} = \frac{-2/3}{-\sqrt{5}/3} = \frac{2\sqrt{5}}{5}$$

3. Given tangent or cotangent: From Equations 6-2, you know either y/x or x/y. Use this known quantity and $x^2 + y^2 = 1$ to find x and y.

EXAMPLE 6-16: For $\cot t = -\frac{5}{4}$ and t in quadrant II, find the values of the other trigonometric functions of t.

Solution: You know that $\cot t = -\frac{5}{4} = x/y$. You are looking for values of x and y that satisfy $x^2 + y^2 = 1$. From $-\frac{5}{4} = x/y$, you get $x = -\frac{5}{4}y$, so

$$\left(-\frac{5}{4}y\right)^2 + y^2 = 1$$

$$\frac{25}{16}y^2 + y^2 = 1$$

$$\frac{41}{16}y^2 = 1$$

$$y^2 = \frac{16}{41}$$

$$y = \pm\frac{4}{\sqrt{41}}$$

In this problem, $x < 0$ and $y > 0$ (t lies in quadrant II), so $y = \frac{4}{\sqrt{41}} = \frac{4\sqrt{41}}{41}$ and

$$x = -\frac{5}{4}y = \left(-\frac{5}{4}\right)\frac{4\sqrt{41}}{41} = \frac{-5\sqrt{41}}{41}$$

Now use Equations 6-2 to complete the problem:

$$\sin t = y = \frac{4\sqrt{41}}{41} \qquad \csc t = \frac{1}{y} = \frac{1}{4/\sqrt{41}} = \frac{\sqrt{41}}{4}$$

$$\cos t = x = \frac{-5\sqrt{41}}{41} \qquad \sec t = \frac{1}{x} = \frac{1}{-5/\sqrt{41}} = \frac{-\sqrt{41}}{5}$$

$$\tan t = \frac{y}{x} = \frac{-4}{5} \qquad \cot t = \frac{x}{y} = \frac{-5}{4}$$

SUMMARY

1. If α is the angle in standard position and $P(x, y)$ lies at distance $r = \sqrt{x^2 + y^2}$ ($r \neq 0$) from the origin on the terminal side of α, the trigonometric ratios are defined as:

$$\sin \alpha = \frac{y}{r} \qquad \csc \alpha = \frac{r}{y}$$

$$\cos \alpha = \frac{x}{r} \qquad \sec \alpha = \frac{r}{x}$$

$$\tan \alpha = \frac{y}{x} \qquad \cot \alpha = \frac{x}{y}$$

where $\csc \alpha$ and $\cot \alpha$ are not defined if $y = 0$ and $\sec \alpha$ and $\tan \alpha$ are not defined if $x = 0$.
2. If x is any real number, the value of the trigonometric function $\sin x$ is equal to the value of the trigonometric ratio $\sin x$, where x is the angle in standard position of measure x radians. The other five trigonometric functions are defined similarly.
3. The wrapping function gives an alternate definition of the trigonometric functions in terms of real numbers.
4. If $P(x, y)$ is a point on the unit circle and t is a real number such that $W(t) = P(x, y)$, then:

$$\sin t = y \qquad \csc t = \frac{1}{y}$$

$$\cos t = x \qquad \sec t = \frac{1}{x}$$

$$\tan t = \frac{y}{x} \qquad \cot t = \frac{x}{y}$$

5. Each trigonometric function maintains a constant sign for all numbers in a quadrant.
6. The values of sine and cosine should be memorized for $0, \frac{1}{6}\pi, \frac{1}{4}\pi, \frac{1}{3}\pi, \frac{1}{2}\pi, \pi, \frac{3}{2}\pi,$ and 2π; these values can be used to deduce the values of the other trigonometric functions.

RAISE YOUR GRADES
Can you...?

☑ find the value of the trigonometric ratios, given a point on the terminal side of an angle
☑ find the value of the trigonometric functions for any real number
☑ give the signs of the trigonometric functions for any quadrant
☑ give the values of the trigonometric functions for angles of $0, \frac{1}{6}\pi,$ $\frac{1}{4}\pi, \frac{1}{3}\pi, \frac{1}{2}\pi, \pi, \frac{3}{2}\pi,$ plus or minus any multiple of 2π
☑ find the values of the other trigonometric functions given the value and quadrant of one of the functions
☑ define the domain and range of the wrapping function

SOLVED PROBLEMS

PROBLEM 6-1 Use the coordinate values of $P(3, 2)$, lying on the terminal side of α in Figure 6-6, to find the values of the six trigonometric ratios of α.

Solution: From the diagram, $r = \sqrt{x^2 + y^2} = \sqrt{3^2 + 2^2} = \sqrt{13}$. Then,

$$\sin \alpha = \frac{y}{r} = \frac{2}{\sqrt{13}} = \frac{2\sqrt{13}}{13} \qquad \csc \alpha = \frac{r}{y} = \frac{\sqrt{13}}{2}$$

$$\cos \alpha = \frac{x}{r} = \frac{3}{\sqrt{13}} = \frac{3\sqrt{13}}{13} \qquad \sec \alpha = \frac{r}{x} = \frac{\sqrt{13}}{3}$$

$$\tan \alpha = \frac{y}{x} = \frac{2}{3} \qquad \cot \alpha = \frac{x}{y} = \frac{3}{2}$$

[See Section 6-1.]

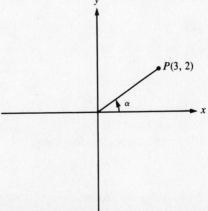

Figure 6-6

PROBLEM 6-2   Let *P* and *Q* be two terminal points on the line $y = 2x$ in the first quadrant. Show that the value of sin α is the same no matter which terminal point is used in computing the value of sin α (see Figure 6-7).

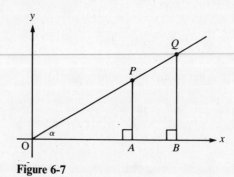

Figure 6-7

Solution: Since triangles *AOP* and *BOQ* contain the same angle α and each has a right angle, the third angle of the two triangles is the same. From your work in geometry, recall that this means that triangles *AOP* and *BOQ* are *similar*. Therefore, $d_{AP}/d_{OP} = d_{BQ}/d_{OQ}$; however, by the definition of the trigonometric functions, you can see that $d_{AP}/d_{OP} = \sin \alpha$ and $d_{BQ}/d_{OQ} = \sin \alpha$. Therefore, you get the same value for sin α no matter which point on the terminal side of α you use to calculate the ratio.   [See Section 6-1.]

PROBLEM 6-3   In which quadrants could the terminal side of angle α lie for: **(a)** sin $\alpha < 0$; **(b)** tan $\alpha < 0$; **(c)** sin α cos $\alpha < 0$?

Solution:

(a) Since $\sin \alpha = y/r$ and $r > 0$, sin $\alpha < 0$ when $y < 0$. These conditions occur only in quadrants III and IV.

(b) Since $\tan \alpha = y/x$, tan $\alpha < 0$ when y and x are of opposite sign; this occurs only in quadrants II and IV.

(c) Since $\sin \alpha \cos \alpha = (y/r)(x/r) = xy/r^2$, sin α cos $\alpha < 0$ when x and y are of opposite sign; this happens only in quadrants II and IV.   [See Section 6-2.]

PROBLEM 6-4   Use the right triangle of Figure 6-8 to express xy in terms of r, α, *and* θ, in two different ways.

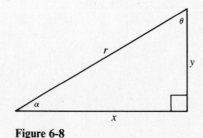

Figure 6-8

Solution: From Figure 6-8, $\sin \alpha = y/r$ so $y = r \sin \alpha$, and $\cos \alpha = x/r$, so $x = r \cos \alpha$. Also, $\sin \theta = x/r$ and $\cos \theta = y/r$ so $x = r \sin \theta$ and $y = r \cos \theta$. Multiplying the first expression for x by the second expression for y, you get

$$xy = r \cos \alpha \, r \cos \theta = r^2 \cos \alpha \cos \theta$$

Multiplying the second expression for x by the first expression for y gives

$$xy = r \sin \theta \, r \sin \alpha = r^2 \sin \theta \sin \alpha$$ [See Section 6-1.]

PROBLEM 6-5 If the terminal side of α lies in quadrant II, and θ equals the acute angle between the terminal side of α and the negative x axis, find expressions for the six trigonometric functions of α in terms of the trigonometric functions of θ (see Figure 6-9).

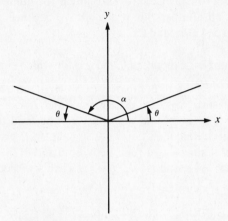

Figure 6-9

Solution: You can apply the definitions of the trigonometric ratios (Equations 6-1) directly to α, but θ is not in standard position. To find the trigonometric ratios corresponding to θ, draw θ in standard position as shown in Figure 6-9. The ratios for θ can then be found. The results are

$$\sin \alpha = \frac{y}{r} = \sin \theta \qquad\qquad \csc \alpha = \frac{r}{y} = \csc \theta$$

$$\cos \alpha = \frac{-x}{r} = -\left(\frac{x}{r}\right) = -\cos \theta \qquad \sec \alpha = \frac{r}{-x} = -\left(\frac{r}{x}\right) = -\sec \theta$$

$$\tan \alpha = \frac{y}{-x} = -\left(\frac{y}{x}\right) = -\tan \theta \qquad \cot \alpha = \frac{-x}{y} = -\left(\frac{x}{y}\right) = -\cot \theta$$

[See Section 6-1 and 6-3C.]

PROBLEM 6-6 Find the radius r of the cone shown in Figure 6-10.

Solution: If you slice the cone by a plane perpendicular to the base and through the center line of the cone, the right triangle shown in Figure 6-10 is formed. In this right triangle, $\tan \alpha = h/r$, so $h = r \tan \alpha$.

[See Section 6-1.]

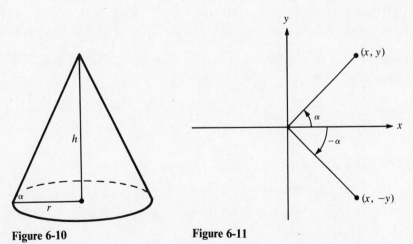

Figure 6-10   **Figure 6-11**

PROBLEM 6-7 Find all numbers between 0 and 2π for which $\sin \alpha = \sin(-\alpha)$.

Solution: If (x, y) is the point at the end of the terminal side of angle α (terminal side of α in quadrant I and measured in radians) and r is the length of the line segment from $(0,0)$ to (x, y), then $\sin \alpha = y/r$. The point on the terminal side of $-\alpha$ is $(x, -y)$ (see Figure 6-11). Thus

$\sin(-\alpha) = -y/r$. Now $\sin \alpha = y/r$ and $\sin(-\alpha) = -y/r$; therefore, $\sin(-\alpha) = -\sin \alpha$. You can use a similar argument to show that $\sin(-\alpha) = -\sin \alpha$ no matter where the terminal side of α lies. You can now restate the problem as: Find all numbers between 0 and 2π for which $\sin \alpha = -\sin \alpha$. Since a number is equal to its negative if and only if the number is zero, you can further restate the problem as: Find all numbers for which $\sin \alpha = 0$. Since $\sin \alpha = y/r$, $\sin \alpha = 0$ only when $y = 0$. This occurs when $\alpha = 0, \pi$, and 2π.   [See Section 6-2.]

PROBLEM 6-8   Find all numbers x (between 0 and 2π) for which $\sin 2x = \pm 1$.

Solution: Whenever you want to find all numbers x between 0 and 2π for which $\sin(kx) = \pm 1$, you must find all solutions of $\sin \alpha = \pm 1$ for α between 0 and $2\pi k$; if α is such a solution, $x = \alpha/k$ is a solution of $\sin(kx) = \pm 1$. The same procedure can be used to solve similar equations involving the other trigonometric functions.

   You must find all solutions of $\sin \alpha = \pm 1$ for α between 0 and $2(2\pi) = 4\pi$. From your knowlege of sine, $\alpha = \frac{1}{2}\pi, \frac{3}{2}\pi, \frac{1}{2}\pi + 2\pi = \frac{5}{2}\pi$, and $\frac{3}{2}\pi + 2\pi = \frac{7}{2}\pi$. Then set $2x$ equal to $\frac{1}{2}\pi, \frac{3}{2}\pi, \frac{5}{2}\pi$, and $\frac{7}{2}\pi$, in turn. The solutions are $x = \frac{1}{4}\pi, \frac{3}{4}\pi, \frac{5}{4}\pi$, and $\frac{7}{4}\pi$.   [See Section 6-2.]

PROBLEM 6-9   Can $|\sin \alpha + \cos \alpha| = 2$ for any angle α?

Solution: Recall that $\sin \alpha$ and $\cos \alpha$ both take values between -1 and 1. In order for the equality to be satisfied there must be an α for which $\sin \alpha = \cos \alpha = +1$ or $\sin \alpha = \cos \alpha = -1$. If you consider Table 6-1, you see that whenever the $\sin \alpha = \pm 1$, $\cos \alpha$ is zero and vice versa. Therefore, $|\sin \alpha + \cos \alpha| < 2$ for all α.   [See Section 6-2.]

PROBLEM 6-10   If $\tan \alpha = a > 0$ with $\sin \alpha < 0$, find $\sec \alpha$ and $\csc \alpha$ in terms of a.

Solution: From the definition of $\tan \alpha$, you know that $(1, a)$ or $(-1, -a)$ could be points on the terminal side of α. Since $\sin \alpha < 0$ you must eliminate $(1, a)$. For $(-1, -a)$, $r = \sqrt{1 + a^2}$. From Equations 6-1,

$$\sec \alpha = \frac{r}{x} = \frac{\sqrt{1 + a^2}}{-1} \quad \text{and} \quad \csc \alpha = \frac{r}{y} = \frac{\sqrt{1 + a^2}}{-a} \qquad \text{[See Section 6-2.]}$$

PROBLEM 6-11   Find all intervals in which α can lie if $0 \leqslant \alpha \leqslant 2\pi$ and $\tan \alpha > 1$.

Solution: Consider the variation of the tangent function. When $\alpha = 0$, $\tan \alpha = 0$. As α increases from 0 to $\frac{1}{2}\pi$, $\tan \alpha$ is positive and increasing. When $\alpha = \frac{1}{4}\pi$, $\tan \alpha = 1$. When α is slightly less than $\frac{1}{2}\pi$, $\tan \alpha$ is very large and positive. This means that $\tan \alpha > 1$ for α in the interval $\left(\frac{1}{4}\pi, \frac{1}{2}\pi\right)$. As α increases from $\frac{1}{2}\pi$ to π, $\tan \alpha$ is negative but $\tan \pi = 0$. As $\tan \alpha$ increases from π to $\frac{3}{2}\pi$, $\tan \alpha$ increases from 0 to large positive values, with $\tan \frac{5}{4}\pi = 1$. This means that $\tan \alpha > 1$ for α in the intervals $\left(\frac{5}{4}\pi, \frac{3}{2}\pi\right)$ and $\left(\frac{1}{4}\pi, \frac{1}{2}\pi\right)$.   [See Sections 6-1 and 6-2.]

PROBLEM 6-12   Given that $(10, 24)$ lies on the terminal side of α, find **(a)** $\sin \alpha + \cos \alpha$; **(b)** $\sin \alpha - \cos \alpha$; **(c)** $\sin \alpha \csc \alpha$.

Solution: Since $r^2 = x^2 + y^2 = 10^2 + 24^2 = 676$, $r = \sqrt{676} = 26$. Then, $\sin \alpha = y/r = \frac{24}{26} = \frac{12}{13}$, $\cos \alpha = x/r = \frac{10}{26} = \frac{5}{13}$, and $\csc \alpha = r/y = \frac{26}{24} = \frac{13}{12}$.

(a) $\sin \alpha + \cos \alpha = \dfrac{12}{13} + \dfrac{5}{13} = \dfrac{17}{13}$

(b) $\sin \alpha - \cos \alpha = \dfrac{12}{13} - \dfrac{5}{13} = \dfrac{7}{13}$

(c) $\sin \alpha \csc \alpha = \dfrac{12}{13} \times \dfrac{13}{12} = 1$ [See Section 6-1.]

PROBLEM 6-13   Use your calculator to determine whether the following statements are true or false: **(a)** $\sin 65° + \sin 25° = \sin 90°$;   **(b)** $\sin 65° - \sin 25° = \sin 40°$;   **(c)** $\sin 80° = 2 \sin 40° \cos 40°$.

Solution: Your calculator produces the following results: $\sin 25° = 0.4226$, $\sin 65° = 0.9063$, $\sin 40° = 0.6428$, $\cos 40° = 0.7660$, and $\sin 80° = 0.9848$.

(a) Because $\sin 65° + \sin 25° = 0.9063 + 0.4226 = 1.3289 \neq 1 = \sin 90°$, statement (a) is false.

(b) Because $\sin 65° - \sin 25° = 0.9063 - 0.4226 = 0.4837 \neq 0.6428 = \sin 40°$, statement (b) is false.

(c) Because $2 \sin 40° \cos 40° = 2(0.6428)(0.7660) = 0.9848 = \sin 80°$, statement (c) is true.

[See Section 6-3.]

PROBLEM 6-14 Figure 6-12 shows angle α in standard position. Distance OP is 108.93 and angle OPB is 53.8°; find $\sin \alpha$ and α.

Solution: Angle α is $90° - OPB = 90° - 53.8° = 36.2°$, and $\sin \alpha = \sin 36.2° = 0.5906$.

[See Section 6-1.]

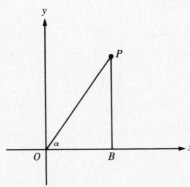

Figure 6-12

PROBLEM 6-15 For $\cos \alpha = 0.6406$ and $\sin \alpha > 0$, find (a) $\tan \alpha$; (b) $\csc \alpha$; (c) $\sec \alpha$.

Solution: Since cosine and sine are both positive, α must lie in quadrant I. Use $\boxed{\cos^{-1}}$ or $\boxed{\arccos}$ on your calculator to find $\alpha = 50.2°$.

(a) $\tan 50.2° = 1.2002$

(b) $\csc 50.2° = \dfrac{1}{\sin 50.2°} = \dfrac{1}{0.7683} = 1.3016$

(c) $\sec \alpha = \dfrac{1}{\cos \alpha} = \dfrac{1}{0.6406} \approx 1.5610$

[See Section 6-3.]

PROBLEM 6-16 Let α and θ be angles that lie between 0 and $\frac{1}{2}\pi$ radians. For $\sin \alpha = \sin \theta$, show that $\alpha = \theta$.

Solution: From the definition of the sine function, you know that the $\sin \alpha$ function is one-to-one if α lies on the interval $[0, \frac{1}{2}\pi]$. From the definition of a one-to-one function (see Chapter 4), $\sin \alpha = \sin \theta$ implies that $\alpha = \theta$.

[See Section 6-1.]

PROBLEM 6-17 A line segment of length 6 with one end fixed at the origin and the other end at point A on the positive x axis is rotated through an angle α in a counterclockwise direction about the origin until the sector formed has an area of $\frac{3}{4}\pi$ square units. Find $\sin \alpha$, $\cos \alpha$, and $\tan \alpha$.

Solution: Recall that the area of a sector equals $\frac{1}{2}r^2\alpha$, where α is the sector angle in radians. Substituting the given quantities into this formula, $\frac{3}{4}\pi = \frac{1}{2}(6^2)\alpha$ or $\alpha = \frac{1}{24}\pi = 0.1309$ radians. Place your calculator in radian mode, enter $\boxed{0.1309}$, and press $\boxed{\sin}$ to get $\sin 0.1309 = 0.1305$. In the same way, $\cos 0.1309 = 0.9914$ and $\tan 0.1309 = 0.1317$.

[See Section 6-1.]

PROBLEM 6-18 Find $W(t)$ for: (a) $t = 1$; (b) $t = \pi$; (c) $t = \frac{1}{12}\pi$.

Solution:

(a) Since $t = 1$ is equivalent to a 1-radian angle, and the terminal point is on the unit circle ($r = 1$), you get

$$\sin 1 = 0.8415 = \frac{y}{r} = \frac{y}{1}, \text{ so } y = 0.8415.$$

Also,

$$\cos 1 = 0.5403 = \frac{x}{r} = \frac{x}{1}, \text{ so } x = 0.5403.$$

Thus, $W(t) = (0.5403, 0.8415)$.

(b) Since $t = \pi$ is the point half-way around the unit circle in a counterclockwise direction from $(1, 0)$, $W(t) = (-1, 0)$. You'll get the same answer if you use your calculator.

(c) Using the procedure in (a), $\sin \frac{1}{12}\pi = y/r = y = 0.2588$ and $\cos \frac{1}{12}\pi = x/r = x = 0.9659$. Thus, $W(t) = (0.9659, 0.2588)$. [See Section 6-2.]

PROBLEM 6-19 Let t be a real number such that $W(t) = \left(\frac{12}{13}, \frac{5}{13}\right)$. Find $\sec t$ and $\cot t$.

Solution: The definition of $W(t)$ implies that $(x, y) = \left(\frac{12}{13}, \frac{5}{13}\right)$. From Equations 6-2,

$$\sec t = \frac{1}{x} = \frac{13}{12}$$

$$\cot t = \frac{x}{y} = \frac{\frac{12}{13}}{\frac{5}{13}} = \frac{12}{5}.$$ [See Section 6-2.]

PROBLEM 6-20 For $\sin t = \frac{4}{5}$ and $\cos t = \frac{3}{5}$, find t.

Solution: From Equations 6-2, $\sin t = x = \frac{4}{5}$ and $\cos t = y = \frac{3}{5}$. So t is the angle in radians whose sin is $\frac{4}{5}$. Place your calculator in radian mode, enter $\boxed{0.8}$ $\boxed{\sin^{-1}}$ to get $t = 0.9273$ radians. [See Section 6-3.]

PROBLEM 6-21 The terminal side of α lies in quadrant II and $\sin \alpha = \frac{1}{2}$; find $\cos \alpha$ and $\tan \alpha$.

Solution: Since you are given $\sin \alpha = y/r = \frac{1}{2}$, use $(x/r)^2 + (y/r)^2 = 1$, so $(x/r)^2 + \left(\frac{1}{2}\right)^2 = 1$. You find that $x/r = -\frac{\sqrt{3}}{2}$ (you choose the minus sign because the terminal side is in quadrant II). Then,

$$\cos \alpha = \frac{x}{r} = -\frac{\sqrt{3}}{2}$$

$$\tan \alpha = \frac{y}{x} = \frac{y/r}{x/r} = \frac{1/2}{-\sqrt{3}/2} = -\frac{\sqrt{3}}{3}$$ [See Section 6-3.]

PROBLEM 6-22 The terminal side of α lies in quadrant III and $\cos \alpha = -\frac{2}{3}$; find $\sin \alpha$ and $\tan \alpha$.

Solution: Since you are given $\cos \alpha = x/r = -\frac{2}{3}$, use $(x/r)^2 + (y/r)^2 = 1$, so $\left(-\frac{2}{3}\right)^2 + (y/r)^2 = 1$. You find that $y/r = -\frac{\sqrt{5}}{3}$ (you choose the minus sign because the terminal side is in quadrant III). Then,

$$\sin \alpha = \frac{y}{r} = -\frac{\sqrt{5}}{3}$$

$$\tan \alpha = \frac{y}{x} = \frac{y/r}{x/r} = \frac{-\sqrt{5}/3}{-2/3} = \frac{\sqrt{5}}{2}$$ [See Section 6-3.]

PROBLEM 6-23 The terminal side of α lies in quadrant IV and $\tan \alpha = -2$; find $\sin \alpha$ and $\cos \alpha$.

Solution: Since you are given $\tan \alpha = y/x = -2$, use $1^2 + (y/x)^2 = (r/x)^2$, so $1^2 + (-2)^2 = (r/x)^2$. You find that $r/x = \sqrt{5}$ (you choose the positive value because the terminal side is in quadrant IV).

Then,

$$\sin\alpha = \frac{y}{r} = \frac{y/x}{r/x} = -\frac{2}{\sqrt{5}} = \frac{-2\sqrt{5}}{5}$$

$$\cos\alpha = \frac{x}{r} = \frac{1}{r/x} = \frac{1}{\sqrt{5}} = \frac{\sqrt{5}}{5} \qquad \text{[See Section 6-3.]}$$

PROBLEM 6-24 The terminal side of α lies in quadrant I and $\cot\alpha = 2$; find $\sin\alpha$ and $\cos\alpha$.

Solution: Since you are given $\cot\alpha = x/y = 2$, use $(x/y)^2 + 1^2 = (r/y)^2$, so $2^2 + 1^2 = (r/y)^2$. You find that $r/y = \sqrt{5}$ (you choose the positive value because the terminal side is in quadrant I). Then,

$$\sin\alpha = \frac{y}{r} = \frac{1}{r/y} = \frac{1}{\sqrt{5}} = \frac{\sqrt{5}}{5}$$

$$\cos\alpha = \frac{x}{r} = \frac{x/y}{r/y} = \frac{2}{\sqrt{5}} = \frac{2\sqrt{5}}{5} \qquad \text{[See Section 6-3.]}$$

PROBLEM 6-25 The terminal side of t lies in quadrant II and $\sec t = -3$; find $\sin t$ and $\tan t$.

Solution: Since $\sec t = r/x = -3 = 1/\cos t$, $\cos t = -\frac{1}{3}$. From Equations 6-2, $\cos t = x$, so $W(t) = (-\frac{1}{3}, y)$. Use the equation of the unit circle to find y:

$$y = \pm\sqrt{1 - x^2} = \pm\sqrt{1 - \left(-\frac{1}{3}\right)^2} = \pm\sqrt{\frac{8}{9}} = \frac{\pm\sqrt{8}}{3}$$

You choose the positive value because t lies in quadrant II. Now use Equations 6-2 to complete the problem:

$$\sin t = y = \frac{\sqrt{8}}{3}$$

$$\tan t = \frac{y}{x} = \frac{\sqrt{8}/3}{-1/3} = -\sqrt{8} \qquad \text{[See Section 6-3.]}$$

Supplementary Exercises

PROBLEM 6-26 For the following points on the terminal side of α, find the six trigonometric functions of α: (**a**) $(3, -4)$; (**b**) $(-4, -3)$; (**c**) $(-5, 12)$; (**d**) $(12, 5)$.

PROBLEM 6-27 Give the values of a between 0 and 2π for which $\cos 2a$ is positive.

PROBLEM 6-28 Give the values of a between 0 and 2π for which $\tan(a/2) = 1$.

PROBLEM 6-29 Angle α is the acute angle between the x axis and the line from the origin to $(18, -7)$; find (**a**) $\sin\alpha$; (**b**) $\tan\alpha$; (**c**) $\sec\alpha$.

PROBLEM 6-30 For $\sin\alpha = -\frac{1}{2}$ and $\cos\alpha > 0$, find $\sec\alpha$.

PROBLEM 6-31 For $\tan\alpha = -1$ and positive $\cos\alpha$, find $\csc\alpha$.

PROBLEM 6-32 The terminal side of angle α in standard position contains the point $(-1, 3)$. Find all of the trigonometric functions of α.

PROBLEM 6-33   For $\cos \alpha = \frac{3}{4}$ with $-\frac{1}{2}\pi < \alpha < 0$, find all of the trigonometric functions of α.

PROBLEM 6-34   Angle α is the angle between the positive x axis and the line from the origin to $(-5, 12)$; find all of the trigonometric functions of α.

PROBLEM 6-35   An angle α in standard position has its terminal side in the fourth quadrant with $\cos \alpha = \frac{\sqrt{3}}{2}$. Find the values of $\csc \alpha$ and $\tan \alpha$.

PROBLEM 6-36   For $\sin \alpha = \frac{1}{\sqrt{2}}$ and $\frac{1}{2}\pi < \alpha < \pi$, find $\tan \alpha$.

PROBLEM 6-37   Find an angle α (in radians) such that $0 < \alpha < \frac{1}{2}\pi$ and $\sin \alpha = -\sin \frac{7}{6}\pi$.

PROBLEM 6-38   Find one positive value of α such that $\cos \alpha = \cos(-20°)$.

PROBLEM 6-39   The terminal side of α (in standard position) contains the point $(33, -56)$. Find all six trigonometric functions of α.

PROBLEM 6-40   Let P and Q be two points in the second quadrant on the line $y = -2x$. Show that the sine ratio is the same no matter which point on $y = -2x$ is used to compute the sine ratio.

PROBLEM 6-41   In which quadrants does the terminal side of α lie for: (a) $\sec \alpha < 0$; (b) $\cot \alpha > 0$; (c) $\sin \alpha \csc \alpha > 0$?

PROBLEM 6-42   Find the six trigonometric functions of α given that P has coordinates: (a) $(3, 7)$; (b) $(-2, 8)$; (c) $(-3, -4)$.

PROBLEM 6-43   In Figure 6-13, the length of line segment OP is 8 and $\alpha = \frac{1}{3}\pi$; find the coordinates of P.

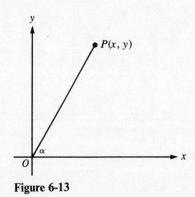

Figure 6-13

PROBLEM 6-44   For $\sin \alpha = 2/\sqrt{u^2 + 4}$ and $\cot \alpha < 0$, find $\cos \alpha$, $\tan \alpha$, $\csc \alpha$, $\sec \alpha$, and $\cot \alpha$ in terms of u.

PROBLEM 6-45   Let P, with coordinates (a, b), be a point on the terminal side of angle α, in standard position. If P' is a point on the terminal side of $-\alpha$, with $d_{OP} = d_{OP'}$, where O is the origin, what are the coordinates of P'?

PROBLEM 6-46   A bug crawls $\frac{25}{4}\pi$ units in a counterclockwise direction from the starting point $(5, 0)$ on a circle of diameter 10 units. Angle α is the angle subtended by the bug's path; find $\sin \alpha$, $\cos \alpha$, and $\tan \alpha$.

PROBLEM 6-47   A line segment of length 6 with one end fixed at the origin, lying on the positive x axis, is rotated through an angle α in a counterclockwise direction. The sector formed has an area of π square units. Find $\sin \alpha$, $\cos \alpha$ and $\tan \alpha$.

PROBLEM 6-48 List the sign of each of the six trigonometric functions of α for the following values of α: (a) $\frac{1}{8}\pi$; (b) $\frac{7}{8}\pi$; (c) $\frac{15}{8}\pi$; (d) $\frac{11}{10}\pi$.

PROBLEM 6-49 Figure 6-14 shows angle α in standard position. Suppose you fix point P so that distance OP is 108.93 and drop perpendicular PB to the horizontal axis so that distance OB is 34; find $\sin\alpha$ and α.

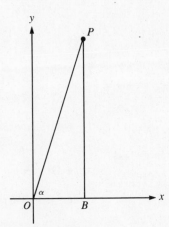

Figure 6-14

PROBLEM 6-50 The terminal side of an angle passes through $(-5, 12)$; find the sine, cosine, and tangent of this angle.

PROBLEM 6-51 A line segment of length 5 with one end fixed at the origin and the other end at point A on the positive x axis is rotated in a counterclockwise direction about the origin until the sector formed has an area of 6π square units. Find $\sec\alpha$, $\csc\alpha$, and $\cot\alpha$.

PROBLEM 6-52 Find the coordinates of the point on the unit circle corresponding to each of the following real numbers: (a) $t = -1$; (b) $t = 3\pi$; (c) $t = -\frac{1}{8}\pi$.

PROBLEM 6-53 Let t be a real number such that $W(t) = \left(-\frac{12}{13}, \frac{5}{13}\right)$. Find $\sec t$ and $\cot t$.

PROBLEM 6-54 Find a value of t such that $\sin t = -\frac{4}{5}$ and $\cos t = \frac{3}{5}$.

Answers to Supplementary Exercises

6-26: (a) $\sin\alpha = -\frac{4}{5}$, $\cos\alpha = \frac{3}{5}$, $\tan\alpha = -\frac{4}{3}$, $\csc\alpha = -\frac{5}{4}$, $\sec\alpha = \frac{5}{3}$, $\cot\alpha = -\frac{3}{4}$; (b) $\sin\alpha = -\frac{3}{5}$, $\cos\alpha = -\frac{4}{5}$, $\tan\alpha = \frac{3}{4}$, $\csc\alpha = -\frac{5}{3}$, $\sec\alpha = -\frac{5}{4}$, $\cot\alpha = \frac{4}{3}$; (c) $\sin\alpha = \frac{12}{13}$, $\cos\alpha = -\frac{5}{13}$, $\tan\alpha = -\frac{12}{5}$, $\csc\alpha = \frac{13}{12}$, $\sec\alpha = -\frac{13}{5}$, $\cot\alpha = -\frac{5}{12}$; (d) $\sin\alpha = \frac{5}{13}$, $\cos\alpha = \frac{12}{13}$, $\tan\alpha = \frac{5}{12}$, $\csc\alpha = \frac{13}{5}$, $\sec\alpha = \frac{13}{12}$, $\cot\alpha = \frac{12}{5}$

6-27: On $[0, \frac{1}{4}\pi]$, $[\frac{3}{4}\pi, \pi]$, $[\pi, \frac{5}{4}\pi]$, and $[\frac{7}{4}\pi, 2\pi]$

6-28: $\frac{1}{2}\pi$

6-29: (a) $\frac{-7\sqrt{373}}{373}$; (b) $-\frac{7}{18}$; (c) $\frac{\sqrt{373}}{18}$

6-30: $\frac{2\sqrt{3}}{3}$

6-31: $-\sqrt{2}$

6-32: $\sin\alpha = \frac{3\sqrt{10}}{10}$, $\cos\alpha = -\frac{\sqrt{10}}{10}$, $\tan\alpha = -3$, $\csc\alpha = \frac{\sqrt{10}}{3}$, $\sec\alpha = -\sqrt{10}$, $\cot\alpha = -\frac{1}{3}$

6-33: $\sin\alpha = -\frac{\sqrt{7}}{4}$, $\cos\alpha = \frac{3}{4}$, $\tan\alpha = -\frac{\sqrt{7}}{3}$, $\csc\alpha = \frac{-4\sqrt{7}}{7}$, $\sec\alpha = \frac{4}{3}$, $\cot\alpha = \frac{-3\sqrt{7}}{7}$

6-34: $\sin\alpha = \frac{12}{13}$, $\cos\alpha = -\frac{5}{13}$, $\tan\alpha = -\frac{12}{5}$, $\csc\alpha = \frac{13}{12}$, $\sec\alpha = -\frac{13}{5}$, $\cot\alpha = -\frac{5}{12}$

6-35: $\csc\alpha = -2$ and $\tan\alpha = \frac{-\sqrt{3}}{3}$

6-36: $\tan\alpha = -1$

6-37: $\frac{1}{6}\pi$

6-38: $\alpha = 20°$

6-39: $\sin\alpha = -\frac{56}{65}$, $\cos\alpha = \frac{33}{65}$, $\tan\alpha = -\frac{56}{33}$, $\csc\alpha = -\frac{65}{56}$, $\sec\alpha = \frac{65}{33}$, $\cot\alpha = -\frac{33}{56}$

6-40: The terminal points lie on the same terminal line; therefore, the sines must be equal.

6-41: (a) II or III; (b) I or III; (c) any quadrant

6-42: (a) $\sin\alpha = \frac{7\sqrt{58}}{58}$, $\cos\alpha = \frac{3\sqrt{58}}{58}$, $\tan\alpha = \frac{7}{3}$, $\csc\alpha = \frac{\sqrt{58}}{7}$, $\sec\alpha = \frac{\sqrt{58}}{3}$, $\cot\alpha = \frac{3}{7}$; (b) $\sin\alpha = \frac{4\sqrt{17}}{17}$, $\cos\alpha = -\frac{\sqrt{17}}{17}$, $\tan\alpha = -4$, $\csc\alpha = \frac{\sqrt{17}}{4}$, $\sec\alpha = -\sqrt{17}$, $\cot\alpha = -\frac{1}{4}$; (c) $\sin\alpha = -\frac{4}{5}$, $\cos\alpha = -\frac{3}{5}$, $\tan\alpha = \frac{4}{3}$, $\csc\alpha = -\frac{5}{4}$, $\sec\alpha = -\frac{5}{3}$, $\cot\alpha = \frac{3}{4}$

6-43: $P = (4, 4\sqrt{3})$

6-44: $\cos\alpha = -u\sqrt{u^2+4}/(u^2+4)$, $\tan\alpha = -2/u$, $\csc\alpha = \sqrt{u^2+4}/2$, $\sec\alpha = -\sqrt{u^2+4}/u$, $\cot\alpha = -u/2$

6-45: $(a, -b)$

6-46: $\sin\alpha = \sqrt{2}/2$, $\cos\alpha = \sqrt{2}/2$, $\tan\alpha = 1$

6-47: $\sin\alpha = 0.1736$, $\cos\alpha = 0.9848$, $\tan\alpha = 0.1763$

6-48:

	$\frac{1}{8}\pi$	$\frac{7}{8}\pi$	$\frac{15}{8}\pi$	$\frac{11}{10}\pi$
$\sin\alpha$	+	+	−	−
$\cos\alpha$	+	−	+	−
$\tan\alpha$	+	−	−	+
$\csc\alpha$	+	+	−	−
$\sec\alpha$	+	−	+	−
$\cot\alpha$	+	−	−	+

6-49: $\sin\alpha = 0.9501$ and $\alpha = 1.25$ rad, or $71.62°$

6-50: $\sin\alpha = \frac{12}{13}$, $\cos\alpha = -\frac{5}{13}$, $\tan\alpha = -\frac{12}{5}$

6-51: $\sec\alpha = 15.92$, $\csc\alpha = 1.002$, $\cot\alpha = 0.063$

6-52: (a) $(0.5403, -0.8415)$; (b) $(-1, 0)$; (c) $(0.924, -0.383)$

6-53: $\sec t = -\frac{13}{12}$, $\cot t = -\frac{12}{5}$

6-54: 5.356 rad, or $306.9°$

7 RELATED ANGLES

THIS CHAPTER IS ABOUT

☑ **Reducing Trigonometric Functions**
☑ **Finding Angles**

7-1. Reducing Trigonometric Functions

Consider the points (x, y), $(-x, y)$ $(-x, -y)$, and $(x, -y)$ on the circle of radius r, where x and y are positive. The trigonometric functions of these four points differ only in sign (see Table 7-1).

TABLE 7-1: Trigonometric functions for related points.

Point	Sine	Cosine	Tangent
(x, y)	y/r	x/r	y/x
$(-x, y)$	y/r	$-x/r$	$y/-x$
$(-x, -y)$	$-y/r$	$-x/r$	$-y/-x$
$(x, -y)$	$-y/r$	x/r	$-y/x$

A. Related angles are acute.

Consider angle α in standard position, with (x, y) a point on the terminal side of α (see Figure 7-1).

Observe that the angles formed with a line drawn

- from the origin to $(-x, y)$ as initial side and the negative x axis as terminal side (see Figure 7-2a),
- from the origin to $(-x, -y)$ as terminal side and the negative x axis as initial side (see Figure 7-2b),
- and from the origin to $(x, -y)$ as initial side and the positive x axis as terminal side (see Figure 7-2c)

all have measure of α. These angles are the **related** or **reference** angles of the three angles whose initial sides are the *positive* x axis and whose terminal sides are the lines from the origin to $(-x, y)$, $(-x, -y)$, and $(x, -y)$, respectively. The measures of these three standard position angles are $\pi - \alpha$, $\pi + \alpha$, and $2\pi - \alpha$ [or $(180 - \alpha)°$, $(180 + \alpha)°$, and $(360 - \alpha)°$], respectively.

From the relationships in Figure 7-2, the related angle of an angle θ between 0 and 2π is

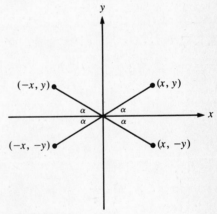

Figure 7-1
Related points.

$$\theta, \qquad \text{if} \quad \theta \text{ is between } 0 \text{ and } \frac{\pi}{2}$$

RELATED ANGLES

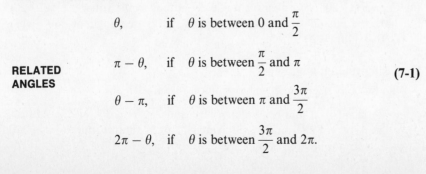

$$\pi - \theta, \quad \text{if} \quad \theta \text{ is between } \frac{\pi}{2} \text{ and } \pi \qquad (7\text{-}1)$$

$$\theta - \pi, \quad \text{if} \quad \theta \text{ is between } \pi \text{ and } \frac{3\pi}{2}$$

$$2\pi - \theta, \quad \text{if} \quad \theta \text{ is between } \frac{3\pi}{2} \text{ and } 2\pi.$$

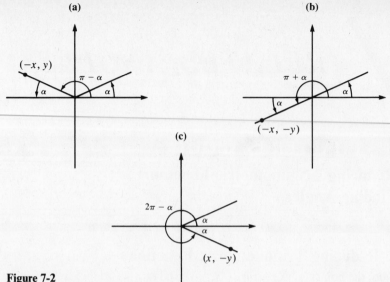

Figure 7-2
Related angles.

A related angle is always an acute angle. To find the related angle corresponding to a *negative* angle, c, between 0 and -2π, find the corresponding positive angle by adding 2π to c; then find the related angle for the positive angle as just described.

Note: Replace $\frac{1}{2}\pi$ by 90°, π by 180°, $\frac{3}{2}\pi$ by 270°, and 2π by 360° when working with angles in degrees.

EXAMPLE 7-1: Find the related angle for: (a) $\frac{4}{5}\pi$; (b) $\frac{5}{4}\pi$; (c) $\frac{21}{24}\pi$; (d) $-\frac{7}{8}\pi$.

Solution:

(a) Because $\frac{4}{5}\pi$ lies between $\frac{1}{2}\pi$ and π, the related angle is $\pi - \frac{4}{5}\pi = \frac{1}{5}\pi$.
(b) Because $\frac{5}{4}\pi$ lies between π and $\frac{3}{2}\pi$, the related angle is $\frac{5}{4}\pi - \pi = \frac{1}{4}\pi$.
(c) Because $\frac{21}{24}\pi$ lies between $\frac{1}{2}\pi$ and π, the related angle is $\pi - \frac{21}{24}\pi = \frac{3}{24}\pi = \frac{1}{8}\pi$.
(d) Because $-\frac{7}{8}\pi$ is negative, the corresponding positive angle is $2\pi + \left(-\frac{7}{8}\pi\right) = \frac{1}{8}\pi$. The related angle of $-\frac{7}{8}\pi$ equals the related angle of $\frac{1}{8}\pi$, which is $\frac{1}{8}\pi$.

EXAMPLE 7-2: Find the related angle for: (a) 315°; (b) 265°; (c) 140°; (d) $-20°$.

Solution:

(a) Because 315° lies between 270° and 360°, the related angle is $(360 - 315)° = 45°$.
(b) Because 265° lies between 180° and 270°, the related angle is $(265 - 180)° = 85°$.
(c) Because 140° is between 90° and 180°, the related angle is $(180 - 140)° = 40°$.
(d) Convert $-20°$ to a positive angle by adding 360°; you get $(360 - 20)° = 340°$; the related angle of $-20°$ equals the related angle of 340°, which is $(360 - 340)° = 20°$.

B. Using related angles

To find any trigonometric function of a number t, you

1. reduce t to a number b between 0 and 2π. You do this by subtracting multiples of 2π from t. If t is given in degrees, subtract multiples of 360° to reduce t to a number b between 0° and 360°.

2. determine the related angle θ of b from the definition of related angles.

3. find the trigonometric function of θ in a table of trigonometric functions or from your calculator.

4. attach the correct sign for the quadrant in which the point lies.

EXAMPLE 7-3: Find the sine, cosine, and tangent of: **(a)** 120°; **(b)** 240°; **(c)** 300°.

Solution:

(a) Because 120° lies between 90° and 180°, its related angle is $(180 - 120)° = 60°$. Recall that $\sin 60° = \frac{\sqrt{3}}{2}$, $\cos 60° = \frac{1}{2}$, $\tan 60° = \sqrt{3}$, so $\sin 120° = \frac{\sqrt{3}}{2}$, $\cos 120° = -\frac{1}{2}$, $\tan 120° = -\sqrt{3}$.

(b) Because 240° lies between 180° and 270°, its related angle is $(240 - 180)° = 60°$, so $\sin 240° = \frac{-\sqrt{3}}{2}$, $\cos 240° = -\frac{1}{2}$, $\tan 240° = \sqrt{3}$.

(c) Because 300° lies between 270° and 360°, its related angle is $(360 - 300)° = 60°$, so $\sin 300° = \frac{-\sqrt{3}}{2}$, $\cos 300° = \frac{1}{2}$, $\tan 300° = -\sqrt{3}$.

EXAMPLE 7-4: Find $\sin \alpha$, $\cos \alpha$, and $\tan \alpha$ for: **(a)** $\alpha = 150°$; **(b)** $\alpha = -45°$; **(c)** $\alpha = 210°$.

Solution:

(a) Because 150° lies between 90° and 180°, the reference angle is $(180 - 150)° = 30°$, so $\cos 150° = -\cos 30° = \frac{-\sqrt{3}}{2}$; $\sin 150° = \sin 30° = \frac{1}{2}$; and $\tan 150° = -\tan 30° = \frac{-\sqrt{3}}{3}$.

(b) You know that $-45°$ equals positive angle $(360 - 45)° = 315°$. Because 315° lies between 270 and 360 degrees, its reference angle is $(360 - 315)° = 45°$, so $\cos(-45°) = \cos 45° = \frac{\sqrt{2}}{2}$; $\sin(-45°) = -\sin 45° = \frac{-\sqrt{2}}{2}$; and $\tan(-45°) = -\tan 45° = -1$.

(c) Note that the reference angle of 210° is $(210 - 180)° = 30°$, so $\cos 210° = -\cos 30° = \frac{-\sqrt{3}}{2}$; $\sin 210° = -\sin 30° = -\frac{1}{2}$; and $\tan 210° = \tan 30° = \frac{\sqrt{3}}{3}$.

EXAMPLE 7-5: Find $\cos \frac{7}{6}\pi$.

Solution: Because $\frac{7}{6}\pi$ lies in the third quadrant, its related angle is $\frac{7}{6}\pi - \pi = \frac{1}{6}\pi$. You know that $\cos \frac{1}{6}\pi = \frac{\sqrt{3}}{2}$, so $\cos \frac{7}{6}\pi = \frac{-\sqrt{3}}{2}$.

EXAMPLE 7-6: Find $\csc \frac{31}{4}\pi$.

Solution: First reduce $\frac{31}{4}\pi$ to an equivalent angle between 0 and 2π: $\frac{31}{4}\pi = \frac{24}{4}\pi + \frac{7}{4}\pi = 3(2\pi) + \frac{7}{4}\pi$. Subtract 6π to find the equivalent angle, $\frac{7}{4}\pi$, which lies in the fourth quadrant. Its related angle is $2\pi - \frac{7}{4}\pi = \frac{1}{4}\pi$, so $\csc \frac{1}{4}\pi = 1/(\sqrt{2}/2) = \sqrt{2}$. Then, since $\frac{7}{4}\pi$ is in the fourth quadrant, $\csc \frac{7}{4}\pi = -\sqrt{2} = \csc \frac{31}{4}\pi$.

EXAMPLE 7-7: Find $\sin 2.4$ and $\tan 2.4$.

Solution: Recall that without the degree sign, 2.4 is measured in radians. There are 2π, or approximately 6.28 rad, on the circumference of a circle, so 2.4 rad lies between $\frac{6.28}{4} = 1.57$ and $\frac{6.28}{2} = 3.14$ rad. This places the angle corresponding to 2.4 rad in the second quadrant. Its reference angle is $\pi - 2.4 = 3.14 - 2.4 = 0.74$ rad. Consult a table of trigonometric functions or use your calculator to find $\sin 0.74 = 0.6743$ and $\tan 0.74 = 0.9131$.

C. Using your calculator

Calculators with trigonometric function keys rapidly provide values of the trigonometric functions. You can find the value of such functions whether the number is positive or negative, or expressed in degrees or radians. To find any trigonometric function of t with a calculator,

1. Place your calculator in the proper mode for radians or degrees. If your calculator has only one mode, and the number is not given in that mode, you will have to convert to the mode of your calculator by using the formulas in Section 3-2D.
2. Enter the number. If the number is too large, you may have to reduce it by subtracting multiples of 2π, or $360°$.
3. Press the appropriate function key. If your calculator doesn't have the desired function key, use the reciprocal relationships to convert the expression into a form that can be entered on your calculator.

EXAMPLE 7-8: Use your calculator to find (**a**) $\sin 536°$; (**b**) $\sec -727°$; (**c**) $\tan 14.3\pi$; (**d**) $\csc -\frac{34}{15}\pi$.

Solution:

(**a**) Set your calculator in degree mode (or convert $536°$ to radians if your calculator has only radian mode) and enter 536. Press $\boxed{\sin}$ to get $\sin 536° = 0.0698$.

(**b**) Set your calculator in degree mode (or convert $-727°$ to radians if your calculator has only radian mode) and enter -727. (This may require that you enter $\boxed{727}$ $\boxed{+/-}$ in order to change the sign.) Press $\boxed{\cos}$ to get $\cos(-727°) = 0.9925$. Finally, press $\boxed{1/x}$ to get $\sec(-727°) = \frac{1}{0.9925} = 1.0075$.

(**c**) Place your calculator in radian mode (or convert 14.3π to degrees if your calculator has only degree mode) and enter $\boxed{14.3}$ $\boxed{\times}$ $\boxed{\pi}$ $\boxed{=}$ $\boxed{\tan}$. If your calculator displays an error message, the number entered is too large for your calculator. Reduce it to a smaller number by subtracting multiples of 2π, in this case, subtract $7(2\pi) = 14\pi$: $14.3\pi - 14\pi = 0.3\pi$. Enter $\boxed{0.3}$ $\boxed{\times}$ $\boxed{\pi}$ $\boxed{=}$ $\boxed{\tan}$ to get $\tan 14.3\pi = \tan 0.3\pi = 1.3764$.

(**d**) Place your calculator in radian mode (or convert $-\frac{34}{15}\pi$ to degrees if your calculator has only degree mode) and enter $\boxed{34}$ $\boxed{+/-}$ $\boxed{\times}$ $\boxed{\pi}$ $\boxed{\div}$ $\boxed{15}$ $\boxed{=}$ $\boxed{\sin}$ to get $\sin(-\frac{34}{15}\pi) = -0.7431$. Press $\boxed{1/x}$ to get $\csc(-\frac{34}{15}\pi) = -1.3456$.

EXAMPLE 7-9: Use your calculator to find (**a**) $\sin 1$; (**b**) $\cos 3.88$; (**c**) $\sec 6$.

Solution: The angles are given in radians, so fix your calculator in radian mode or convert 1, 3.88, and 6 to degrees (you get 57.30, 444.62, and 687.55 degrees, respectively). Then use your calculator as described in Example 7-8 to get

(**a**) $\sin 1 = 0.8415$
(**b**) $\cos 3.88 = -0.7395$
(**c**) $\sec 6 = 1/\cos 6 = 1/0.9602 = 1.0415$

7-2. Finding Angles

A. Using related angles

If you are given a value for one of the trigonometric functions, then you can find all numbers (that is, angles in radians) that produce the given value of the trigonometric function by finding one such angle and using the periodicity of the trigonometric functions to find all other angles. The following examples illustrate the procedure by which you can find angles between 0 and 2π corresponding to a given value of a trigonometric function.

EXAMPLE 7-10: Find α between 0 and 2π for which $\sin \alpha = \frac{\sqrt{2}}{2}$ and $\cos \alpha = \frac{\sqrt{2}}{2}$.

Solution: You know that $\sin \alpha$ is positive in the first and second quadrants and $\cos \alpha$ is positive in the first and fourth quadrants. Since both sine and cosine take positive values, α is in the first quadrant and, from Table 6-1, the acute angle with sine and cosine equal to $\frac{\sqrt{2}}{2}$ is $\frac{1}{4}\pi$.

EXAMPLE 7-11: For $\csc \alpha = -2$ and $\tan \alpha = \frac{\sqrt{3}}{3}$, find α.

Solution: Since $\csc \alpha$ is negative in the third and fourth quadrants and $\tan \alpha$ is positive in the first and third quadrants, α must lie in the third quadrant. From Table 6-1, the acute angle, θ, with $\tan \theta = \frac{\sqrt{3}}{3}$ and $\csc \theta = 2$ is $\theta = \frac{1}{6}\pi$. The solution is the angle in quadrant III whose related angle is $\frac{1}{6}\pi$, or $\alpha = \pi + \frac{1}{6}\pi = \frac{7}{6}\pi$.

EXAMPLE 7-12: Angle α is between $90°$ and $180°$ and $\cos \alpha = -0.9643$; find α in degrees.

Solution: By using a table of trigonometric functions or your calculator, you find that an acute angle whose cosine is $+0.9634$ is $15.55°$. The desired angle is a second-quadrant angle; therefore, from the reference-angle relationship, $\alpha = (180 - 15.55)° = 164.45°$.

B. Using your calculator

If you are given the value of a trigonometric function, you can use your calculator to find the corresponding angle. Enter the given value and use the inverse trigonometric functions, that is, the functions that give *the angle whose sine* (or *cosine*, or *tangent*) *is*. Values of the inverse trigonometric functions are found by first pressing the $\boxed{\text{INV}}$ key and then pressing $\boxed{\text{sin}}$, $\boxed{\text{cos}}$, or $\boxed{\text{tan}}$. Some calculators possess $\boxed{\text{arc sin}}$ or $\boxed{\text{sin}^{-1}}$ keys. You can find further information on inverse functions in Chapter 13. You should consult the directions supplied with your calculator for key arrangements that differ from the above.

EXAMPLE 7-13: Find α in degrees such that $\sin \alpha = -\frac{1}{2}$.

Solution: Place your calculator in degree mode and enter $\boxed{1}\boxed{+/-}\boxed{\div}$ $\boxed{2}\boxed{=}\boxed{\text{INV}}\boxed{\text{sin}}$. The display will show an angle in degrees whose sine is $-\frac{1}{2}$, in this case -30, standing for negative $30°$.

EXAMPLE 7-14: Find α: (a) in radians, such that $\tan \alpha = 3$; (b) in the third quadrant and in degrees, such that $\tan \alpha = 3$.

Solution:

(a) Place your calculator in radian mode. Enter $\boxed{3}\boxed{\text{INV}}\boxed{\text{tan}}$. The result is 1.2490, that is, an angle of 1.2490 rad.
(b) Convert 1.2490 radians to degrees: $1.2490 \text{ rad} = 1.2490(180)/\pi = 71.56°$. To find the angle in the third quadrant with tangent equal to 3, find the angle in quadrant III whose related angle is $71.56°$, that is, $\alpha = (180 + 71.56)° = 251.56°$.

EXAMPLE 7-15: Find α in radians if $-\pi < \alpha < 0$ and $\cos \alpha = 0.6$.

Solution: Fix your calculator in radian mode. Enter $\boxed{0.6}\boxed{\text{INV}}\boxed{\text{cos}}$. The result is 0.9273, that is, 0.9273 rad. However, 0.9273 rad lies between 0 and $\frac{1}{2}\pi$ radians. Use $\cos \alpha = \cos(-\alpha)$ to get $\alpha = -0.9273$ rad.

SUMMARY

1. The related angle of angle α is

θ if θ is between 0 and $\frac{1}{2}\pi$

$\pi - \theta$ if θ is between $\frac{1}{2}\pi$ and π

$\theta - \pi$ if θ is between π and $\frac{3}{2}\pi$

$2\pi - \theta$ if θ is between $\frac{3}{2}\pi$ and 2π

2. To find the values of the trigonometric functions for any angle by the related-angle method, find a related angle, find the trigonometric functions of the related angle, and attach the sign according to the quadrant of the given angle.
3. To find the values of the trigonometric functions directly from your calculator, enter the number and press the appropriate trigonometric function key.
4. To find the number (angle in radians) that corresponds to the value of a trigonometric function, find an acute angle that has the given value (without sign). This is a related angle of the desired angle. Use the relationships between angles and related angles to complete the solution.
5. To find the number (angle in radians) that corresponds to the value of a trigonometric function directly from your calculator, enter the number and press the appropriate inverse trigonometric function, or arc function, key.

RAISE YOUR GRADES

Can you...?

☑ find the reference angle for a given angle
☑ find an angle given the related angle
☑ find the values of the trigonometric functions for any real number using related angles
☑ find the values of the trigonometric functions for any real number using your calculator
☑ find an angle (using related angles) given the value of one or more of the trigonometric functions for that angle
☑ find an angle (using your calculator) given the value of one or more of the trigonometric functions for that angle

SOLVED PROBLEMS

PROBLEM 7-1 Use related angles to find $\sin \frac{5}{6}\pi$ and $\cos \frac{5}{6}\pi$.

Solution: The related angle for $\frac{5}{6}\pi$ is $\pi - \frac{5}{6}\pi = \frac{1}{6}\pi$. From Table 6-1, $\sin \frac{1}{6}\pi = \frac{1}{2}$ and $\cos \frac{1}{6}\pi = \frac{\sqrt{3}}{2}$. Since $\frac{5}{6}\pi$ is a quadrant-II angle, $\sin \frac{5}{6}\pi = \frac{1}{2}$ and $\cos \frac{5}{6}\pi = \frac{-\sqrt{3}}{2}$.　　　　　　[See Section 7-1.]

PROBLEM 7-2 Angle α is an angle in radians in the third quadrant. Find α if: **(a)** $\sin \alpha = -\frac{1}{2}$; **(b)** $\cos \alpha = -\frac{1}{2}$. Express your answer in radians.

Solution:

(a) Recall that $\sin \frac{1}{6}\pi = \frac{1}{2}$. A quadrant-III angle related to $\frac{1}{6}\pi$ is $\pi + \frac{1}{6}\pi = \frac{7}{6}\pi$. Therefore, $\sin \frac{7}{6}\pi = -\frac{1}{2}$.
(b) Remember that $\cos \frac{1}{3}\pi = \frac{1}{2}$. A quadrant-III angle related to $\frac{1}{3}\pi$ is $\pi + \frac{1}{3}\pi = \frac{4}{3}\pi$. Thus, $\cos \frac{4}{3}\pi = -\frac{1}{2}$.　　　　　　[See Section 7-2.]

PROBLEM 7-3 Angle α is between 0 and 2π, $\sin \alpha = \frac{-\sqrt{3}}{2}$, and $\cos \alpha = \frac{1}{2}$; find α (in radians).

Solution: Since sine is negative in quadrants III and IV and cosine is positive in quadrants I and IV, α must lie in quadrant IV. From Table 6-1, $\sin \frac{1}{3}\pi = \frac{\sqrt{3}}{2}$ and $\cos \frac{1}{3}\pi = \frac{1}{2}$. Therefore, the angle between 0 and 2π related to $\frac{1}{3}\pi$ in quadrant IV is $2\pi - \frac{1}{3}\pi = \frac{5}{3}\pi$.　　　　　　[See Section 7-2.]

PROBLEM 7-4 Find $\cos\alpha$, $\tan\alpha$, $\cot\alpha$, $\sec\alpha$, and $\csc\alpha$ if: **(a)** $\sin\alpha = \frac{3}{5}$ and $0 < \alpha < \frac{1}{2}\pi$; **(b)** $\sin\alpha = \frac{3}{5}$ and $\frac{1}{2}\pi < \alpha < \pi$.

Solution:

(a) You want to find a point (x, y) in the first quadrant for which $\sin\alpha = y/r = \frac{3}{5}$. If you let $y = 3$ and $r = 5$, then $x^2 = r^2 - y^2 = 5^2 - 3^2 = 25 - 9 = 16$, so $x = +4$ (you choose the plus sign because $0 < \alpha < \frac{1}{2}\pi$). Then, $\cos\alpha = x/r = \frac{4}{5}$, $\tan\alpha = y/x = \frac{3}{4}$, $\csc\alpha = r/y = \frac{5}{3}$, $\sec\alpha = r/x = \frac{5}{4}$, and $\cot\alpha = x/y = \frac{4}{3}$.

(b) By definition, $\sin\alpha = y/r = \frac{3}{5}$. If you let $y = 3$ and $r = 5$, then $x^2 = r^2 - y^2 = 5^2 - 3^2 = 25 - 9 = 16$, so $x = -4$ (you choose the minus sign because $\cos\alpha < 0$ for $\frac{1}{2}\pi < \alpha < \pi$). Then, $\cos\alpha = x/r = -\frac{4}{5}$, $\tan\alpha = y/x = -\frac{3}{4}$, $\csc\alpha = r/y = \frac{5}{3}$, $\sec\alpha = r/x = -\frac{5}{4}$, and $\cot\alpha = x/y = -\frac{4}{3}$. [See Section 7-1.]

PROBLEM 7-5 What are the values of $\sin\alpha$ and $\cos\alpha$ if: **(a)** $\tan\alpha = 1$; **(b)** $\tan\alpha = -1$?

Solution:

(a) From Table 6-1, $\frac{1}{4}\pi$ is an angle with tangent equal to 1. Since $\frac{1}{4}\pi$ is a related angle of $\frac{5}{4}\pi$ $\left(\frac{5}{4}\pi - \pi = \frac{1}{4}\pi\right)$, $\tan\frac{5}{4}\pi = 1$. Then, from your calculator or a table of trigonometric functions, you find $\sin\frac{1}{4}\pi = \cos\frac{1}{4}\pi = \frac{\sqrt{2}}{2}$, or $\sin\frac{5}{4}\pi = \cos\frac{5}{4}\pi = \frac{-\sqrt{2}}{2}$.

(b) From Table 6-1, $-\frac{1}{4}\pi$ is an angle with tangent equal to -1. The positive angles with tangent equal to -1 are $\frac{7}{4}\pi(2\pi - \frac{1}{4}\pi)$ and $\frac{3}{4}\pi(\pi - \frac{1}{4}\pi)$. Then, from your calculator or from a table of trigonometric functions, you find $\sin\frac{7}{4}\pi = \frac{-\sqrt{2}}{2}$, $\cos\frac{7}{4}\pi = \frac{\sqrt{2}}{2}$, $\sin\frac{3}{4}\pi = \frac{\sqrt{2}}{2}$, and $\cos\frac{3}{4}\pi = \frac{-\sqrt{2}}{2}$.

PROBLEM 7-6 Approximate the values of α between 0 and 360° for which: **(a)** $\sin\alpha = \frac{3}{5}$; **(b)** $\sin\alpha = \frac{12}{13}$.

Solution:

(a) The acute angle α for which $\sin\alpha = \frac{3}{5}$ is found to be 36.9° on your calculator. The sine function is positive only in quadrants I and II. Thus, the only other angle that satisfies $\sin\alpha = \frac{3}{5}$ is $(180 - 36.9)° = 143.1°$ (the quadrant-II angle whose related angle is 36.9°).

(b) The quadrant-I value of α for which $\sin\alpha = \frac{12}{13}$ is 67.4°. The only other angle that satisfies $\sin\alpha = \frac{12}{13}$ is $(180 - 67.4)° = 112.6°$ (the quadrant-II angle whose related angle is 67.4°). [See Section 7-2.]

PROBLEM 7-7 The terminal side of angle α passes through point (a, b); for which numbers a and b will $\sin\alpha = b/a$?

Solution: Begin by sketching the ray from the origin passing through (a, b) as in Figure 7-3. Any α in standard position with this ray as terminal side satisfies $\sin\alpha = b/\sqrt{a^2 + b^2}$, and this equals b/a only if $b = 0$. The only angles for which $\sin\alpha = b/a$ are those with $b = 0$, that is, all angles whose terminal side coincides with either the positive or negative x axis. [See Section 7-1.]

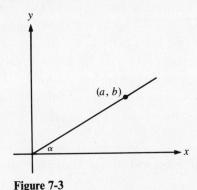

Figure 7-3

PROBLEM 7-8 Tell whether the cosine function is an even function, an odd function, or neither. Tell whether the sine function is even, odd, or neither.

Solution: Recall from Chapter 4 that f is an even function if $f(t) = f(-t)$ for any t and an odd function if $f(-t) = -f(t)$ for any t. You can show that $\cos(-\alpha) = \cos\alpha$. The argument for α, a quadrant-I angle, is as follows: Since $\cos\alpha = x/r$ and $\cos(-\alpha) = x/r$, then $\cos(-\alpha) = \cos\alpha$ (see Figure 7-4). Similarly, $\sin(-\alpha) = -\sin\alpha$ since $\sin\alpha = y/r$ and $\sin(-\alpha) = -y/r$ (see Figure 7-4). You should verify that these identities, $\sin(-\alpha) = -\sin\alpha$ and $\cos(-\alpha) = \cos\alpha$, are true for any α. Thus, cosine is an even function and sine is an odd function.

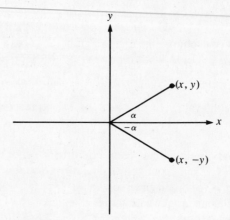

Figure 7-4

PROBLEM 7-9 Find α such that $0 < \alpha < 90°$ and: **(a)** $\cos\alpha = \cos 190°$; **(b)** $\sin\alpha = -\sin(-95°)$; **(c)** $\sin\alpha = \sin 890°$.

Solution:

(a) Because cosine is positive in quadrant I and negative in quadrant III, no α satisfies the given condition.

(b) For any α, $\sin(-\alpha) = -\sin\alpha$, so $-\sin(-95°) = \sin 95°$. Then you can restate the problem as: Find α, $0 < \alpha < 90°$, for which $\sin\alpha = \sin 95°$. The solution is the related angle of $95°$ or, $(180 - 95)° = 85°$.

(c) Because the sine function repeats its values for each change of $360°$ in the variable, $\sin 890° = \sin(720 + 170)° = \sin 170°$. Then you can restate the problem as: Find α in quadrant I for which $\sin\alpha = \sin 170°$. The solution is the related angle of $170°$, or $(180° - 170°) = 10°$.

[See Section 7-2.]

PROBLEM 7-10 Find a number x such that $0 < x < \frac{1}{2}\pi$ and: **(a)** $\sin x = \sin 2$; **(b)** $\tan x = \tan(-4)$; **(c)** $\cos x = \cos\frac{27}{4}\pi$.

Solution:

(a) Since $\sin 2 = 0.9093$, x is the number whose sine is 0.9093. To find such a number between 0 and $\frac{1}{2}\pi$, fix your calculator in radian mode, enter $\boxed{0.9093}\ \boxed{\text{INV}}\ \boxed{\sin}$. The result is $x = 1.1416$ rad. This number is the solution since it is between 0 and $\frac{1}{2}\pi$ and $\sin 1.1416 = 0.9093 = \sin 2$.
Keep your calculator in radian mode for parts **(b)** and **(c)**.

(b) Enter $\boxed{4}\ \boxed{+/-}\ \boxed{\tan}$ to get $\tan(-4) = -1.1578$. Since the tangent function is nonnegative for $0 < x < \frac{1}{2}\pi$, there is no x that satisfies both conditions.

(c) Enter $\boxed{27}\ \boxed{\times}\ \boxed{\pi}\ \boxed{\div}\ \boxed{4}\ \boxed{=}\ \boxed{\cos}$ to get $\cos\frac{27}{4}\pi = -0.7071$. Again, since the cosine function is nonnegative for $0 < x < \frac{1}{2}\pi$, there is no x that satisfies both conditions. [See Section 7-2.]

PROBLEM 7-11 Find α if: **(a)** α is a quadrant-II angle and $\sin\alpha = \frac{\sqrt{2}}{2}$; **(b)** $-\frac{1}{2}\pi < \alpha < 0$ and $\cos\alpha = \frac{1}{2}$; **(c)** α is a quadrant-III angle and $\tan\alpha = 1$; **(d)** α is a quadrant-IV angle and $\sin\alpha = \frac{-\sqrt{3}}{2}$.

Solution:

(a) Since $\sin\frac{1}{4}\pi = \frac{\sqrt{2}}{2}$, the solution is the quadrant-II angle whose related angle is $\frac{1}{4}\pi$: $\pi - \frac{1}{4}\pi = \frac{3}{4}\pi$.

(b) Since $\cos(-\frac{1}{3}\pi) = \cos\frac{1}{3}\pi = \frac{1}{2}$, the solution is the quadrant-IV angle $-\frac{1}{3}\pi$.

(c) Since $\tan \alpha = 1$, the solution is the quadrant-III angle whose related angle is $\frac{1}{4}\pi$: $\pi + \frac{1}{4}\pi = \frac{5}{4}\pi$.

(d) Since $\sin \frac{1}{3}\pi = \frac{\sqrt{3}}{2}$, the solution is the quadrant-IV angle whose related angle is $\frac{1}{3}\pi$: $2\pi - \frac{1}{3}\pi = \frac{5}{3}\pi$.

[See Section 7-2.]

PROBLEM 7-12 Angle $\alpha = \frac{7}{6}\pi$; find $\sin \alpha$, $\cos \alpha$, $\tan \alpha$, $\cot \alpha$, $\csc \alpha$, and $\cot \alpha$ without using your calculator.

Solution: The related angle is $\frac{7}{6}\pi - \pi = \frac{1}{6}\pi$. From Table 6-1, $\sin \frac{1}{6}\pi = \frac{1}{2}$, $\cos \frac{1}{6}\pi = \frac{\sqrt{3}}{2}$, $\tan \frac{1}{6}\pi = \frac{1}{\sqrt{3}}$, or $\frac{\sqrt{3}}{3}$, $\csc \frac{1}{6}\pi = 2$, $\sec \frac{1}{6}\pi = \frac{2\sqrt{3}}{3}$, and $\cot \frac{1}{6}\pi = \sqrt{3}$. The values of these functions at $\frac{7}{6}\pi$ differ only in sign. They are: $\sin \frac{7}{6}\pi = -\frac{1}{2}$, $\cos \frac{7}{6}\pi = \frac{-\sqrt{3}}{2}$, $\tan \frac{7}{6}\pi = \frac{\sqrt{3}}{3}$, $\csc \frac{7}{6}\pi = -2$, $\sec \frac{7}{6}\pi = \frac{-2\sqrt{3}}{3}$, and $\cot \frac{7}{6}\pi = \sqrt{3}$.

[See Section 7-1.]

PROBLEM 7-13 Find the values of $\sin \alpha$, $\cos \alpha$, and $\tan \alpha$ if: (a) $\sec \alpha = 2$; (b) $\sec \alpha = -2$.

Solution:

(a) From the reciprocal relationships, $1/\cos \alpha = \sec \alpha = 2$, so $\cos \alpha = \frac{1}{2}$. Let $x = 1$ and $r = 2$. From $r^2 = x^2 + y^2$, $2^2 = 1^2 + y^2$ or $y^2 = 3$. Then y is either $\sqrt{3}$ or $-\sqrt{3}$. Therefore, $\sin \alpha = y/r = \frac{\sqrt{3}}{2}$ or $\frac{-\sqrt{3}}{2}$ and $\tan \alpha = y/x = \sqrt{3}$ or $-\sqrt{3}$.

(b) As in part (a), $1/\cos \alpha = \sec \alpha = -2$, so $\cos \alpha = -\frac{1}{2}$. As in part (a), let $x = -1$ and $r = 2$. Then, $y = \sqrt{3}$ or $-\sqrt{3}$. So, $\sin \alpha = y/r = \frac{\sqrt{3}}{2}$ or $\frac{-\sqrt{3}}{2}$ and $\tan \alpha = y/x = \sqrt{3}$ or $-\sqrt{3}$.

[See Section 7-1.]

PROBLEM 7-14 Angle $\alpha = \frac{1}{5}\pi$; find $\sin \alpha$ and $\cos(\frac{1}{2}\pi - \alpha)$.

Solution: Using your calculator, $\sin \frac{1}{5}\pi = 0.5878$, and $\cos(\frac{1}{2}\pi - \frac{1}{5}\pi) = \cos \frac{3}{10}\pi = 0.5878$. These two results are equal because $\cos(\frac{1}{2}\pi - \alpha) = \sin \alpha$ for α an acute angle.

[See Sections 5-1D and 7-1.]

PROBLEM 7-15 Angle $\alpha = \frac{3}{5}\pi$; find $\cos \alpha/2$ and $+\sqrt{(1 + \cos \alpha)/2}$.

Solution: Substituting $\frac{3}{5}\pi$ for α, you get $\cos(\frac{3}{5}\pi/2) = \cos \frac{3}{10}\pi = 0.5878$ and $+\sqrt{(1 + \cos \frac{3}{5}\pi)/2} = +\sqrt{(1 + (-0.3090))/2} = 0.5878$.

PROBLEM 7-16 For $\cos \alpha = \frac{\sqrt{3}}{2}$, find $(\sec \alpha)^2$, or $\sec^2 \alpha$, and $(\tan \alpha)^2$, or $\tan^2 \alpha$.

Solution: From the reciprocal relations, $\sec \alpha = 1/\cos \alpha = \frac{2}{\sqrt{3}} = r/x$. Let $r = 2$ and $x = \sqrt{3}$. Then $y^2 = r^2 - x^2 = 2^2 - (\sqrt{3})^2 = 1$, so $y = +1$ or -1. You get $\tan \alpha = y/x = \frac{1}{\sqrt{3}}$ or $\frac{-1}{\sqrt{3}}$. Also, $\sec^2 \alpha = (\frac{2}{\sqrt{3}})^2 = \frac{4}{3}$ and $\tan^2 \alpha = (\frac{\pm 1}{\sqrt{3}})^2 = \frac{1}{3}$. You have shown that $1 + \tan^2 \alpha = \sec^2 \alpha$ for $\cos \alpha = \frac{\sqrt{3}}{2}$. This trigonometric identity is true for any α (see Chapter 9).

[See Section 7-1.]

PROBLEM 7-17 Find $\sin 8$ and $\cos 8$.

Solution: First express 8 rad as an a multiple of 2π plus an angle between 0 and 2π: $8 = 6.2832 + 1.7168 = 2\pi + 1.7168$. Then use the fact that the values of the sine and cosine functions repeat over each interval of 2π radians, that is, $\sin 8 = \sin(2\pi + 1.7168) = \sin 1.7168$ and $\cos 8 = \cos(2\pi + 1.7168) = \cos 1.7168$. The values for $\sin 1.7168$ and $\cos 1.7168$ are found by placing your calculator in radian mode, entering 1.7168, and pressing the appropriate key: $\sin 1.7168 = 0.9894$ and $\cos 1.7168 = -0.1455$.

[See Section 7-1.]

Supplementary Exercises

PROBLEM 7-18 Find the values of $\sin \frac{5}{4}\pi$ and $\cos \frac{5}{4}\pi$.

PROBLEM 7-19 Find the values of $\sin \frac{127}{6}\pi$ and $\cos \frac{127}{6}\pi$.

PROBLEM 7-20 Find the values of $\sin 240°$ and $\cos 240°$.

PROBLEM 7-21 Find an angle α (in radians) in the fourth quadrant for which: (a) $\sin \alpha = \frac{-\sqrt{2}}{2}$; (b) $\cos \alpha = \frac{\sqrt{2}}{2}$.

PROBLEM 7-22 Find an angle α (in radians) in the second quadrant for which: (a) $\sin \alpha = \frac{\sqrt{3}}{2}$; (b) $\cos \alpha = \frac{-\sqrt{3}}{2}$.

PROBLEM 7-23 Angle α is between 0 and 360°, $\sin \alpha = \frac{1}{3}$, and $\cos \alpha = \frac{-\sqrt{8}}{3}$; find α.

PROBLEM 7-24 For $\sin \alpha = \frac{2}{3}$ and $\cos \alpha = \frac{-\sqrt{5}}{3}$, find $\tan \alpha$.

PROBLEM 7-25 Find all values of α (in radians) for: (a) $\tan \alpha > 0$; (b) $\tan \alpha < 0$.

PROBLEM 7-26 Find the reference, or related, angle for each angle in the given unit: (a) 225°; (b) 540°; (c) $-270°$; (d) $\frac{29}{12}\pi$ rad; (e) $\frac{24}{7}\pi$ rad; (f) $-\frac{15}{4}\pi$ rad; (g) $\frac{837}{3}°$; (h) 179 rad; (i) $73\pi°$.

PROBLEM 7-27 Find the sine of each of the angles in Problem 7-26.

PROBLEM 7-28 Evaluate $\tan \alpha$ for: (a) $\alpha = 0.46\pi$; (b) 0.47π; (c) 0.54π.

PROBLEM 7-29 Find a number x such that $\frac{1}{2}\pi < x < \pi$ and: (a) $\sin x = \sin 2$; (b) $\tan x = \tan(-4)$; (c) $\cos x = \cos \frac{27}{4}\pi$.

PROBLEM 7-30 Use your calculator to find α between 0° and 360° for: (a) α in quadrant II and $\sin \alpha = \frac{3}{4}$; (b) α in quadrant III and $\sin \alpha = -\frac{3}{4}$; (c) α in quadrant IV and $\sin \alpha = -\frac{3}{4}$; (d) α in quadrant IV and $\cos \alpha = 0.1351$; (e) α in quadrant III and $\tan \alpha = 2.5106$.

PROBLEM 7-31 In each of the following find two angles in degrees with positive measure less than 360° for which: (a) $\tan \alpha = \frac{\sqrt{3}}{3}$; (b) $\sin \alpha = -\frac{1}{2}$; (c) $\cos \alpha = \frac{-\sqrt{2}}{2}$; (d) $\tan \alpha = -1$; (e) $\sin \alpha = \frac{-\sqrt{2}}{2}$; (f) $\cos \alpha = \frac{1}{2}$.

PROBLEM 7-32 Use your calculator to find the value of: (a) $\sec 100°$; (b) $\csc 305°$; (c) $\cot 200°$.

PROBLEM 7-33 Use your calculator to find the value of: (a) $(\sec 2)^2 - (\tan 2)^2$; (b) $(\csc 42°)^2 - (\cot 42°)^2$.

PROBLEM 7-34 Use your calculator to find the value of α between 0° and 360° for: (a) $\sec \alpha = 3.1421$ and α in quadrant IV; (b) $\cot \alpha = 0.4215$ and α in quadrant III; (c) $\sec \alpha = -1.2145$ and α in quadrant III; (d) $\csc \alpha = 5.3100$ and α in quadrant II; (e) $\cot \alpha = -8.1025$ and α in quadrant IV; (f) $\csc \alpha = -3.1857$ and α in quadrant III.

PROBLEM 7-35 Find a value of α between 0 and 2π for which $\sin \alpha = -\frac{1}{2}$ and $\cos \alpha < 0$.

PROBLEM 7-36 Find a value of α between 2π and 4π for which $\cos \alpha = \frac{1}{2}$ and $\sin \alpha < 0$.

PROBLEM 7-37 Find a value of α between $-\pi$ and -2π for which $\sec \alpha = 3$ and $\cot \alpha > 0$.

PROBLEM 7-38 Find the values of $\sin \alpha$, $\cos \alpha$, and $\tan \alpha$ for $0 < \alpha < \frac{1}{2}\pi$ and: (a) $\csc \alpha = 3$; (b) $\cot \alpha = 3$.

PROBLEM 7-39 For $\sin \alpha < 0$ and $\tan \alpha = 4.8$, find α between 0 and 2π.

PROBLEM 7-40 For $\sin \alpha > 0$ and $\cot \alpha = -0.7$, find α between 2π and 8π.

PROBLEM 7-41 For $\tan \alpha < 0$ and $\cos \alpha = -0.38$, find α between 0° and 360°.

PROBLEM 7-42 For $\cot \alpha > 0$ and $\sin \alpha = -0.53$, find α between $\frac{3}{2}\pi$ and $\frac{11}{2}\pi$.

PROBLEM 7-43   Angle $\alpha = \frac{3}{7}\pi$; find $\tan \alpha$ and $\cot(\frac{1}{2}\pi - \alpha)$.

PROBLEM 7-44   Angle $\alpha = \frac{3}{5}\pi$; find $\sin \frac{1}{2}\alpha$ and $+\sqrt{(1 - \cos\alpha)/2}$.

PROBLEM 7-45   For $\cos \alpha = \frac{-\sqrt{3}}{2}$, find $\csc^2\alpha$ and $\cot^2\alpha$.

Answers to Supplementary Exercises

7-18: $\frac{-\sqrt{2}}{2}$ and $\frac{-\sqrt{2}}{2}$

7-19: $-\frac{1}{2}$ and $\frac{-\sqrt{3}}{2}$

7-20: $\frac{-\sqrt{3}}{2}$ and $-\frac{1}{2}$

7-21: **(a)** $\frac{7}{4}\pi$; **(b)** $\frac{7}{4}\pi$

7-22: **(a)** $\frac{2}{3}\pi$; **(b)** $\frac{5}{6}\pi$

7-23: $\alpha = 160.5°$

7-24: $\tan \alpha = \frac{-2\sqrt{5}}{5}$

7-25: **(a)** $\alpha = n\pi + a$ where n is any integer and $0 < a < \frac{1}{2}\pi$; **(b)** $\alpha = n\pi + a$ where n is any integer and $\frac{1}{2}\pi < a < \pi$.

7-26: **(a)** 45°; **(b)** 0°; **(c)** 90°; **(d)** $\frac{5}{12}\pi$; **(e)** $\frac{3}{7}\pi$; **(f)** $\frac{1}{4}\pi$; **(g)** 81°; **(h)** 0.071 rad; **(i)** 49.34°

7-27: **(a)** $\frac{-\sqrt{2}}{2}$; **(b)** 0; **(c)** 1; **(d)** 0.966; **(e)** −.975; **(f)** $\frac{\sqrt{2}}{2}$; **(g)** −0.988; **(h)** 0.071; **(i)** −0.758

7-28: **(a)** 7.92; **(b)** 10.58; **(c)** −7.92

7-29: **(a)** 2; **(b)** 2.28; **(c)** $\frac{3}{4}\pi$

7-30: **(a)** 131.4°; **(b)** 228.6°; **(c)** 311.4°; **(d)** 277.8°; **(e)** 248.3°

7-31: **(a)** 30° and 210°; **(b)** 210° and 330°; **(c)** 135° and 225°; **(d)** 135° and 315°; **(e)** 135° and 315°; **(f)** 60° and 300°

7-32: **(a)** −5.76; **(b)** −1.22; **(c)** 2.75

7-33: **(a)** 1; **(b)** 1

7-34: **(a)** 288.56°; **(b)** 247.14°; **(c)** 214.58°; **(d)** 169.15°; **(e)** 352.96°; **(f)** 198.29°

7-35: $\frac{7}{6}\pi$

7-36: $\frac{11}{3}\pi$

7-37: −5.05

7-38: **(a)** $\sin \alpha = \frac{1}{3}$, $\cos \alpha = \frac{2\sqrt{2}}{3}$, $\tan \alpha = \frac{\sqrt{2}}{4}$; **(b)** $\sin \alpha = \frac{\sqrt{10}}{10}$, $\cos \alpha = \frac{3\sqrt{10}}{10}$, $\tan \alpha = \frac{1}{3}$

7-39: 4.51 rad

7-40: 8.46, 14.75, 21.03, and 27.31

7-41: 112.33°

7-42: 9.98 rad and 16.27 rad

7-43: Both equal 4.38

7-44: Both equal 0.809

7-45: $\csc^2\alpha = 4$ and $\cot^2\alpha = 3$

8 GRAPHS OF THE TRIGONOMETRIC FUNCTIONS

THIS CHAPTER IS ABOUT

☑ **Graphs of the Trigonometric Functions**
☑ **Period, Amplitude, and Phase**
☑ **Graphing $y = K\sin(Lx + M) + C$**
☑ **Graphing by Addition of Ordinates**
☑ **Introduction to Harmonic Motion**

8-1. Graphs of the Trigonometric Functions

Selected values assumed by the trigonometric functions appear in Table 6-1. You can use these values to sketch the graphs of the trigonometric functions by plotting the angle in radians or degrees on the horizontal axis (value of x) and the value of the trigonometric function on the vertical axis (value of y). Use values of x from 0 to 2π for sine and cosine, from $-\pi$ to π for the others. When you join the points with a smooth curve, you'll get graphs like those in Figure 8-1, defined to be the *standard* graphs of the trigonometric functions.

You can use the graphs of Figure 8-1 to determine the domains and ranges of the trigonometric functions:

- You find the domain by finding the set of all values of x for which a value of y is defined; if a vertical line passing through any x intersects the graph, that value of x is in the domain. The domains of the trigonometric functions exclude those values of x at which vertical asymptotes occur.
- You find the range by finding the set of all values of y assumed by the function; if a horizontal line passing through any y intersects the graph, that value of y is in the range.

TABLE 8-1.
Domain and range of the trigonometric functions.

Function	Domain	Range
sine	$\{x \mid x \text{ is real}\}$	$\{y \mid y \text{ is real and } -1 \leqslant y \leqslant 1\}$
cosine	$\{x \mid x \text{ is real}\}$	$\{y \mid y \text{ is real and } -1 \leqslant y \leqslant 1\}$
tangent	$\left\{x \mid x \text{ is real and } x \neq \dfrac{\pi}{2} + n\pi, \; n \text{ any integer}\right\}$	$\{y \mid y \text{ is real}\}$
cosecant	$\left\{x \mid x \text{ is real and } x \neq n\pi, \; n \text{ any integer}\right\}$	$\{y \mid y \text{ is real and } y \leqslant -1 \text{ or } y \geqslant 1\}$
secant	$\left\{x \mid x \text{ is real and } x \neq \dfrac{\pi}{2} + n\pi, \; n \text{ any integer}\right\}$	$\{y \mid y \text{ is real and } y \leqslant -1 \text{ or } y \geqslant 1\}$
cotangent	$\left\{x \mid x \text{ is real and } x \neq n\pi, \; n \text{ any integer}\right\}$	$\{y \mid y \text{ is real}\}$

By inspecting the graphs of the trigonometric functions in Figure 8-1, you see that the domains and ranges of the trigonometric functions are as given in Table 8-1.

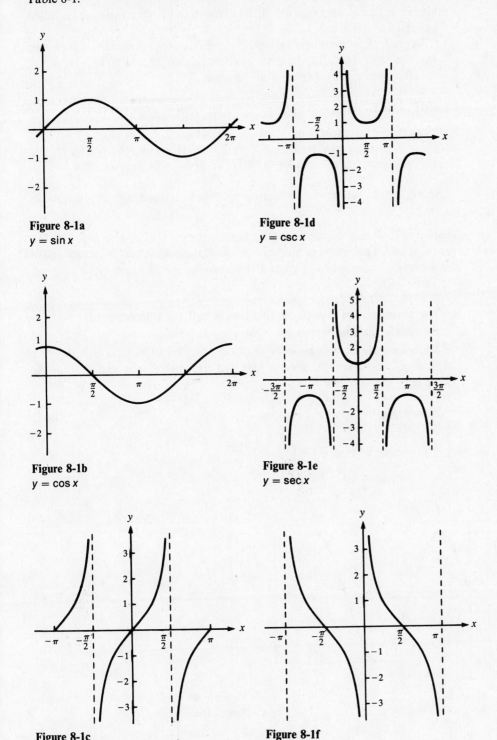

Figure 8-1a
$y = \sin x$

Figure 8-1d
$y = \csc x$

Figure 8-1b
$y = \cos x$

Figure 8-1e
$y = \sec x$

Figure 8-1c
$y = \tan x$

Figure 8-1f
$y = \cot x$

8-2. Period, Amplitude, and Phase

A. Periodic functions repeat over fixed intervals.

In considering the graphs of the trigonometric functions in Figure 8-1, note that the graphs are 'smooth'; that is, the graphs do not have sharp corners or breaks within their domains. This behavior is called **continuity**. You can say that the trigonometric functions are **continuous** wherever they are defined.

Note also that the curves of the trigonometric functions repeat at fixed intervals. This is due to the repetition of the trigonometric ratios and is called **periodicity**.

In general, a function is **periodic** if there is a positive number p such that $f(x + p) = f(x)$ for all x. If p is the smallest number for which $f(x + p) = f(x)$ for all x, then p is the **period of the function**.

EXAMPLE 8-1: What is the period of the function in Figure 8-2?

Solution: There are many numbers, p, for which $f(x + p) = f(x)$ for all x, including 2, 4, 6, or any positive even number. However, the period of this function is 2 since 2 is the smallest p for which $f(x + p) = f(x)$ for all x.

EXAMPLE 8-2: What is the period of each trigonometric function in Figure 8-1?

Solution: Observe the graphs of the functions in Figure 8-1 and the definition of period. The period of sine, cosine, cosecant, and secant is 2π radians, or 360 degrees; the period of tangent and cotangent is π radians, or 180 degrees (see Table 8-2).

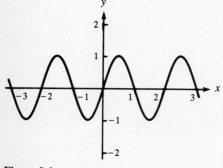

Figure 8-2
$y = \sin \pi x$

B. The amplitude of a periodic function is half the difference of the maximum and minimum values.

The **amplitude** of any periodic function $y = f(x)$ is half of the difference between the maximum y value (M) and the minimum y value (m) in the range of the function:

AMPLITUDE $$A = \frac{1}{2}(M - m) \tag{8-1}$$

Amplitudes are illustrated in Figure 8-3.

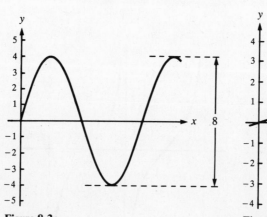

Figure 8-3a
Amplitude $= \frac{1}{2}(8) = 4$

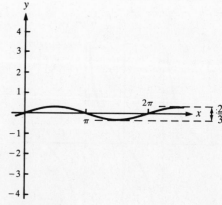

Figure 8-3b
Amplitude $= \frac{1}{2}(\frac{2}{3}) = \frac{1}{3}$

EXAMPLE 8-3: What is the amplitude of each trigonometric function in Figure 8-1?

Solution: If you observe the graphs of the trigonometric functions in Figure 8-1, you see that the amplitude for sine and cosine is 1, while the other trigonometric functions take values ranging from arbitrarily large negative numbers to arbitrarily large positive numbers. The amplitude for such functions is said to be undefined or infinite (see Table 8-2).

TABLE 8-2: Periods, Amplitudes, and Asymptotes of the Trigonometric Functions.

Function	Period in Radians	Period in Degrees	Amplitude	Vertical Asymptotes
sine	2π	360	1	None
cosine	2π	360	1	None
tangent	π	180	infinite	$\dfrac{\pi}{2} + n\pi$
cosecant	2π	360	infinite	$n\pi$
secant	2π	360	infinite	$\dfrac{\pi}{2} + n\pi$
cotangent	π	180	infinite	$n\pi$

C. The phase shift depends on the interval of repetition.

The two graphs shown in Figure 8-4 have period 2π and amplitude 1; they also have the same shape. If you shift the graph of $y = f(x)$ to the right by $\frac{1}{2}\pi$, it coincides with the graph of $y = g(x)$. If you shift the graph of $y = g(x)$ to the left by $\frac{1}{2}\pi$, it coincides with the graph of $y = f(x)$. The only difference in the graphs is this shift, called a **phase shift**. You can write $g(x) = f(x - \frac{1}{2}\pi)$ or $f(x) = g(x + \frac{1}{2}\pi)$. You say that $g(x) = f(x - \frac{1}{2}\pi)$ has a phase shift of $\frac{1}{2}\pi$ units to the right of the graph of $f(x)$ (called a phase shift of $+\frac{1}{2}\pi$ units), or you say that $f(x) = g(x + \frac{1}{2}\pi)$ has a phase shift of $\frac{1}{2}\pi$ units to the left of the graph of $g(x)$ (called a phase shift of $-\frac{1}{2}\pi$ units).

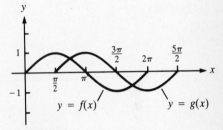

Figure 8-4
Phase shift.

Note: Don't be confused by the sign and direction of the phase shift. When the phase shift is a positive number, the graph of the function is shifted right by that number of units. In Figure 8-4, let $f(x) = \sin x$; it follows that $g(x) = \sin(x - \frac{1}{2}\pi)$. You may want to try a few representative values of x to convince yourself of this relationship. If we agree that $f(x) = \sin x$ is the standard graph, then the phase shift of $g(x) = \sin(x - \frac{1}{2}\pi)$ is $+\frac{1}{2}\pi$, that is, $\frac{1}{2}\pi$ units to the right of $f(x)$. See Section 8-3D for further discussion of this point.

8-3. Graphing $y = K \sin(Lx + M) + C$

The following discussion concentrates on the sine function. Everything applies, with appropriate changes, to the other trigonometric functions.

A. The amplitude is $|K|$.

Recall that $y = \sin x$ has a period of 2π radians, or 360 degrees, and amplitude of 1. For $y = K \sin x$, where K is any constant, each value of y is multiplied by K; so the maximum value of $K \sin x$ is $|K|$ and the minimum value is $-|K|$. Then the amplitude is

$$\frac{|K| - (-|K|)}{2} = \frac{2|K|}{2} = |K|$$

instead of 1. The period of the function is not changed.

B. The horizontal axis depends on C.

If $\sin x_0 = n$, then $C + \sin x_0 = C + n$, for any constant C and any number x_0. Therefore, the graph of $y = C + \sin x$ is translated by C units, in an upward or positive direction if $C > 0$ and in a downward or negative direction if $C < 0$. The axis of the function is then the line $y = C$ instead of

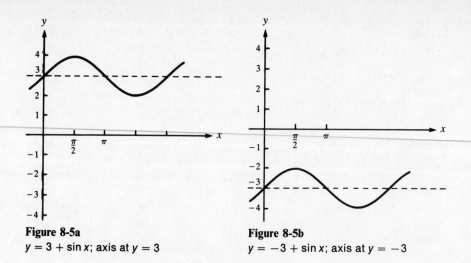

Figure 8-5a
$y = 3 + \sin x$; axis at $y = 3$

Figure 8-5b
$y = -3 + \sin x$; axis at $y = -3$

$y = 0$; the period and amplitude of the function are not affected. This is shown in Figure 8-5.

C. The period depends on *L*.

The period of each of the trigonometric functions is given in Table 8-2. The period of $\sin x$ is 2π, or 360 degrees. What is the period of $\sin 2x$? If you let $y = 2x$, then as x increases from 0 to π, y increases from 0 to 2π radians, a complete period of the sine function. Because $\sin 2x$ completes one cycle or period as x increases from 0 to π, the period of $y = \sin 2x$ is π.

What is the period of $\sin \frac{1}{3}x$? If you let $y = \frac{1}{3}x$, then as x increases from 0 to 6π, y increases from 0 to 2π radians, one period of the sine function. Therefore the period of $\sin \frac{1}{3}x$ is 6π.

These two examples suggest the following principle concerning the period of the sine function: as Lx moves through 2π radians, or 360 degrees, $\sin Lx$ completes one cycle or period; however x only moves through $2\pi/|L|$ radians, or $360/|L|$ degrees. This means that the period of $\sin Lx$ is $2\pi/|L|$ radians, or $360/|L|$ degrees. Graphs are shown in Figure 8-6.

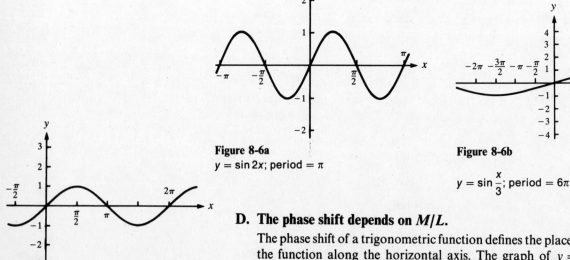

Figure 8-6a
$y = \sin 2x$; period $= \pi$

Figure 8-6b
$y = \sin \dfrac{x}{3}$; period $= 6\pi$

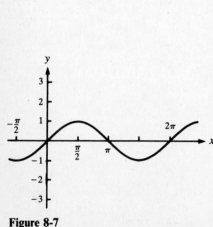

Figure 8-7
$y = \sin x$

D. The phase shift depends on *M/L*.

The phase shift of a trigonometric function defines the placement of values of the function along the horizontal axis. The graph of $y = \sin x$ (shown in Figure 8-7) is considered the standard sine graph. The function $y = \sin x$ takes the value zero when $x = 0$ and is zero after completing one cycle at $x = 2\pi$.

Now consider the curve in Figure 8-8. The curve $y = \sin(x + \frac{1}{4}\pi)$ has the same period, amplitude, and axis as $y = \sin x$; however, the addition of $\frac{1}{4}\pi$ to the argument has shifted the graph so that $y = 0$ when $x = -\frac{1}{4}\pi$, and one period is completed at $x = -\frac{1}{4}\pi + 2\pi = \frac{7}{4}\pi$. If $\sin x = y$ at $x = x_0$, then $\sin(x + \frac{1}{4}\pi) = y$ at $x = x_0 - \frac{1}{4}\pi$. You say that a phase shift to the left by $\frac{1}{4}\pi$ radians has occurred. The function $y = \sin(x - \frac{1}{4}\pi) = \sin(x + (-\frac{1}{4}\pi))$ has the same period, amplitude, and axis as $y = \sin x$; however, $-\frac{1}{4}\pi$ in the argument has caused a shift of the graph so that $y = 0$ when $x = \frac{1}{4}\pi$, and one period is completed at $x = \frac{1}{4}\pi + 2\pi = \frac{9}{4}\pi$. If $\sin x = y$ at $x = x_0$, then $\sin(x - \frac{1}{4}\pi) = y$ at $x = x_0 + \frac{1}{4}\pi$. You say that a phase shift to the right by $\frac{1}{4}\pi$ radians has occurred.

You could have found the phase shift by computing the value of x at which $x + M = 0$. In general, the phase shift of $y = \sin(Lx + M)$ is found by solving $Lx + M = 0$ for x, that is, $x = -M/L$; the phase shift is to the right by $-M/L$ units if $-M/L$ is positive, and to the left by M/L units if $-M/L$ is negative.

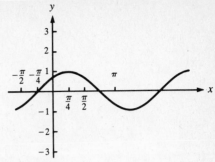

Figure 8-8

$$y = \sin\left(x + \frac{\pi}{4}\right)$$

E. Summary

1. Multiplying any of the trigonometric functions by a constant K, e.g., $K \sin x$, $K \cos x$, $K \cot x$, etc., multiplies the amplitude of that function by K.

2. The amplitude of $K \cdot f$, where f is any of the trigonometric functions, is $|K|$ times the amplitude of f. For example, the amplitude of $K \sin x$ is $|K|$ times the amplitude of $\sin x$, or $|K| \cdot 1 = |K|$.

3. The axis of $C + f$, where f is any of the trigonometric functions, is the line $y = C$. For example, the axis of $C + \sin x$ is the line $y = C$.

4. The period of $f(Lx)$, where f is any of the trigonometric functions, equals the period of f divided by L. For example, the period of $\sin Lx$ is $2\pi/|L|$.

5. The phase shift of $f(Lx + M)$, where f is any of the trigonometric functions, is right by $-M/L$ units if $-M/L > 0$ and left by M/L units if $-M/L < 0$. For example, the phase shift of $\sin(Lx + M)$ is right by $-M/L$ units if $-M/L > 0$ and left by M/L units if $-M/L < 0$.

The following graphing examples illustrate the key points of this section.

EXAMPLE 8-5: Graph $y = 2 + 3\sin(0.5x - 30°)$.

Solution: The amplitude is 3, the axis is the line $y = 2$, the period is $360°/0.5 = 720°$, and phase is shifted right by $-M/L = -(-30°)/0.5 = 60°$. The graph is shown in Figure 8-9.

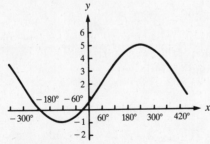

Figure 8-9

$$y = 2 + 3\sin(0.5x - 30°)$$

EXAMPLE 8-6: Graph $y = -4 + 3\cos(2x + \frac{2}{3}\pi)$.

Solution: The amplitude is 3, the axis is the line $y = -4$, the period is $2(\frac{1}{2}\pi) = \pi$, and phase is shifted left by $\frac{1}{3}\pi$ because $-M/L = -\frac{2}{3}\pi/2 = -\frac{1}{3}\pi$. The graph is shown in Figure 8-10.

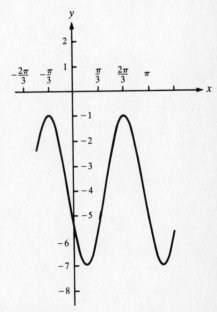

Figure 8-10

$$y = -4 + 3\cos\left[2\left(x + \frac{\pi}{3}\right)\right]$$

EXAMPLE 8-7: Graph $y = 3 - \tan\frac{1}{2}x$.

Solution: The amplitude is unchanged, the axis is the line $y = 3$, and the period is 2π. Subtracting $\tan\frac{1}{2}x$ from 3 has the same effect as adding $-1(\tan\frac{1}{2}x)$ to 3. Multiplying the tangent function by -1 changes the phase by $90°$, so the graph looks like the graph of the cotangent function. See Figure 8-11 for the graph.

EXAMPLE 8-8: Graph $y = \sin(2x + \frac{1}{3}\pi)$.

Solution: The amplitude is unchanged, the horizontal axis remains $y = 0$, the period is $2(\frac{1}{2}\pi) = \pi$, and, since $-M/L = -(\frac{1}{3}\pi)/2 = -\frac{1}{6}\pi$, the phase is shifted left by $\frac{1}{6}\pi$. See Figure 8-12 for the graph.

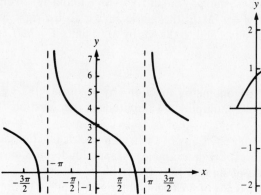

Figure 8-11

$$y = 3 - \tan\frac{x}{2}$$

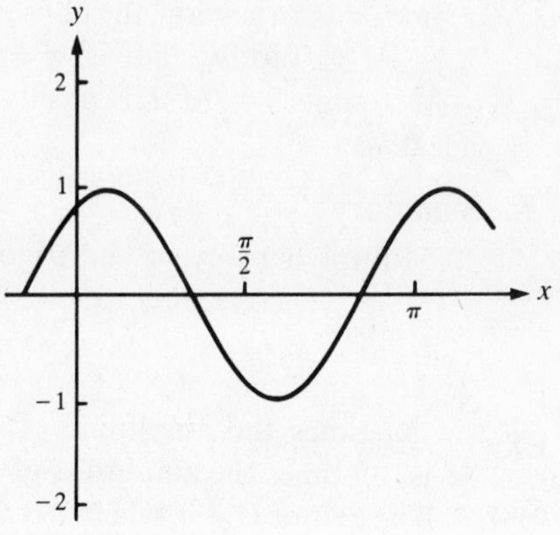

Figure 8-12

$$y = \sin\left(2x + \frac{\pi}{3}\right)$$

8-4. Graphing by Addition of Ordinates

When you are required to graph a function expressible as the sum of two or more functions, you can often simplify your task by using **addition of ordinates** as illustrated in the following examples.

EXAMPLE 8-9: Sketch the graph of $y = f(\theta) = \sin\theta + \cos\theta$.

Solution: The graph of y can be sketched from knowledge of the graphs of $\sin\theta$ and $\cos\theta$:

1. Since both $\sin\theta$ and $\cos\theta$ are of period 2π, $f(\theta)$ has period 2π. This follows from

$$f(\theta) = \sin(\theta + 2\pi) + \cos(\theta + 2\pi) = \sin\theta + \cos\theta = f(\theta)$$

2. Sketch the graphs of $\sin\theta$ and $\cos\theta$ on the same set of axes.
3. For any θ, the value of the given $f(\theta)$ is the sum of two y values; $y_1 = \cos\theta$ and $y_2 = \sin\theta$ (see Figure 8-13).

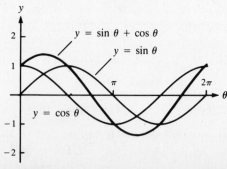

Figure 8-13

$$y = \sin\theta + \cos\theta$$

EXAMPLE 8-10: Sketch the graph of $y = f(\theta) = \theta + \sin\theta$.

Solution: Sketch the graphs of $y = \theta$ and $y = \sin\theta$ on the same axes. For each θ, the corresponding value of y is the sum of the values of y for $\sin\theta$ and θ. The result is shown in Figure 8-14.

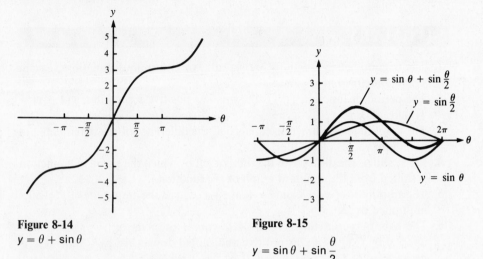

Figure 8-14
$y = \theta + \sin\theta$

Figure 8-15
$$y = \sin\theta + \sin\frac{\theta}{2}$$

EXAMPLE 8-11: Sketch the graph of $y = f(\theta) = \sin\theta + \sin\frac{1}{2}\theta$.

Solution: Reproduce the graphs of $\sin\theta$ and $\sin\frac{1}{2}\theta$ on the same set of axes. For each θ, the value of y for the function is the sum of the values of y for $\sin\theta$ and $\sin\frac{1}{2}\theta$. This is shown in Figure 8-15.

8-5. Introduction to Harmonic Motion

If you view a rotating wheel (e.g. a Ferris wheel) from the edge of the wheel and focus on one point of the wheel (or one car on the Ferris wheel), the rotary motion is not apparent; the point (or the car) appears to oscillate up and down in a regular, or periodic, fashion between a highest point and a lowest point. This suggests that the motion of that point might be described by a sine or cosine function.

Let x be the coordinate of a point on a line L, that is, think of L as a coordinate axis with a zero point and a positive direction. If the point is moving, think of x as a function of time, that is, as $x = f(t)$. If the point is oscillating in a periodic manner between some smallest and largest value, then you can say that the motion is **harmonic**. It can be described by the equation $x = a\sin(bt + c)$ for some constants $a, b,$ and c. Whenever an object moves according to such an equation, you say that the motion is **simple harmonic motion**.

EXAMPLE 8-12: The position of a point on a line is given by $x = 2\sin(t + \frac{1}{4}\pi)$. Describe the motion of the point.

Solution: The initial position of the point at $t = 0$ is $x = 2\sin\frac{1}{4}\pi = \frac{2\sqrt{2}}{2} = \sqrt{2}$. As t increases, $\sin(t + \frac{1}{4}\pi)$ increases until $t = \frac{1}{4}\pi$ and $x = 2\sin\frac{1}{2}\pi = 2$. As t continues to increase, $\sin(t + \frac{1}{4}\pi)$ will begin to decrease, becoming 0 at $t = \frac{3}{4}\pi$, and continuing until $t = \frac{5}{4}\pi$ and $x = 2\sin(\frac{5}{4}\pi + \frac{1}{4}\pi) = 2\sin\frac{3}{2}\pi = -2$. Now $\sin(t + \frac{1}{4}\pi)$ begins to increase again until $t = 2\pi$, at which time the point is back to its starting position. The cycle will again repeat with increasing t. Figure 8-16 illustrates the motion of the point.

The function $x = 2\sin(t + \frac{1}{4}\pi)$, which defines the harmonic motion of the point, has an amplitude of 2, a period of $2\pi/1 = 2\pi$, and a phase shift of $-\frac{1}{4}\pi$. Note these values from Figure 8-16.

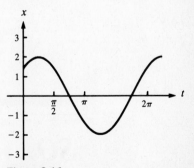

Figure 8-16

$$y = 2\sin\left(t + \frac{\pi}{4}\right)$$

EXAMPLE 8-13: Describe the motion of an object whose equation is $x = \frac{1}{2}\sin 2t$, where t is measured in seconds.

Solution: This equation describes the motion of a point that moves back and forth between $x = -\frac{1}{2}$ and $x = \frac{1}{2}$ over time intervals of length $2(\frac{1}{2}\pi) = \pi$ seconds. The motion starts at $x = 0$, proceeds to $x = \frac{1}{2}$, reverses and proceeds through $x = 0$ to $x = -\frac{1}{2}$, reverses and moves back to the right through zero to $\frac{1}{2}$, etc.

SUMMARY

1. The trigonometric functions are smooth, or continuous, functions, except where undefined.
2. The period of function $f(x)$ is the smallest positive p such that $f(x + p) = f(x)$ for all x.
3. The amplitude of a function equals half the difference of the maximum and minimum values of the function.
4. Phase shift is the distance to the right or left by which the graph of a function is shifted from the standard graph of the function.
5. The amplitude of $y = K \sin(Lx + M) + C$ equals the amplitude of $y = \sin x$ multiplied by $|K|$.
6. The axis of $y = K \sin(Lx + M) + C$ equals the axis of $y = \sin x$ shifted upward by C units if C is positive and shifted downward by C units if C is negative.
7. The period of $y = K \sin(Lx + M) + C$ equals the period of $\sin x$ divided by $|L|$.
8. The phase shift of $y = K \sin(Lx + M) + C$ equals $-M/L$.
9. Graphing by addition of ordinates is carried out by plotting graphs of the functions that are being summed on the same axes and adding corresponding y values for each x value.
10. Harmonic motion of an object is expressed by $x = a \sin(bt + c)$, where x is the position of the object, t is time, and a, b, and c are motion constants.

RAISE YOUR GRADES

Can you ...?

☑ plot graphs of the six trigonometric functions
☑ determine the domain and range of each trigonometric function
☑ find the period given the graph of a periodic function
☑ find the amplitude given the graph of a periodic function
☑ determine the phase shift given the graph of a periodic function
☑ find the amplitude, period, axis, and phase shift of
$y = K \sin(Lx + M) + C$; find the same if sine is replaced by any of the other trigonometric functions
☑ graph by the addition-of-ordinates technique
☑ plot the graph of the harmonic motion of an object and find the equation of harmonic motion

SOLVED PROBLEMS

Determine the period, amplitude, and phase shift for the functions in Problems 8-1 through 8-10.

[See Sections 8-1 and 8-3E.]

PROBLEM 8-1 $y = 4 \cos x$

Solution:

Amplitude: 4(amplitude of cosine) = 4(1) = 4

Axis: axis of cosine is $y = 0$

Period: period of cosine = 2π

Phase shift: 0

PROBLEM 8-2 $y = \cos 4x$

Solution:

Amplitude: amplitude of cosine = 1

Axis: axis of cosine is $y = 0$

Period: (period of cosine)/4 = $2(\frac{1}{4}\pi) = \frac{1}{2}\pi$

Phase shift: 0

PROBLEM 8-3 $y = \frac{1}{4}\cos 4x$

Solution:

Amplitude: (amplitude of cosine)$(\frac{1}{4})$ = $(1)(\frac{1}{4}) = \frac{1}{4}$

Axis: axis of cosine is $y = 0$

Period: (period of cosine)/4 = $2(\frac{1}{4}\pi) = \frac{1}{2}\pi$

Phase shift: 0

PROBLEM 8-4 $y = \frac{1}{4}\left[4 + \cos(\frac{1}{4}x)\right]$

Solution:

Amplitude: (amplitude of cosine)$(\frac{1}{4})$ = $(1)(\frac{1}{4}) = \frac{1}{4}$

Axis: (axis of cosine) + $\frac{1}{4}(4)$ is $y = 0 + 1 = 1$

Period: (period of cosine)/$\frac{1}{4}$ = $2\pi/\frac{1}{4} = 8\pi$

Phase shift: 0

PROBLEM 8-5 $y = 4 + \cos x$

Solution:

Amplitude: amplitude of cosine = 1

Axis: (axis of cosine) + 4 is $y = 0 + 4 = 4$

Period: period of cosine = 2π

Phase shift: 0

PROBLEM 8-6 $y = \sec \frac{1}{2}x$

Solution:

Amplitude: amplitude of secant is undefined or infinite

Axis: axis of secant is $y = 0$

Period: (period of secant)/$\frac{1}{2}$ = $2\pi/\frac{1}{2} = 4\pi$

Phase shift: 0

PROBLEM 8-7 $y = \frac{1}{2}\sec x$

Solution:

Amplitude: (amplitude of secant)$(\frac{1}{2})$ = undefined or infinite

Axis: axis of secant is $y = 0$

Period: period of secant = 2π

Phase shift: 0

PROBLEM 8-8 $y = \cot 3x$

Solution:

Amplitude: amplitude of cotangent is undefined or infinite

Axis: axis of cotangent is $y = 0$

Period: (period of cotangent)/3 = $\frac{1}{3}\pi$

Phase shift: 0

PROBLEM 8-9 $y = \cot(x - 3)$

Solution:

Amplitude: amplitude of cotangent is undefined or infinite

Axis: axis of cotangent is $y = 0$

Period: period of cotangent = π

Phase shift: $-(-3) = 3$ units to the right

PROBLEM 8-10 $y = \sin[2(x - \frac{1}{2})]$

Solution:

Amplitude: amplitude of sine = 1

Axis: axis of sine is $y = 0$

Period: (period of sine)/2 = $2\pi/2 = \pi$

Phase shift: $-(-\frac{1}{2}) = \frac{1}{2}$ units to the right

Find the period, axis, amplitude, and phase shift for the functions in problems 8-11 through 8-19. Sketch a graph of the function showing at least one period. [See Sections 8-1 and 8-3E.]

PROBLEM 8-11 $y = 2 - \cos x$

Solution:

Amplitude: amplitude of cosine = 1

Axis: (axis of cosine) + 2 is $y = 0 + 2 = 2$

Period: period of cosine = 2π

Phase shift: π units to the right

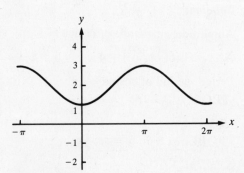

Figure 8-17
$y = 2 - \cos x$

Note: The phase shift is a result of subtracting $\cos x$ from 2. It is equivalent to multiplying the y values by -1. You can sketch the graphs of $\cos x$ and $-1 \cos x$ to verify this.

PROBLEM 8-12 $y = -2\cos x$

Solution:

Amplitude: (amplitude of cosine)(2) = 2

Axis: axis of cosine is $y = 0$

Period: period of cosine = 2π

Phase shift: π units to the right

   (See note for Problem 8-11.)

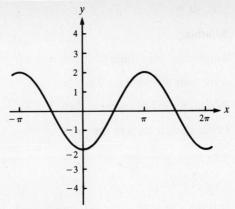

Figure 8-18
$y = -2\cos x$

PROBLEM 8-13 $y = \cos(x - \pi)$

Solution:

Amplitude: amplitude of cosine = 1

Axis: axis of cosine is $y = 0$

Period: period of cosine = 2π

Phase shift: $-(-\pi) = \pi$ units to the right

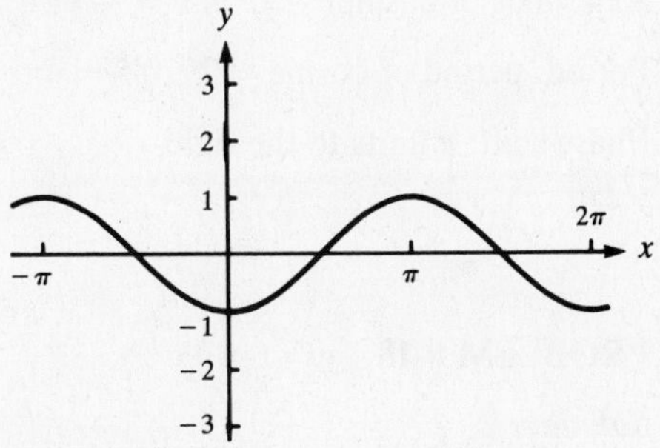

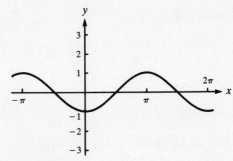

Figure 8-19
$y = \cos(x - \pi)$

PROBLEM 8-14 $y = \cos(\pi - x)$

Solution:

Amplitude: amplitude of cosine = 1

Axis: axis of cosine is $y = 0$

Period: period of cosine = 2π

Phase shift: π units to the right

   (See note for Problem 8-11.)

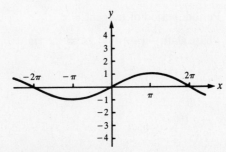

Figure 8-20
$y = \cos(\pi - x)$

PROBLEM 8-15 $y = \cos[0.5(x - \pi)]$

Solution:

Amplitude: amplitude of cosine = 1

Axis: axis of cosine is $y = 0$

Period: period of cosine = $2\pi/0.5 = 4\pi$

Phase shift: π units to the right

Figure 8-21
$y = \cos[0.5(x - \pi)]$

PROBLEM 8-16 $y = 2 + \cos[0.5(x - \pi)]$

Solution:

Amplitude: amplitude of cosine = 1

Axis: (axis of cosine) + 2 is $y = 2$

Period: period of cosine = $2\pi/0.5 = 4\pi$

Phase shift: π units to the right

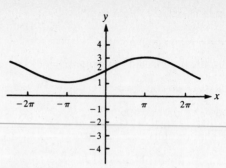

Figure 8-22

$y = 2 + \cos[0.5(x - \pi)]$

PROBLEM 8-17 $y = \cos[0.5(x - \pi)] - \frac{1}{2}\pi$

Solution:

Amplitude: amplitude of cosine = 1

Axis: (axis of cosine) $- \frac{1}{2}\pi$ is $y = -\frac{1}{2}\pi$

Period: period of cosine = $2\pi/0.5 = 4\pi$

Phase shift: π units to the right

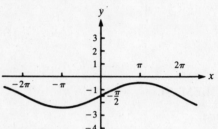

Figure 8-23

$y = \cos[0.5(x - \pi)] - \dfrac{\pi}{2}$

PROBLEM 8-18 $y = \tan\frac{1}{3}x$

Solution:

Amplitude: amplitude of tangent is undefined

Axis: axis of tangent is $y = 0$

Period: (period of tangent)$/\frac{1}{3} = \pi/\frac{1}{3} = 3\pi$

Phase shift: 0

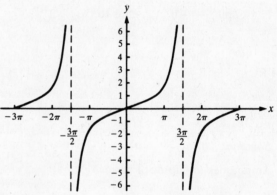

Figure 8-24

$y = \tan\dfrac{x}{3}$

PROBLEM 8-19 $y = \tan\left(x - \frac{1}{4}\pi\right)$

Solution:

Amplitude: amplitude of tangent is undefined

Axis: axis of tangent is $y = 0$

Period: period of tangent = π

Phase shift: $-(-\frac{1}{4}\pi) = \frac{1}{4}\pi$ to the right

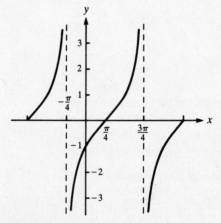

Figure 8-25

$y = \tan\left(x - \dfrac{\pi}{4}\right)$

Use the addition-of-ordinates technique to sketch graphs of the functions in Problems 8-20 through 8-22. [See Section 8-5.]

PROBLEM 8-20 $y = 2\sin x + \cos x$

Solution: Plot graphs of $2\sin x$ and $\cos x$ on the same axes. For each x, the y value of $2\sin x + \cos x$ is the sum of the y values for $2\sin x$ and $\cos x$. The completed graph is shown in Figure 8-26.

PROBLEM 8-21 $y = \sin x + \cos\frac{1}{2}x$

Solution: Plot graphs of $\sin x$ and $\cos\frac{1}{2}x$ on the same axes. For each x, the y value of $\sin x + \cos\frac{1}{2}x$ is the sum of the y values for $\sin x$ and $\cos\frac{1}{2}x$. The completed graph is shown in Figure 8-27.

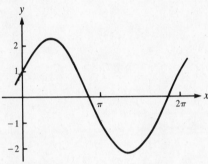

Figure 8-26

$y = 2\sin x + \cos x$

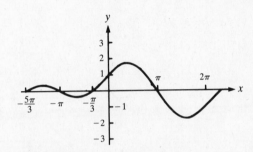

Figure 8-27

$y = \sin x + \cos\dfrac{x}{2}$

PROBLEM 8-22 $y = \cos\frac{1}{2}x + \sin 2x$

Solution: Plot graphs of $\cos\frac{1}{2}x$ and $\sin 2x$ on the same axes. For each x, the y value of $\cos\frac{1}{2}x + \sin 2x$ is the sum of the y values for $\cos\frac{1}{2}x$ and $\sin 2x$. The completed graph is shown in Figure 8-28.

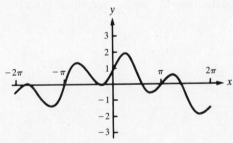

Figure 8-28

$y = \cos\dfrac{x}{2} + \sin 2x$

PROBLEM 8-23 An object is moving according to the equation $y = \sqrt{2}\sin(8\pi t - \frac{1}{4}\pi)$, where t is time measured in minutes. What is the period, how many cycles are completed in 10 minutes, and what is the distance of the object from the origin when $t = 0$?

Solution: The period of the function is $2\pi/8\pi = \frac{1}{4}$ minute. A period equal to $\frac{1}{4}$ minute means that 4 cycles are completed in one minute, or 40 cycles are completed in 10 minutes. The distance of the object from the origin when $t = 0$ is

$$\left|\sqrt{2}\sin\left(8\pi(0) - \frac{\pi}{4}\right)\right| = \left|\sqrt{2}\sin\left(-\frac{\pi}{4}\right)\right| = \left|\sqrt{2}\left(\frac{-\sqrt{2}}{2}\right)\right| = |-1| = 1 \text{ unit}$$

[See Section 8-5.]

PROBLEM 8-24 Write an equation to describe an object moving in harmonic motion with a period of 3 seconds and an amplitude of 5 units.

Solution: The equation of harmonic motion is $x = a\sin(bt + c)$, where a, b, and c are constants. From your work with the cosine function, you know that b controls the period and a controls the amplitude of the function. Since the amplitude is 5 and the period is 3, take $a = 5$, $b = \frac{2}{3}\pi$, and $c = 0$. The resulting function, $x = 5\sin\frac{2}{3}\pi t$, has a period of 3 seconds and amplitude of 5 units, if t is measured in seconds. [See Section 8-5.]

PROBLEM 8-25 Show graphically that $x > \sin x$ for $0 < x < \frac{1}{2}\pi$.

Solution: Select values of x between 0 and $\frac{1}{2}\pi$. Use your calculator to find $\sin x$ for each value. Very carefully plot the points $(x, \sin x)$ and sketch a smooth curve joining these points. Also make a sketch of the graph of $y = x$ on the same axes. From Figure 8-29 you see that the line $y = x$ remains above the graph of $\sin x$ for x between 0 and $\frac{1}{2}\pi$. [See Section 8-1.]

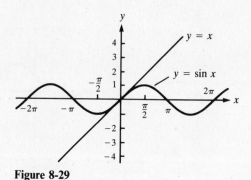

Figure 8-29

Supplementary Exercises

For the following functions determine the period, amplitude, and phase shift.

PROBLEM 8-26 $y = 2 - \sin\left[2\left(x - \frac{1}{2}\right)\right]$

PROBLEM 8-27 $y = -3\tan(3 - x)$

PROBLEM 8-28 $y = 3\tan\frac{1}{3}x$

PROBLEM 8-29 $y = 2 + \sec 2x$

PROBLEM 8-30 $y = 2 - \sec 2x$

PROBLEM 8-31 $y = 2 - \sec(-2x)$

PROBLEM 8-32 $y = 2 + 3\sec(x - \pi)$

PROBLEM 8-33 $y = 2 + 3\sec[\pi(x - \pi)]$

PROBLEM 8-34 $y = 2 + 3\sec[(1/\pi)(x - \pi)]$

PROBLEM 8-35 $y = \frac{3}{4}\sin\left[3\left(x - \frac{4}{3}\right)\right]$

Find the period, amplitude, and phase shift, and sketch a graph of each of the following functions showing at least one period of the function.

PROBLEM 8-36 $y = \tan[2(x - \pi)]$

PROBLEM 8-37 $y = \sec\frac{1}{4}x$

PROBLEM 8-38 $y = \sec[2(x - \pi)]$

PROBLEM 8-39 $y = \sec(x - \frac{1}{4}\pi)$

Use the addition-of-ordinates technique to sketch graphs of the following functions.

PROBLEM 8-40 $y = 2\cos\frac{1}{2}x + 4\sin 2x$

PROBLEM 8-41 $y = \sin x - 3\cos x$

PROBLEM 8-42 $y = \sin(x - \frac{1}{4}\pi) + \sin(x + \frac{1}{4}\pi)$

PROBLEM 8-43 An object is moving according to the equation $y = 3\sin(\frac{1}{4}t + 2)$, where t is time measured in seconds. What is the period, how many cycles are completed in 4 seconds, and what is the distance of the object from the origin when $t = 0$?

PROBLEM 8-44 An object moving in harmonic motion has a period of 2 minutes and an amplitude of 10 units; write the equation of motion for the object.

PROBLEM 8-45 Use graphs of the functions to find an interval that contains a value of x for which the graphs of $y = \cos x$ and $y = x$ have a point of intersection.

Answers to Supplementary Exercises

8-26: amplitude is 1, axis is $y = 2$, period is π, phase shift is $\frac{1}{2}$ unit to the right

8-27: amplitude is undefined, axis is $y = 0$, period is π, phase shift is 3

8-28: amplitude is undefined, axis is $y = 0$, period is 3π, phase shift is 0

8-29: amplitude is undefined, axis is $y = 2$, period is π, phase shift is 0

8-30: amplitude is undefined, axis is $y = 2$, period is π, phase shift is 0

8-31: amplitude is undefined, axis is $y = 2$, period is π, phase shift is 0

8-32: amplitude is undefined, axis is $y = 2$, period is 2π, phase shift is π

8-33: amplitude is undefined, axis is $y = 2$, period is 2, phase shift is π

8-34: amplitude is undefined, axis is $y = 2$, period is $2\pi^2$, phase shift is π

8-35: amplitude is $\frac{3}{4}$, axis is $y = 0$, period is $\frac{2}{3}\pi$, phase shift is $\frac{4}{3}$

8-36: amplitude is undefined, axis is $y = 0$, period is $\frac{1}{2}\pi$, phase shift is π units to the right

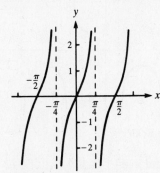

Figure 8-30
$y = \tan[2(x - \pi)]$

8-37: amplitude is undefined, axis is $y = 0$, period is 8π, phase shift is 0

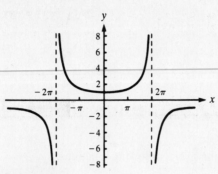

Figure 8-31

$$y = \sec \frac{x}{4}$$

8-38: amplitude is undefined, axis is $y = 0$, period is π, phase shift is π units to the right

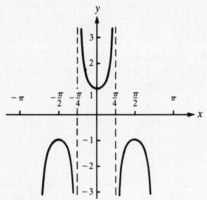

Figure 8-32
$$y = \sec[2(x - \pi)]$$

8-39: amplitude is undefined, axis is $y = 0$, period is 2π, phase shift is $\frac{1}{4}\pi$ units to the right

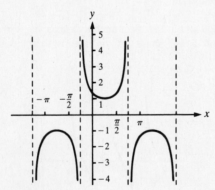

Figure 8-33

$$y = \sec\left(x - \frac{\pi}{4}\right)$$

8-40:

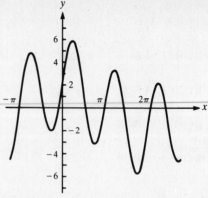

Figure 8-34

$$y = 2\cos \frac{x}{2} + 4\sin 2x$$

8-41:

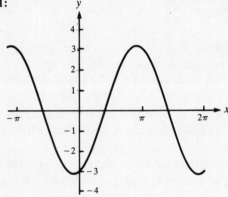

Figure 8-35
$$y = \sin x - 3\cos x$$

8-42:

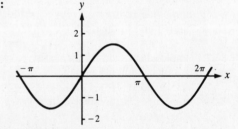

Figure 8-36

$$y = \sin\left(x - \frac{\pi}{4}\right) + \sin\left(x + \frac{\pi}{4}\right)$$

8-43: period is $2\pi/\frac{1}{4} = 8\pi$; $4/8\pi$ cycles are completed in 4 seconds, distance at $t = 0$ is $3\sin 2$ units

8-44: $y = 10\sin \pi t$ where t is in minutes

8-45: $\left[0, \frac{1}{2}\pi\right]$

9 TRIGONOMETRIC IDENTITIES

THIS CHAPTER IS ABOUT

☑ **Definition of an Identity**
☑ **Trigonometric Identities**
☑ **Verifying Trigonometric Identities**

9-1. Definition of an Identity

An **identity** is an equation that is satisfied by all allowable values of the variable.

EXAMPLE 9-1: Write three algebraic identities.

Solution: The following are algebraic identities:

(a) $x^2 - 1 = (x - 1)(x + 1)$, for all x
(b) $x + 1 + 2x - 3 = 3x - 2$, for all x

(c) $\dfrac{1}{(1/x)} = x$, for all $x \neq 0$

You can see that these algebraic identities derive from the principles of algebraic addition and multiplication.

EXAMPLE 9-2: Rewrite the identities of Example 9-1 with trigonometric functions for x.

Solution: Any trigonometric function can be used in place of x. We have chosen $\cos \theta$ in (a), $\tan \theta$ in (b), and $\sin \theta$ in (c).

(a) $\cos^2 \theta - 1 = (\cos \theta - 1)(\cos \theta + 1)$, for all θ
(b) $\tan \theta + 1 + 2 \tan \theta - 3 = 3 \tan \theta - 2$, for all θ for which $\tan \theta$ is defined

(c) $\dfrac{1}{1/\sin \theta} = \sin \theta$, for all θ for which $\sin \theta \neq 0$

Even though these identities involve trigonometric functions, their validity stems from algebraic principles rather than properties of the trigonometric functions. Therefore, they are also considered algebraic identities.

9-2. Trigonometric Identities

In contrast to the identities in Example 9-1, yet similar in nature, are the identity relationships that exist among the trigonometric functions; they result from the way in which the trigonometric functions are defined.

A. Trigonometric identities derived on arbitrary circles or right triangles

Let (x, y) be any point except the origin on the terminal side of angle θ, at a distance $r = \sqrt{x^2 + y^2}$ from the origin (see Figure 9-1). Note that a circle of radius r and a right triangle with one angle denoted by θ are formed. For the triangle and circle shown in Figure 9-1, recall the definitions from Chapter 5:

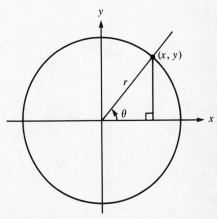

Figure 9-1
Relationships among a circle, right triangle, and angle.

$\sin \theta = y/r$; $\csc \theta = r/y$, for $y \neq 0$; $\cos \theta = x/r$; $\sec \theta = r/x$, for $x \neq 0$; $\tan \theta = y/x$, for $x \neq 0$; $\cot \theta = x/y$, for $y \neq 0$. The following identities are consequences of these definitions.

1. Reciprocal identities

$$\frac{1}{\sin \theta} = \frac{1}{y/r} = \frac{r}{y} = \csc \theta \qquad \sin \theta \csc \theta = 1$$

RECIPROCAL IDENTITIES
$$\frac{1}{\cos \theta} = \frac{1}{x/r} = \frac{r}{x} = \sec \theta \qquad \cos \theta \sec \theta = 1 \qquad \textbf{(9-1)}$$

$$\frac{1}{\tan \theta} = \frac{1}{y/x} = \frac{x}{y} = \cot \theta \qquad \tan \theta \cot \theta = 1$$

2. Pythagorean identities
Because (x, y) satisfies the equation $x^2 + y^2 = r^2$, you can divide both sides by r^2: $(x/r)^2 + (y/r)^2 = 1$, which produces the identity $\cos^2\theta + \sin^2\theta = 1$. If you divide $x^2 + y^2 = r^2$ by x^2 you get $1 + (y/x)^2 = (r/x)^2$, which produces the identity $1 + \tan^2\theta = \sec^2\theta$. If you divide $x^2 + y^2 = r^2$ by y^2 you get $(x/y)^2 + 1 = (r/y)^2$, which produces the identity $\cot^2\theta + 1 = \csc^2\theta$. These are the Pythagorean identities:

$$\cos^2\theta + \sin^2\theta = 1$$

PYTHAGOREAN IDENTITIES
$$1 + \tan^2\theta = \sec^2\theta \qquad \textbf{(9-2)}$$

$$1 + \cot^2\theta = \csc^2\theta$$

3. Ratio identities
Included among the identities that follow from the definitions of Chapter 5 are the ratio identities:

$$\tan \theta = \frac{y}{x} = \frac{y/r}{x/r} = \frac{\sin \theta}{\cos \theta}$$

RATIO IDENTITIES
$$\qquad \textbf{(9-3)}$$

$$\cot \theta = \frac{x}{y} = \frac{x/r}{y/r} = \frac{\cos \theta}{\sin \theta}$$

B. Trigonometric identities on the unit circle

Consider the definitions of the trigonometric functions on the unit circle (see Chapter 6). You can also derive the trigonometric identities of Section **A** by using unit-circle definitions. This indicates that the two sets of definitions of the trigonometric functions are equivalent.

If (x, y) is a point on the unit circle (see Figure 9-2) at a distance $t > 0$ measured in a counterclockwise direction on the circumference of the unit circle from the point $(1, 0)$, then the following alternative definitions of the trigonometric functions apply: $\sin t = y$; $\csc t = 1/y$, for $y \neq 0$; $\cos t = x$; $\sec t = 1/x$, for $x \neq 0$; $\tan t = y/x$, for $x \neq 0$; $\cot t = x/y$, for $y \neq 0$. The reciprocal, Pythagorean, and ratio identities take the following forms:

1. Reciprocal identities

$$\csc t = \frac{1}{y} = \frac{1}{\sin t} \qquad \sin t = \frac{1}{\csc t} \qquad \sin t \csc t = 1$$

$$\sec t = \frac{1}{x} = \frac{1}{\cos t} \qquad \cos t = \frac{1}{\sec t} \qquad \cos t \sec t = 1 \qquad \textbf{(9-4)}$$

$$\cot t = \frac{x}{y} = \frac{1}{\tan t} \qquad \tan t = \frac{1}{\cot t} \qquad \tan t \cot t = 1$$

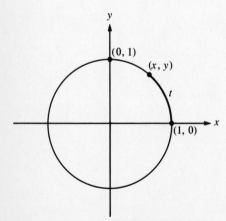

Figure 9-2
Point on unit circle and related arc of the circle.

2. Pythagorean identities

$$\cos^2 t + \sin^2 t = 1$$

$$1 + \tan^2 t = \sec^2 t \qquad \textbf{(9-5)}$$

$$\cot^2 t + 1 = \csc^2 t$$

3. Ratio identities

$$\tan t = \frac{y}{x} = \frac{\sin t}{\cos t}$$

$$\cot t = \frac{x}{y} = \frac{\cos t}{\sin t} \qquad \textbf{(9-6)}$$

9-3. Verifying Trigonometric Identities

When you apply algebraic processes such as factorization, addition, multiplication, and simplification of fractions to the fundamental identities, you can verify trigonometric identities or derive new ones. Let's begin by looking at some examples that demonstrate techniques used in verifying identities. You'll usually find that performing the indicated operations will aid you in carrying out the verification.

EXAMPLE 9-3: Simplify $(\cos\theta/\sin\theta)^2 - (1/\sin^2\theta)$.

Solution: Square the quantity inside the left-most parentheses and subtract the two fractions:

$$\left(\frac{\cos\theta}{\sin\theta}\right)^2 - \frac{1}{\sin^2\theta} = \frac{\cos^2\theta}{\sin^2\theta} - \frac{1}{\sin^2\theta}$$

$$= \frac{\cos^2\theta - 1}{\sin^2\theta}$$

Rewrite $\cos^2\theta + \sin^2\theta = 1$ as $\cos^2\theta - 1 = -\sin^2\theta$ and substitute:

$$\frac{\cos^2\theta - 1}{\sin^2\theta} = \frac{-\sin^2\theta}{\sin^2\theta} = -1$$

Factoring expressions will help you to find identities.

EXAMPLE 9-4: Simplify $\sin^4\theta - \cos^4\theta$.

Solution: Factor as the difference of two squares and use $\sin^2\theta + \cos^2\theta = 1$:

$$\sin^4\theta - \cos^4\theta = (\sin^2\theta - \cos^2\theta)(\sin^2\theta + \cos^2\theta)$$

$$= (\sin^2\theta - \cos^2\theta)(1)$$

$$= 1 - \cos^2\theta - \cos^2\theta$$

$$= 1 - 2\cos^2\theta$$

EXAMPLE 9-5: Simplify $(\cot x \sec x + \csc x)\sin x$.

Solution: Rewrite the expression in terms of $\sin x$ and $\cos x$:

$$(\cot x \sec x + \csc x)\sin x = \left[\left(\frac{\cos x}{\sin x}\right)\left(\frac{1}{\cos x}\right) + \frac{1}{\sin x}\right]\sin x$$

Do the multiplication and clear fractions:

$$\left(\frac{\cos x}{\sin x}\right)\left(\frac{\sin x}{\cos x}\right) + \frac{\sin x}{\sin x} = 1 + 1 = 2$$

In verifying identities, you can

1. Begin with the expression on the left-hand side of the equals sign and use algebraic principles and trigonometric identities to show that it equals the expression on the right-hand side of the equals sign.
2. Begin with the expression on the right-hand side and show that it equals the expression on the left-hand side.
3. Reduce the expressions on both sides of the equal sign to some common expression.

It is usually best to begin with the more complex expression and work toward the simpler expression. The following examples illustrate all three approaches.

EXAMPLE 9-6: Show that $\sin^2\theta/(1 - \cos\theta) = 1 + \cos\theta$.

Solution:
$$\frac{\sin^2\theta}{1 - \cos\theta} = 1 + \cos\theta$$

Substitute $1 - \cos^2\theta$ for $\sin^2\theta$:
$$\frac{1 - \cos^2\theta}{1 - \cos\theta} =$$

Factor:
$$\frac{(1 + \cos\theta)(1 - \cos\theta)}{1 - \cos\theta} =$$

Simplify the fraction:
$$1 + \cos\theta = 1 + \cos\theta$$

EXAMPLE 9-7: Show that $\cos^4\theta - \sin^4\theta = 2\cos^2\theta - 1$.

Solution:
$$\cos^4\theta - \sin^4\theta = 2\cos^2\theta - 1$$

Factor:
$$(\cos^2\theta + \sin^2\theta)(\cos^2\theta - \sin^2\theta) =$$

Use $\cos^2\theta + \sin^2\theta = 1$:
$$(1)(\cos^2\theta - \sin^2\theta) =$$

Substitute $1 - \cos^2\theta$ for $\sin^2\theta$:
$$\cos^2\theta - (1 - \cos^2\theta) =$$

Simplify:
$$2\cos^2\theta - 1 = 2\cos^2\theta - 1$$

EXAMPLE 9-8: Show that $\sec\theta\csc\theta - 2\cos\theta\csc\theta = \tan\theta - \cot\theta$.

Solution:
$$\sec\theta\csc\theta - 2\cos\theta\csc\theta = \tan\theta - \cot\theta$$

Use the reciprocal identities:
$$\frac{1}{\sin\theta}\left(\frac{1}{\cos\theta}\right) - \frac{2\cos\theta}{\sin\theta} =$$

Clear fractions:
$$\frac{1 - 2\cos^2\theta}{\sin\theta\cos\theta} =$$

Rewrite $-2\cos^2\theta$ as $-\cos^2\theta - \cos^2\theta$:
$$\frac{1 - \cos^2\theta - \cos^2\theta}{\sin\theta\cos\theta} =$$

Substitute $\sin^2\theta$ for $1 - \cos^2\theta$:
$$\frac{\sin^2\theta - \cos^2\theta}{\sin\theta\cos\theta} =$$

Rewrite as two fractions:
$$\frac{\sin^2\theta}{\sin\theta\cos\theta} - \frac{\cos^2\theta}{\sin\theta\cos\theta} =$$

Simplify the fractions:
$$\frac{\sin\theta}{\cos\theta} - \frac{\cos\theta}{\sin\theta} =$$

Use the ratio identities:
$$\tan\theta - \cot\theta = \tan\theta - \cot\theta$$

EXAMPLE 9-9: Show that $(1 + \sin x)/\cos x = \cos x/(1 - \sin x)$.

Solution: In this solution begin with the quantity on the right side of the equals sign and perform algebraic operations until you reach the expression on the left side of the equals sign.

$$\frac{1 + \sin x}{\cos x} = \frac{\cos x}{1 - \sin x}$$

Multiply numerator and denominator by $1 + \sin x$:

$$= \cos x \left(\frac{1 + \sin x}{1 - \sin^2 x} \right)$$

Use $1 - \sin^2 x = \cos^2 x$:

$$= \cos x \left(\frac{1 + \sin x}{\cos^2 x} \right)$$

Cancel $\cos x$:

$$\frac{1 + \sin x}{\cos x} = \frac{1 + \sin x}{\cos x}$$

EXAMPLE 9-10: Show that $(\tan \theta - 1)/(\tan \theta + 1) = (1 - \cot \theta)/(1 + \cot \theta)$.

Solution: For this example reduce both sides of the identity to the same quantity.

$$\frac{\tan \theta - 1}{\tan \theta + 1} = \frac{1 - \cot \theta}{1 + \cot \theta}$$

Use $\tan \theta = \sin \theta / \cos \theta$:

$$\frac{(\sin \theta / \cos \theta) - 1}{(\sin \theta / \cos \theta) + 1} =$$

Multiply numerator and denominator by $\cos \theta$:

$$\frac{\sin \theta - \cos \theta}{\sin \theta + \cos \theta} = \frac{1 - \cot \theta}{1 + \cot \theta}$$

Use $\cot \theta = \cos \theta / \sin \theta$:

$$= \frac{1 - (\cos \theta / \sin \theta)}{1 + (\cos \theta / \sin \theta)}$$

Multiply numerator and denominator by $\sin \theta$:

$$\frac{\sin \theta - \cos \theta}{\sin \theta + \cos \theta} = \frac{\sin \theta - \cos \theta}{\sin \theta + \cos \theta}$$

Now you have shown that the quantities on the right and left sides of the original expression are equal, so the identity is established.

SUMMARY

1. An identity is an equation satisfied by all allowable values of the variables.
2. Trigonometric identities consist of definitions of the trigonometric functions, reciprocal identities, Pythagorean identities, and ratio identities.
3. The reciprocal identities are

$$\csc \theta = \frac{1}{\sin \theta} \quad \text{or} \quad \sin \theta = \frac{1}{\csc \theta}$$

$$\sec \theta = \frac{1}{\cos \theta} \quad \text{or} \quad \cos \theta = \frac{1}{\sec \theta}$$

$$\cot \theta = \frac{1}{\tan \theta} \quad \text{or} \quad \tan \theta = \frac{1}{\cot \theta}$$

4. The Pythagorean identities are

$$\sin^2 \theta + \cos^2 \theta = 1$$

$$1 + \tan^2 \theta = \sec^2 \theta$$

$$\cot^2 \theta + 1 = \csc^2 \theta$$

5. The ratio identities are

$$\tan \theta = \frac{\sin \theta}{\cos \theta}$$

$$\cot \theta = \frac{\cos \theta}{\sin \theta}$$

6. Apply algebraic procedures to the fundamental identities to derive new identities.

7. In verifying identities, it is usually best to begin with the more complex expression and reduce it to the simpler expression on the opposite side of the equals sign.

8. In verifying identities, it is often helpful to express everything in terms of sine and cosine.

RAISE YOUR GRADES

Can you...?

☑ identify algebraic and trigonometric identities
☑ state the definitions of the trigonometric functions
☑ state the reciprocal identities
☑ state the Pythagorean identities
☑ state the ratio identities
☑ verify trigonometric identities involving the above identities

SOLVED PROBLEMS

All of the solved problems in this chapter use the fundamental identities and techniques outlined in Section 9-3.

PROBLEM 9-1 Simplify $(\sin x + \cos x)(\sec x \csc x)$. Express your answer in terms of $\sin x$ and $\cos x$.

Solution: Use the reciprocal identities and multiply:

$$(\sin x + \cos x)(\sec x \csc x) = (\sin x + \cos x)\left(\frac{1}{\cos x}\right)\left(\frac{1}{\sin x}\right)$$

$$= \frac{\cos x + \sin x}{\sin x \cos x}$$

PROBLEM 9-2 Simplify $(\sin x + \csc x)^2$. Express your answer in terms of $\sin^2 x$.

Solution: Use $\csc x = 1/\sin x$ and square:

$$(\sin x + \csc x)^2 = \left(\sin x + \frac{1}{\sin x}\right)^2 = \left(\frac{\sin^2 x + 1}{\sin x}\right)^2$$

$$= \frac{\sin^4 x + 2\sin^2 x + 1}{\sin^2 x}$$

$$= \sin^2 x + 2 + \frac{1}{\sin^2 x}$$

PROBLEM 9-3   Simplify $\cos^3\theta - \sin^3\theta$. Express your answer in terms of $\sin\theta$ and $\cos\theta$.

Solution: Factor and use $\sin^2\theta + \cos^2\theta = 1$:

$$\cos^3\theta - \sin^3\theta = (\cos\theta - \sin\theta)(\cos^2\theta + \sin\theta\cos\theta + \sin^2\theta)$$

$$= (\cos\theta - \sin\theta)(\sin\theta\cos\theta + 1)$$

PROBLEM 9-4   Simplify $\left(\dfrac{4\sin^4 x}{\cos^2 x}\right)\left(\dfrac{\cos^3 x}{2\sin^2 x}\right)^2$. Express your answer in terms of $\cos x$.

Solution:

$$\left(\frac{4\sin^4 x}{\cos^2 x}\right)\left(\frac{\cos^3 x}{2\sin^2 x}\right)^2 = \left(\frac{4\sin^4 x}{\cos^2 x}\right)\left(\frac{\cos^6 x}{4\sin^4 x}\right)$$

$$= \cos^4 x$$

PROBLEM 9-5   In the algebraic expression $\sqrt{4 + x^2}$, let $x = 2\tan\theta$. Simplify and express in terms of $\sec\theta$.

Solution: $\qquad \sqrt{4 + x^2} = \sqrt{4 + (2\tan\theta)^2} = \sqrt{4 + 4\tan^2\theta} = 2\sqrt{1 + \tan^2\theta}$

Use $1 + \tan^2\theta = \sec^2\theta$: $\qquad 2\sqrt{1 + \tan^2\theta} = 2\sqrt{\sec^2\theta} = 2\,|\sec\theta|$

PROBLEM 9-6   In the algebraic expression $\sqrt{25 - x^2}$, let $x = 5\sin\theta$. Simplify and express in terms of $\cos\theta$.

Solution: $\qquad \sqrt{25 - x^2} = \sqrt{25 - (5\sin\theta)^2} = \sqrt{25 - 25\sin^2\theta} = 5\sqrt{1 - \sin^2\theta}$

Use $1 - \sin^2\theta = \cos^2\theta$: $\qquad 5\sqrt{1 - \sin^2\theta} = 5\sqrt{\cos^2\theta} = 5\,|\cos\theta|$

PROBLEM 9-7   Show that $1/(\tan\theta\cos\theta) = \csc\theta$.

Solution:

$$\frac{1}{(\tan\theta\cos\theta)} = \csc\theta$$

Use $\tan\theta = \sin\theta/\cos\theta$:

$$\frac{1}{\left(\dfrac{\sin\theta}{\cos\theta}\right)\cos\theta} =$$

Simplify:

$$\frac{1}{\sin\theta} =$$

Use $\csc\theta = 1/\sin\theta$: $\qquad\qquad \csc\theta = \csc\theta$

Note: This problem illustrates the use of $\sin\theta$ and $\cos\theta$ to replace $\tan\theta$ and $\csc\theta$. Many of the problems you'll encounter in your study of trigonometry will require these substitutions, or substitutions of $\cos\theta/(\sin\theta)$ for $\cot\theta$ and $1/\cos\theta$ for $\sec\theta$.

PROBLEM 9-8   Simplify $\sec^2\theta - \tan^2\theta$.

Solution: Since $\sec\theta = 1/\cos\theta$ and $\tan\theta = \sin\theta/\cos\theta$,

$$\sec^2\theta - \tan^2\theta = \frac{1}{\cos^2\theta} - \frac{\sin^2\theta}{\cos^2\theta}$$

$$= \frac{1 - \sin^2\theta}{\cos^2\theta}$$

Use $1 - \sin^2\theta = \cos^2\theta$: $\qquad \dfrac{1 - \sin^2\theta}{\cos^2\theta} = \dfrac{\cos^2\theta}{\cos^2\theta} = 1$

An alternate solution uses $\sec^2\theta = 1 + \tan^2\theta$; the answer follows immediately.

PROBLEM 9-9   Show that $\sin^2\theta + \cos^2\theta = \sec^2\theta - \tan^2\theta$.

Solution:
$$\sin^2\theta + \cos^2\theta = \sec^2\theta - \tan^2\theta$$

Use $\sin^2\theta + \cos^2\theta = 1$:
$$1 = \sec^2\theta - \tan^2\theta$$

From Problem 9-8, $\sec^2\theta - \tan^2\theta = 1$:
$$1 = 1$$

You see that both sides of the original expression equal 1; the identity follows.

PROBLEM 9-10 Show that $1 - \cos\theta = \sin^2\theta/(1 + \cos\theta)$.

Solution:
$$1 - \cos\theta = \frac{\sin^2\theta}{1 + \cos\theta}$$

Multiply by $\dfrac{1 + \cos\theta}{1 + \cos\theta}$:
$$\frac{(1 - \cos\theta)(1 + \cos\theta)}{(1 + \cos\theta)} =$$

Simplify:
$$\frac{1 - \cos^2\theta}{1 + \cos\theta} =$$

Use $1 - \cos^2\theta = \sin^2\theta$:
$$\frac{\sin^2\theta}{1 + \cos\theta} = \frac{\sin^2\theta}{1 + \cos\theta}$$

You can also verify this identity by working with the expression on the right side of the equals sign:

$$1 - \cos\theta = \frac{\sin^2\theta}{1 + \cos\theta}$$

Multiply by $\dfrac{1 - \cos\theta}{1 - \cos\theta}$:
$$= \frac{\sin^2\theta(1 - \cos\theta)}{1 - \cos^2\theta}$$

Use $1 - \cos^2\theta = \sin^2\theta$:
$$= \frac{\sin^2\theta(1 - \cos\theta)}{\sin^2\theta}$$

Simplify:
$$1 - \cos\theta = 1 - \cos\theta$$

PROBLEM 9-11 Show that $\left(1 - \dfrac{\cos\theta}{\tan\theta/\sin\theta}\right)\cot\theta = \dfrac{1}{\csc\theta\sec\theta}$.

$$\left(1 - \frac{\cos\theta}{\tan\theta/\sin\theta}\right)\cot\theta = \frac{1}{\csc\theta\sec\theta}$$

Use the ratio identities:
$$\left(1 - \frac{\cos\theta}{\dfrac{\dfrac{\sin\theta}{\cos\theta}}{\sin\theta}}\right)\left(\frac{\cos\theta}{\sin\theta}\right) =$$

Simplify:
$$(1 - \cos^2\theta)\left(\frac{\cos\theta}{\sin\theta}\right) =$$

Use $1 - \cos^2\theta = \sin^2\theta$:
$$\sin^2\theta\left(\frac{\cos\theta}{\sin\theta}\right) =$$

$$\sin\theta\cos\theta =$$

Use the reciprocal identities:
$$\frac{1}{\csc\theta\sec\theta} = \frac{1}{\csc\theta\sec\theta}$$

PROBLEM 9-12 Show that $1 - 2\cos^2\theta = 2\sin^2\theta - 1$.

Solution:
$$1 - 2\cos^2\theta = 2\sin^2\theta - 1$$

Use $\cos^2\theta = 1 - \sin^2\theta$:
$$1 - 2(1 - \sin^2\theta) =$$

Simplify:
$$1 - 2 + 2\sin^2\theta =$$

$$2\sin^2\theta - 1 = 2\sin^2\theta - 1$$

PROBLEM 9-13 Show that $\dfrac{\sin^2\theta - \cos^2\theta}{\sin\theta + \cos\theta} = \dfrac{1}{\csc\theta} - \dfrac{1}{\sec\theta}$.

Solution:
$$\frac{\sin^2\theta - \cos^2\theta}{\sin\theta + \cos\theta} = \frac{1}{\csc\theta} - \frac{1}{\sec\theta}$$

Factor:
$$\frac{(\sin\theta - \cos\theta)(\sin\theta + \cos\theta)}{\sin\theta + \cos\theta} =$$

Simplify:
$$\sin\theta - \cos\theta =$$

Use the reciprocal identities:
$$\frac{1}{\csc\theta} - \frac{1}{\sec\theta} = \frac{1}{\csc\theta} - \frac{1}{\sec\theta}$$

PROBLEM 9-14 Show that $\dfrac{1}{1 + \cos\theta} + \dfrac{1}{1 - \cos\theta} = 2\csc^2\theta$.

Solution:
$$\frac{1}{1 + \cos\theta} + \frac{1}{1 - \cos\theta} = 2\csc^2\theta$$

Combine fractions:
$$\frac{(1 - \cos\theta) + (1 + \cos\theta)}{1 - \cos^2\theta} =$$

Simplify and use $1 - \cos^2\theta = \sin^2\theta$:
$$\frac{2}{\sin^2\theta} =$$

Use $\csc\theta = 1/\sin\theta$:
$$2\csc^2\theta = 2\csc^2\theta$$

PROBLEM 9-15 Express $\sec\theta + \csc\theta$ in terms of $\sin\theta$ and $\cos\theta$. Simplify the result.

Solution:
$$\sec\theta + \csc\theta = \frac{1}{\cos\theta} + \frac{1}{\sin\theta}$$
$$= \frac{\sin\theta + \cos\theta}{\sin\theta\cos\theta}$$

PROBLEM 9-16 Show that $\dfrac{\sec\theta + \csc\theta}{\tan\theta + \cot\theta} = \sin\theta + \cos\theta$.

Solution:
$$\frac{\sec\theta + \csc\theta}{\tan\theta + \cot\theta} = \sin\theta + \cos\theta$$

Use the reciprocal and ratio identities:
$$\frac{\dfrac{1}{\cos\theta} + \dfrac{1}{\sin\theta}}{\dfrac{\sin\theta}{\cos\theta} + \dfrac{\cos\theta}{\sin\theta}} =$$

Clear fractions:
$$\frac{\dfrac{\sin\theta + \cos\theta}{\sin\theta\cos\theta}}{\dfrac{\sin^2\theta + \cos^2\theta}{\cos\theta\sin\theta}} =$$

Simplify:
$$\frac{\sin\theta + \cos\theta}{\sin^2\theta + \cos^2\theta} =$$

Use $\sin^2\theta + \cos^2\theta = 1$:
$$\sin\theta + \cos\theta = \sin\theta + \cos\theta$$

PROBLEM 9-17 Show that $\dfrac{\cos\theta}{1 + \sin\theta} + \tan\theta = \sec\theta$.

Solution:
$$\frac{\cos\theta}{1+\sin\theta} + \tan\theta = \sec\theta$$

Use $\tan\theta = \sin\theta/\cos\theta$:
$$\frac{\cos\theta}{1+\sin\theta} + \frac{\sin\theta}{\cos\theta} =$$

Combine fractions:
$$\frac{\cos^2\theta + \sin\theta + \sin^2\theta}{\cos\theta(1+\sin\theta)} =$$

Use $\cos^2\theta + \sin^2\theta = 1$:
$$\frac{\sin\theta + 1}{\cos\theta(1+\sin\theta)} =$$

$$\frac{1}{\cos\theta} =$$

$$\sec\theta = \sec\theta$$

PROBLEM 9-18   Show that $\dfrac{\sin^3\theta}{\sec\theta} + \dfrac{\cos^3\theta}{\csc\theta} = \sin\theta\cos\theta$.

Solution:
$$\frac{\sin^3\theta}{\sec\theta} + \frac{\cos^3\theta}{\csc\theta} = \sin\theta\cos\theta$$

Use the reciprocal identities:
$$\sin^3\theta\cos\theta + \cos^3\theta\sin\theta =$$

Factor:
$$(\sin\theta\cos\theta)(\sin^2\theta + \cos^2\theta) =$$

Use $\sin^2\theta + \cos^2\theta = 1$:
$$\sin\theta\cos\theta = \sin\theta\cos\theta$$

PROBLEM 9-19   Show that $\sin\theta\cos\theta\tan\theta\cot\theta\sec\theta\csc\theta = 1$.

Solution:
$$\sin\theta\cos\theta\tan\theta\cot\theta\sec\theta\csc\theta = 1$$

Use the reciprocal and ratio identities:
$$\sin\theta\cos\theta\,\frac{\sin\theta}{\cos\theta}\,\frac{\cos\theta}{\sin\theta}\,\frac{1}{\cos\theta}\,\frac{1}{\sin\theta} = 1$$

Cancel common terms:
$$1 = 1$$

PROBLEM 9-20   Show that $\dfrac{1}{\sin\theta} + \dfrac{1}{\cos\theta} + \dfrac{1}{\tan\theta} + \dfrac{1}{\cot\theta} + \dfrac{1}{\sec\theta} + \dfrac{1}{\csc\theta} =$

$\sin\theta + \cos\theta + \tan\theta + \cot\theta + \sec\theta + \csc\theta$.

Solution: The result follows directly from the reciprocal identities.

PROBLEM 9-21   Show that $\sin\theta + \cos\theta + \tan\theta + \cot\theta + \sec\theta + \csc\theta =$

$\dfrac{2 + \cos\theta - \cos^2\theta}{\sin\theta} + \dfrac{2 + \sin\theta - \sin^2\theta}{\cos\theta}$.

Solution: Work with the left-hand side: $\sin\theta + \cos\theta + \tan\theta + \cot\theta + \sec\theta + \csc\theta$

Use the ratio and reciprocal identities:
$$\sin\theta + \cos\theta + \frac{\sin\theta}{\cos\theta} + \frac{\cos\theta}{\sin\theta} + \frac{1}{\cos\theta} + \frac{1}{\sin\theta}$$

Rearrange the terms:
$$\left(\sin\theta + \frac{\cos\theta}{\sin\theta} + \frac{1}{\sin\theta}\right) + \left(\cos\theta + \frac{\sin\theta}{\cos\theta} + \frac{1}{\cos\theta}\right)$$

Combine the collected terms:
$$\left(\frac{\sin^2\theta + \cos\theta + 1}{\sin\theta}\right) + \left(\frac{\cos^2\theta + \sin\theta + 1}{\cos\theta}\right)$$

Use $\sin^2\theta = 1 - \cos^2\theta$ and $\cos^2\theta = 1 - \sin^2\theta$:
$$\frac{1 - \cos^2\theta + \cos\theta + 1}{\sin\theta} + \frac{1 - \sin^2\theta + \sin\theta + 1}{\cos\theta}$$

Combine terms:
$$\frac{2 + \cos\theta - \cos^2\theta}{\sin\theta} + \frac{2 + \sin\theta - \sin^2\theta}{\cos\theta}$$

PROBLEM 9-22 Show that $\sin^2 x + 2\sin x\cos y + \cos^2 y = (\sin x + \cos y)^2$.

Solution: Use $a^2 + 2ab + b^2 = (a + b)^2$ with $a = \sin x$ and $b = \cos y$. The result follows directly.

PROBLEM 9-23 Show that $\sec^2\theta(1 - \sin^2\theta) = 1$.

Solution:

$$\sec^2\theta(1 - \sin^2\theta) = 1$$

Use $1 - \sin^2\theta = \cos^2\theta$

$$\sec^2\theta\cos^2\theta = 1$$

Use $\cos\theta = 1/\sec\theta$

$$\sec^2\theta\,\frac{1}{\sec^2\theta} = 1$$

$$1 = 1$$

PROBLEM 9-24 Show that $\dfrac{\sin x\cos y - \cos x\sin y}{\cos x\cos y + \sin x\sin y} = \dfrac{\tan x - \tan y}{1 + \tan x\tan y}$.

Solution:

Use the ratio identities:

$$= \frac{\dfrac{\sin x}{\cos x} - \dfrac{\sin y}{\cos y}}{1 + \dfrac{\sin x}{\cos x}\dfrac{\sin y}{\cos y}}$$

Combine fractions:

$$= \frac{\dfrac{\sin x\cos y - \cos x\sin y}{\cos x\cos y}}{\dfrac{\cos x\cos y + \sin x\sin y}{\cos x\cos y}}$$

Clear fractions:

$$\frac{\sin x\cos y - \cos x\sin y}{\cos x\cos y + \sin x\sin y} = \frac{\sin x\cos y - \cos x\sin y}{\cos x\cos y + \sin x\sin y}$$

PROBLEM 9-25 Show that $\dfrac{1 - 2\sin^2\theta}{\sin\theta\cos\theta} = \cot\theta - \tan\theta$.

Solution:

Use the ratio identities:

$$= \frac{\cos\theta}{\sin\theta} - \frac{\sin\theta}{\cos\theta}$$

Combine fractions:

$$= \frac{\cos^2\theta - \sin^2\theta}{\sin\theta\cos\theta}$$

Use $\cos^2\theta = 1 - \sin^2\theta$:

$$\frac{1 - 2\sin^2\theta}{\sin\theta\cos\theta} = \frac{1 - 2\sin^2\theta}{\sin\theta\cos\theta}$$

PROBLEM 9-26 Show that $\dfrac{1}{\sin^2\theta} + \dfrac{1}{\cos^2\theta} = \dfrac{1 + \tan^2\theta}{\sin^2\theta}$.

Solution:

Rewrite:

$$= \frac{1}{\sin^2\theta} + \frac{\tan^2\theta}{\sin^2\theta}$$

Use $\tan^2\theta = \sin^2\theta/\cos^2\theta$:

$$= \frac{1}{\sin^2\theta} + \frac{\dfrac{\sin^2\theta}{\cos^2\theta}}{\sin^2\theta}$$

Simplify:

$$\frac{1}{\sin^2\theta} + \frac{1}{\cos^2\theta} = \frac{1}{\sin^2\theta} + \frac{1}{\cos^2\theta}$$

PROBLEM 9-27 Show that $\dfrac{1}{1 + \cos \theta} + \dfrac{1}{1 - \cos \theta} = 2 \csc^2\theta$.

Solution:

Combine fractions: $\dfrac{1 - \cos \theta + 1 + \cos \theta}{(1 - \cos \theta)(1 + \cos \theta)} =$

Simplify: $\dfrac{2}{1 - \cos^2\theta} =$

Use $1 - \cos^2\theta = \sin^2\theta$: $\dfrac{2}{\sin^2\theta} =$

Use $\csc \theta = 1/\sin \theta$: $2 \csc^2\theta = 2 \csc^2\theta$

PROBLEM 9-28 Show that $\cos \theta \csc \theta + \cot \theta = 2 \cot \theta$.

Solution:

$$\cos \theta \csc \theta + \cot \theta = \cos \theta \, \frac{1}{\sin \theta} + \frac{\cos \theta}{\sin \theta} = \frac{2 \cos \theta}{\sin \theta} = 2 \cot \theta$$

PROBLEM 9-29 Use identities to find the value of: (**a**) $\csc x$ when $\sin x = \frac{1}{4}$; (**b**) $\sec x$ when $\cos x = -\frac{2}{3}$; (**c**) $\cot x$ when $\tan x = \frac{9}{2}$; (**d**) $\cos x$ when $\sin x = \frac{2}{5}$ and $\frac{1}{2}\pi < x < \pi$; (**e**) $\sin x$ when $\cos x = -\frac{3}{4}$ and $\pi < x < \frac{3}{2}\pi$.

Solution:

(**a**) $\csc x = \dfrac{1}{\sin x} = \dfrac{1}{1/4} = 4$

(**b**) $\sec x = \dfrac{1}{\cos x} = -\dfrac{1}{2/3} = -\dfrac{3}{2}$

(**c**) $\cot x = \dfrac{1}{\tan x} = \dfrac{1}{9/2} = \dfrac{2}{9}$

(**d**) Since $\cos^2 x + \sin^2 x = 1$, $\cos^2 x + \left(\frac{2}{5}\right)^2 = 1$. Solving for $\cos^2 x$, you get $\cos^2 x = \frac{21}{25}$ or $\cos x = \pm\sqrt{\frac{21}{25}} = \pm\frac{\sqrt{21}}{5}$. Choose the negative sign because x lies in quadrant II. Then $\cos x = \frac{-\sqrt{21}}{5}$.

(**e**) Since $\cos^2 x + \sin^2 x = 1$, $\left(-\frac{3}{4}\right)^2 + \sin^2 x = 1$. Solving for $\sin^2 x$, $\sin^2 x = \frac{7}{16}$ or $\sin = \pm\frac{\sqrt{7}}{4}$. Choose the negative sign because x lies in quadrant III. Then $\sin x = \frac{-\sqrt{7}}{4}$.

PROBLEM 9-30 Tell what is wrong with the statement "If the terminal side of α lies in quadrant I and $\tan \alpha = \frac{2}{5}$, then (since $\tan \alpha = \sin \alpha/\cos \alpha$) it follows that $\sin \alpha = 2$ and $\cos \alpha = 5$."

Solution: It is true that $\tan \alpha$ can be $\frac{2}{5}$, but the sine and cosine functions can never take values greater than $+1$ nor less than -1.

PROBLEM 9-31 Tell what is wrong with the statement "If $\tan \alpha = 3$ and the terminal side of α lies in quadrant III, then (since $1 + \tan^2\alpha = \sec^2\alpha$) $\sec \alpha = \sqrt{1 + 9} = \sqrt{10}$."

Solution: From the given identity, $\sec^2\alpha = 10$, but since the terminal side of α lies in quadrant III, the secant cannot be positive.

PROBLEM 9-32 Tell what is wrong with the statement "If $\sin \alpha = \frac{3}{5}$ and the terminal side of α lies in quadrant II, then (since $\sin^2\alpha + \cos^2\alpha = 1$) $\cos \alpha = \sqrt{1 - \frac{9}{25}} = \sqrt{\frac{16}{25}} = \frac{4}{5}$."

Solution: As in Problem 9-31, the cosine cannot be positive in quadrant II.

PROBLEM 9-33 Show that $(\sec x + \tan x)(1 - \sin x) = \cos x$.

Solution: $(\sec x + \tan x)(1 - \sin x) = \left(\dfrac{1}{\cos x} + \dfrac{\sin x}{\cos x}\right)(1 - \sin x) = \dfrac{1 + \sin x}{\cos x}(1 - \sin x)$

$$= \frac{1 - \sin^2 x}{\cos x} = \frac{\cos^2 x}{\cos x} = \cos x$$

PROBLEM 9-34 Show that $\sin^4 x - \cos^4 x + \cos^2 x = \sin^2 x$.

Solution: $\quad \sin^4 x - \cos^4 x + \cos^2 x = (\sin^2 x - \cos^2 x)(\sin^2 x + \cos^2 x) + \cos^2 x$
$$= (\sin^2 x - \cos^2 x)(1) + \cos^2 x$$
$$= \sin^2 x$$

PROBLEM 9-35 Show that $\dfrac{1 + \tan^2 x}{\tan^2 x} = \csc^2 x$.

Solution: $\quad \dfrac{1 + \tan^2 x}{\tan^2 x} = \dfrac{1}{\tan^2 x} + 1 = \dfrac{1}{\dfrac{\sin^2 x}{\cos^2 x}} + 1 = \dfrac{\cos^2 x}{\sin^2 x} + 1 = \dfrac{\cos^2 x + \sin^2 x}{\sin^2 x}$

$$= \frac{1}{\sin^2 x} = \csc^2 x$$

An alternate solution uses $\cot^2 x + 1 = \csc^2 x$; the answer follows readily.

PROBLEM 9-36 Show that $\dfrac{\sec x}{\tan x + \cot x} = \sin x$.

Solution:

$$\frac{\sec x}{\tan x + \cot x} = \frac{\dfrac{1}{\cos x}}{\dfrac{\sin x}{\cos x} + \dfrac{\cos x}{\sin x}} = \frac{\dfrac{1}{\cos x}}{\dfrac{\sin^2 x + \cos^2 x}{\cos x \sin x}}$$

$$= \frac{\dfrac{1}{\cos x}}{\dfrac{1}{\cos x \sin x}} = \sin x$$

PROBLEM 9-37 Show that $\cos^4 x - \sin^4 x = \cos^2 x - \sin^2 x$.

Solution: $\quad \cos^4 x - \sin^4 x = (\cos^2 x + \sin^2 x)(\cos^2 x - \sin^2 x) = (1)(\cos^2 x - \sin^2 x)$

PROBLEM 9-38 Show that $\dfrac{1 + \sec x}{\tan x + \sin x} = \csc x$.

Solution:

$$\frac{1 + \sec x}{\tan x + \sin x} = \frac{1 + \dfrac{1}{\cos x}}{\dfrac{\sin x}{\cos x} + \sin x} = \frac{\dfrac{\cos x + 1}{\cos x}}{\dfrac{\sin x + \sin x \cos x}{\cos x}}$$

$$= \frac{\cos x + 1}{\sin x(1 + \cos x)} = \frac{1}{\sin x} = \csc x$$

Supplementary Exercises

PROBLEM 9-39   Factor $\sin^3\alpha + 8$.

PROBLEM 9-40   Simplify $\dfrac{1}{\cos^2 x - \sin^2 x} - \dfrac{2}{\cos x + \sin x}$.

PROBLEM 9-41   Simplify $\dfrac{\dfrac{\sin\alpha}{\cos\alpha}}{\dfrac{\sin^2\alpha}{\sec\alpha}}$.

PROBLEM 9-42   In the algebraic expression $\sqrt{36 - 16x^2}$, let $x = \left(\frac{3}{2}\right)\cos\alpha$. Simplify the result.

PROBLEM 9-43   In the algebraic expression $\sqrt{x^2 - 4}/x$, let $x = 2\sec\alpha$. Simplify the result.

PROBLEM 9-44   In the algebraic expression $x^2\sqrt{4 + 9x^2}$, let $x = \left(\frac{2}{3}\right)\tan\alpha$. Simplify the result.

PROBLEM 9-45   Express $(\tan x + \cot x)^2$ in terms of sines and cosines. Simplify your answer.

PROBLEM 9-46   Express $\sec^2 x + \csc^2 x$ in terms of sines and cosines. Simplify your answer.

PROBLEM 9-47   Show that $(\tan x + \cot x)^2 = \sec^2 x + \csc^2 x$.

PROBLEM 9-48   Show that $\dfrac{\sin x + \tan x}{1 + \cos x} = \tan x$.

PROBLEM 9-49   Show that $\dfrac{\cos x + \sin x}{\sin x} = 1 + \dfrac{1}{\tan x}$.

PROBLEM 9-50   Show that $\dfrac{1 + \sin x}{\cos x} = \dfrac{\cos x}{1 - \sin x}$.

PROBLEM 9-51   Express $\dfrac{1 + \tan^2 x}{\tan^2 x}$ in terms of sines and cosines. Simplify your answer.

PROBLEM 9-52   Show that $\dfrac{1 + \cot^2 x}{\cot^2 x} = \sec^2 x$.

PROBLEM 9-53   Show that $\dfrac{1 + \csc x}{\csc x} = \dfrac{\cos^2 x}{1 - \sin x}$.

PROBLEM 9-54   Show that $\tan^2 y - \sin^2 y = \tan^2 y \sin^2 y$.

PROBLEM 9-55   Show that $\sin x + \tan x = \tan x(1 + \cos x)$.

PROBLEM 9-56   Show that $\cos^2 x \sec^2 x + \cos^2 x \csc^2 x = \csc^2 x$.

PROBLEM 9-57   Show that $\dfrac{1 + \sin\alpha}{1 - \sin\alpha} = (\sec\alpha + \tan\alpha)^2$.

PROBLEM 9-58 Show that $\dfrac{\cos \alpha}{1 + \sin \alpha} + \tan \alpha = \sec \alpha$.

PROBLEM 9-59 Show that $\dfrac{\sin^3\alpha - \sin \alpha + \cos \alpha}{\sin \alpha} = \cot \alpha - \cos^2\alpha$.

PROBLEM 9-60 Express $\csc x - \cot x \csc x$ in terms of sines and cosines. Simplify your answer.

PROBLEM 9-61 Show that $\dfrac{1 - \cot^4 x}{\csc^2 x(\sin x + \cos x)} = \csc x - \cot x \csc x$.

PROBLEM 9-62 Express $\dfrac{\tan x + \sec x}{\sec x - \tan x}$ in terms of sines and cosines. Simplify your answer.

PROBLEM 9-63 Show that $\dfrac{\tan x + \sec x}{\sec x - \cos x + \tan x} = \csc x$.

PROBLEM 9-64 Show that $\dfrac{\tan \alpha - \sin \alpha}{\sin^3\alpha} = \dfrac{\sec \alpha}{1 + \cos \alpha}$.

PROBLEM 9-65 Show that $\dfrac{1 + \sin x}{\cos x} = \dfrac{\cos x}{1 - \sin x}$.

PROBLEM 9-66 Show that $\dfrac{\tan x - \cos x \cot x}{\csc x} = \dfrac{\sin x}{\cot x} - \dfrac{\cos x}{\sec x}$.

PROBLEM 9-67 Show that $\dfrac{1 + \csc x}{\sec x} = \cos x + \cot x$.

PROBLEM 9-68 Show that $\sqrt{\dfrac{\sec x - 1}{\sec x + 1}} = \dfrac{|\sin x|}{1 + \cos x}$.

PROBLEM 9-69 Show that $\sqrt{\dfrac{1 - \sin x}{1 + \sin x}} = \dfrac{1 - \sin x}{|\cos x|}$.

PROBLEM 9-70 Show that $1 + \cos \alpha + \sin \alpha = \dfrac{2 \cos \alpha \sin \alpha}{\cos \alpha + \sin \alpha - 1}$.

Answers to Supplementary Exercises

9-39: $\sin^3\alpha + 8 = \sin^3\alpha + 2^3 =$
$(\sin \alpha + 2)(\sin^2\alpha - 2 \sin \alpha + 4)$

9-40: $\dfrac{1}{\cos^2 x - \sin^2 x} - \dfrac{2}{\cos x + \sin x} =$
$\dfrac{1 - 2(\cos x - \sin x)}{\cos^2 x - \sin^2 x}$

9-41: $\dfrac{\dfrac{\sin \alpha}{\cos \alpha}}{\dfrac{\sin^2\alpha}{\sec \alpha}} = \left(\dfrac{\sin \alpha}{\cos \alpha}\right)\dfrac{\sec \alpha}{\sin^2\alpha} = \dfrac{1}{\cos^2\alpha \sin \alpha}$

9-42: $\sqrt{36 - 16x^2} = \sqrt{36 - 16(\tfrac{3}{2}\cos \alpha)^2} =$
$\sqrt{36 - 36 \cos^2\alpha} = 6\sqrt{1 - \cos^2\alpha} = 6|\sin \alpha|$

9-43: $\dfrac{\sqrt{x^2 - 4}}{x} = \dfrac{\sqrt{(2 \sec \alpha)^2 - 4}}{2 \sec \alpha} =$
$\dfrac{\sqrt{4 \sec^2\alpha - 4}}{2 \sec \alpha} = \dfrac{2\sqrt{\sec^2\alpha - 1}}{2 \sec \alpha} =$
$\dfrac{2\sqrt{\tan^2\alpha}}{2 \sec \alpha} = \dfrac{|\tan \alpha|}{\sec \alpha} = |\tan \alpha|\cos \alpha$

9-44: $x^2\sqrt{4 + 9x^2} = (\tfrac{2}{3}\tan \alpha)^2\sqrt{4 + 9(\tfrac{2}{3}\tan \alpha)^2} =$
$(\tfrac{4}{9}\tan^2\alpha)\sqrt{4 + 4\tan^2\alpha} = \tfrac{8}{9}\tan^2\alpha|\sec \alpha|$

9-45: $\left(\dfrac{\sin x}{\cos x} + \dfrac{\cos x}{\sin x}\right)^2 = \left(\dfrac{\sin^2 x + \cos^2 x}{\sin x \cos x}\right)^2 =$
$\left(\dfrac{1}{\sin x \cos x}\right)^2 = \dfrac{1}{\sin^2 x \cos^2 x}$

9-46: $\sec^2 x + \csc^2 x = \left(\dfrac{1}{\cos x}\right)^2 + \left(\dfrac{1}{\sin x}\right)^2 =$

$\dfrac{\sin^2 x + \cos^2 x}{\sin^2 x \cos^2 x} = \dfrac{1}{\sin^2 x \cos^2 x}$

9-47: $(\tan x + \cot x)^2 = \dfrac{\sin^2 x}{\cos^2 x} + 2 + \dfrac{\cos^2 x}{\sin^2 x} =$

$(\tan^2 x + 1) + (1 + \cot^2 x) = \sec^2 x + \csc^2 x$

9-48: $\dfrac{\sin x + \tan x}{1 + \cos x} = \dfrac{\sin x + \dfrac{\sin x}{\cos x}}{1 + \cos x} =$

$\dfrac{\dfrac{\sin x \cos x + \sin x}{\cos x}}{1 + \cos x} = \dfrac{\dfrac{\sin x(1 + \cos x)}{\cos x}}{1 + \cos x} =$

$\dfrac{\sin x}{\cos x} = \tan x$

9-49: $1 + \dfrac{1}{\tan x} = 1 + \dfrac{1}{\dfrac{\sin x}{\cos x}} = 1 + \dfrac{\cos x}{\sin x} =$

$\dfrac{\sin x + \cos x}{\sin x}$

9-50: $\dfrac{1 + \sin x}{\cos x} = \left(\dfrac{1 + \sin x}{\cos x}\right)\left(\dfrac{1 - \sin x}{1 - \sin x}\right) =$

$\dfrac{1 - \sin^2 x}{\cos x(1 - \sin x)} = \dfrac{\cos^2 x}{\cos x(1 - \sin x)} =$

$\dfrac{\cos x}{1 - \sin x}$

9-51: $\dfrac{1 + \left(\dfrac{\sin x}{\cos x}\right)^2}{\left(\dfrac{\sin x}{\cos x}\right)^2} =$

$\left(\dfrac{\cos^2 x + \sin^2 x}{\cos^2 x}\right)\left(\dfrac{\cos^2 x}{\sin^2 x}\right) = \dfrac{1}{\sin^2 x} =$

$\csc^2 x$

9-52: $\dfrac{1 + \cot^2 x}{\cot^2 x} = \dfrac{1}{\cot^2 x} + 1 = \tan^2 x + 1 =$

$\sec^2 x$

9-53: $\dfrac{1 + \csc x}{\csc x} = \dfrac{1}{\csc x} + 1 = \sin x + 1 =$

$1 + \sin x = (1 + \sin x)\left(\dfrac{1 - \sin x}{1 - \sin x}\right) = \dfrac{1 - \sin^2 x}{1 - \sin x} =$

$\dfrac{\cos^2 x}{1 - \sin x}$

9-54: $\tan^2 y - \sin^2 y = \tan^2 y\left(1 - \dfrac{\sin^2 y}{\tan^2 y}\right) =$

$\tan^2 y(1 - \cos^2 y) = \tan^2 y \sin^2 y$

9-55: $\sin x + \tan x = \tan x\left(\dfrac{\sin x}{\tan x} + 1\right) =$

$\tan x\left(\dfrac{\sin x}{\dfrac{\sin x}{\cos x}} + 1\right) = \tan x(\cos x + 1)$

9-56: $\cos^2 x \sec^2 x + \cos^2 x \csc^2 x =$

$1 + \dfrac{\cos^2 x}{\sin^2 x} = \dfrac{\sin^2 x + \cos^2 x}{\sin^2 x} = \dfrac{1}{\sin^2 x} =$

$\csc^2 x$

9-57: $(\sec \alpha + \tan \alpha)^2 = \left(\dfrac{1}{\cos \alpha} + \dfrac{\sin \alpha}{\cos \alpha}\right)^2 =$

$\dfrac{(1 + \sin \alpha)^2}{\cos^2 \alpha} = \dfrac{(1 + \sin \alpha)^2}{1 - \sin^2 \alpha} =$

$\left(\dfrac{(1 + \sin \alpha)^2}{(1 - \sin \alpha)(1 + \sin \alpha)}\right)(1 + \sin \alpha)$

$\dfrac{1 + \sin \alpha}{1 - \sin \alpha}$

9-58: $\dfrac{\cos \alpha}{1 + \sin \alpha} + \tan \alpha = \dfrac{\cos \alpha}{1 + \sin \alpha} + \dfrac{\sin \alpha}{\cos \alpha} =$

$\dfrac{\cos^2 \alpha}{\cos \alpha(1 + \sin \alpha)} + \dfrac{\sin \alpha + \sin^2 \alpha}{\cos \alpha(1 + \sin \alpha)} =$

$\dfrac{\cos^2 \alpha + \sin \alpha + \sin^2 \alpha}{\cos \alpha(1 + \sin \alpha)} = \dfrac{1 + \sin \alpha}{\cos \alpha(1 + \sin \alpha)} =$

$\dfrac{1}{\cos \alpha} = \sec \alpha$

9-59: $\dfrac{\sin^3 \alpha - \sin \alpha + \cos \alpha}{\sin \alpha} =$

$\sin^2 \alpha - 1 + \dfrac{\cos \alpha}{\sin \alpha} = -(1 - \sin^2 \alpha) + \dfrac{\cos \alpha}{\sin \alpha} =$

$-\cos^2 \alpha + \cot \alpha$

9-60: $\csc x - \cot x \csc x =$

$\dfrac{1}{\sin x} - \left(\dfrac{\cos x}{\sin x}\right)\left(\dfrac{1}{\sin x}\right) = \left(\dfrac{1}{\sin x}\right)\left(1 - \dfrac{\cos x}{\sin x}\right) =$

$\left(\dfrac{1}{\sin^2 x}\right)(\sin x - \cos x)$

9-61: $\dfrac{1 - \cot^4 x}{\csc^2 x(\sin x + \cos x)} =$

$\dfrac{(1 - \cot^2 x)(1 + \cot^2 x)}{\csc^2 x(\sin x + \cos x)} =$

$\dfrac{1 - \cot^2 x}{\sin x + \cos x} = \dfrac{1 - \left(\dfrac{\cos x}{\sin x}\right)^2}{\sin x + \cos x} =$

$\dfrac{\sin^2 x - \cos^2 x}{\sin^2 x(\sin x + \cos x)} =$

$\dfrac{\sin x - \cos x}{\sin^2 x} = \csc x - \cot x \csc x$

9-62: $\dfrac{\tan x + \sec x}{\sec x - \tan x} = \dfrac{\dfrac{\sin x}{\cos x} + \dfrac{1}{\cos x}}{\dfrac{1}{\cos x} - \dfrac{\sin x}{\cos x}} =$

$\dfrac{\dfrac{\sin x + 1}{\cos x}}{\dfrac{1 - \sin x}{\cos x}} = \dfrac{1 + \sin x}{1 - \sin x}$

9-63: $\dfrac{\tan x + \sec x}{\sec x - \cos x + \tan x} =$

$\dfrac{\dfrac{\sin x}{\cos x} + \dfrac{1}{\cos x}}{\dfrac{1}{\cos x} - \cos x + \dfrac{\sin x}{\cos x}} = \dfrac{\sin x + 1}{1 - \cos^2 x + \sin x} =$

$\dfrac{\sin x + 1}{\sin^2 x + \sin x} = \dfrac{1}{\sin x} = \csc x$

9-64: $\dfrac{\tan \alpha - \sin \alpha}{\sin^3 \alpha} = \dfrac{\dfrac{\sin \alpha}{\cos \alpha} - \sin \alpha}{\sin^3 \alpha} =$

$\dfrac{\dfrac{1}{\cos \alpha} - 1}{\sin^2 \alpha} = \dfrac{\dfrac{1 - \cos \alpha}{\cos \alpha}}{\sin^2 \alpha} =$

$\left(\dfrac{\dfrac{1 - \cos \alpha}{\cos \alpha}}{\sin^2 \alpha} \right) \times \left(\dfrac{1 + \cos \alpha}{1 + \cos \alpha} \right) =$

$\dfrac{1 - \cos^2 \alpha}{\cos \alpha \sin^2 \alpha (1 + \cos \alpha)} = \dfrac{1}{\cos \alpha (1 + \cos \alpha)} =$

$\dfrac{\sec \alpha}{1 + \cos \alpha}$

9-65: $\dfrac{1 + \sin x}{\cos x} = \left(\dfrac{1 + \sin x}{\cos x} \right) \dfrac{1 - \sin x}{1 - \sin x} =$

$\dfrac{1 - \sin^2 x}{\cos x (1 - \sin x)} = \dfrac{\cos^2 x}{\cos x (1 - \sin x)} =$

$\dfrac{\cos x}{1 - \sin x}$

9-66: $\dfrac{\tan x - \cos x \cot x}{\csc x} =$

$\dfrac{\tan x}{\csc x} - \dfrac{\cos x \cot x}{\csc x} = \dfrac{\sin^2 x}{\cos x} - \cos^2 x =$

$\dfrac{\sin x}{\cot x} - \dfrac{\cos x}{\sec x}$

9-67: $\dfrac{1 + \csc x}{\sec x} = \dfrac{1}{\sec x} + \dfrac{\csc x}{\sec x} =$

$\cos x + \dfrac{\cos x}{\sin x} = \cos x + \cot x$

9-68: $\sqrt{\dfrac{\sec x - 1}{\sec x + 1}} = \sqrt{\dfrac{\dfrac{1}{\cos x} - 1}{\dfrac{1}{\cos x} + 1}} =$

$\left(\sqrt{\dfrac{1 - \cos x}{1 + \cos x}} \right)\left(\sqrt{\dfrac{1 + \cos x}{1 + \cos x}} \right) = \sqrt{\dfrac{(1 - \cos^2 x)}{(1 + \cos x)^2}} =$

$\sqrt{\dfrac{\sin^2 x}{(1 + \cos x)^2}} = \dfrac{|\sin x|}{1 + \cos x}$

9-69: $\sqrt{\dfrac{1 - \sin x}{1 + \sin x}} =$

$\left(\sqrt{\dfrac{1 - \sin x}{1 + \sin x}} \right)\left(\sqrt{\dfrac{1 - \sin x}{1 - \sin x}} \right) = \dfrac{1 - \sin x}{\sqrt{1 - \sin^2 x}} =$

$\dfrac{1 - \sin x}{|\cos x|}$

9-70: $\dfrac{2 \cos \alpha \sin \alpha}{\cos \alpha + \sin \alpha - 1} =$

$\left(\dfrac{2 \cos \alpha \sin \alpha}{\cos \alpha + \sin \alpha - 1} \right)\left(\dfrac{\cos \alpha + \sin \alpha + 1}{\cos \alpha + \sin \alpha + 1} \right) =$

$(1 + \cos \alpha + \sin \alpha)$

$\times \left(\dfrac{2 \cos \alpha \sin \alpha}{\cos^2 \alpha + 2 \sin \alpha \cos \alpha + \sin^2 \alpha - 1} \right) =$

$(1 + \cos \alpha + \sin \alpha)\left(\dfrac{2 \cos \alpha \sin \alpha}{1 + 2 \sin \alpha \cos \alpha - 1} \right) =$

$1 + \cos \alpha + \sin \alpha$

10 TWO-ANGLE IDENTITIES

THIS CHAPTER IS ABOUT

☑ **Sum and Difference Identities**
☑ **Double-Angle Identities**
☑ **Half-Angle Identities**

10-1. Sum and Difference Identities

In most trigonometry textbooks, you will find that the difference identity is first derived for cosines, extended to the difference identity for sines, extended to the sum identities for cosines and sines, and finally extensions of these are made to sum and difference identities for tangents. Sum and difference formulas for the other trigonometric functions are used only in special cases and will appear only in selected examples and problems.

A. Difference identity for cosine

The derivation of the difference identity for cosine appears in most trigonometry books and will not be repeated here.

COSINE DIFFERENCE IDENTITY
$$\cos(\alpha - \beta) = \cos \alpha \cos \beta + \sin \alpha \sin \beta \qquad (10\text{-}1)$$

EXAMPLE 10-1: Use the difference formula for cosines to simplify (a) $\cos(x - \pi)$; (b) $\cos(\tfrac{1}{2}\pi - x)$; (c) $\cos(x - \tfrac{1}{4}\pi)$.

Solution:

(a) $\cos(x - \pi) = \cos x \cos \pi + \sin x \sin \pi$
$$= \cos x(-1) + \sin x(0) = -\cos x$$

(b) $\cos\left(\dfrac{\pi}{2} - x\right) = \cos\dfrac{\pi}{2}\cos x + \sin\dfrac{\pi}{2}\sin x$
$$= (0)\cos x + (1)\sin x = \sin x$$

Note: This cofunction result was established in Chapter 5 using the definitions of the trigonometric functions. We'll use it to establish the difference identity for the sine function.

(c) $\cos\left(x - \dfrac{\pi}{4}\right) = \cos x \cos\dfrac{\pi}{4} + \sin x \sin\dfrac{\pi}{4}$

$$= \cos x \dfrac{\sqrt{2}}{2} + \sin x \dfrac{\sqrt{2}}{2}$$

$$= \dfrac{\sqrt{2}}{2}(\cos x + \sin x)$$

EXAMPLE 10-2: Find the value of k for which
(a) $\cos 3x \cos 2x + \sin 3x \sin 2x = \cos kx$; (b) $\cos x \cos x + \sin x \sin x = \cos kx$.

Solution:

(a) By comparing the given expression with the right hand side of Equation 10-1, it is easy to see that $\alpha = 3x$ and $\beta = 2x$. Then $\cos kx = \cos(\alpha - \beta) = \cos(3x - 2x) = \cos 1x$, or $k = 1$.

(b) Again, compare the given expression with the right hand side of Equation 10-1. Then $\alpha = x$ and $\beta = x$, or $\cos kx = \cos(\alpha - \beta) = \cos(x - x) = \cos 0$. Thus $k = 0$. From identities in Chapter 9, you know that this is the correct answer because $\sin^2 x + \cos^2 x = 1 = \cos 0$.

B. Difference identity for sine

Let's begin with the identity $\cos(\frac{1}{2}\pi - x) = \sin x$ (see Example 10-1), replacing x with $\alpha - \beta$:

$$\sin(\alpha - \beta) = \cos\left[\frac{\pi}{2} - (\alpha - \beta)\right] = \cos\left[\left(\frac{\pi}{2} - \alpha\right) - (-\beta)\right]$$

Now use Equation 10-1 on the right-hand side of the equation:

$$\cos\left[\left(\frac{\pi}{2} - \alpha\right) - (-\beta)\right] = \cos\left(\frac{\pi}{2} - \alpha\right)\cos(-\beta) + \sin\left(\frac{\pi}{2} - \alpha\right)\sin(-\beta)$$

Apply the cofunction identities, $\cos(\frac{1}{2}\pi - \alpha) = \sin \alpha$, and $\sin(\frac{1}{2}\pi - \alpha) = \cos \alpha$ to the cosine and sine of $\frac{1}{2}\pi - \alpha$:

$$\sin(\alpha - \beta) = \sin \alpha \cos(-\beta) + \cos \alpha \sin(-\beta)$$

Finally, use the negative angle identities, $\sin(-\beta) = -\sin \beta$ and $\cos(-\beta) = \cos \beta$, to produce the difference identity for sines:

SINE DIFFERENCE IDENTITY $\qquad \sin(\alpha - \beta) = \sin \alpha \cos \beta - \cos \alpha \sin \beta \qquad$ **(10-2)**

EXAMPLE 10-3: Verify the cofunction identity, $\sin(\frac{1}{2}\pi - x) = \cos x$, using the difference identity for sine.

Solution:

$$\sin(\tfrac{1}{2}\pi - x) = \sin \tfrac{1}{2}\pi \cos x - \cos \tfrac{1}{2}\pi \sin x$$
$$= 1 \cos x - 0 \sin x = \cos x$$

EXAMPLE 10-4: Use the difference identities for sine and cosine to find the value of $\sin \frac{1}{12}\pi$ and $\cos \frac{1}{12}\pi$.

Solution: Because $\frac{1}{12}\pi = \frac{1}{3}\pi - \frac{1}{4}\pi$,

$$\sin \frac{\pi}{12} = \sin\left(\frac{\pi}{3} - \frac{\pi}{4}\right) = \sin \frac{\pi}{3} \cos \frac{\pi}{4} - \cos \frac{\pi}{3} \sin \frac{\pi}{4}$$

$$= \frac{\sqrt{3}}{2}\left(\frac{\sqrt{2}}{2}\right) - \frac{1}{2}\left(\frac{\sqrt{2}}{2}\right) = \frac{\sqrt{6} - \sqrt{2}}{4}$$

Using the same method and the formula for the cosine of a difference, you get

$$\cos \frac{\pi}{12} = \cos\left(\frac{\pi}{3} - \frac{\pi}{4}\right) = \cos \frac{\pi}{3} \cos \frac{\pi}{4} + \sin \frac{\pi}{3} \sin \frac{\pi}{4}$$

$$= \frac{1}{2}\left(\frac{\sqrt{2}}{2}\right) + \frac{\sqrt{3}}{2}\left(\frac{\sqrt{2}}{2}\right) = \frac{\sqrt{2} + \sqrt{6}}{4}$$

C. Sum identities for cosine and sine

If you use $\sin(-x) = -\sin x$ and $\cos(-x) = \cos x$, and the difference identities for sine and cosine, you can proceed as follows:

$$\cos(\alpha + \beta) = \cos(\alpha - (-\beta)) = \cos \alpha \cos(-\beta) + \sin \alpha \sin(-\beta)$$

which leads to the sum identity for cosines:

COSINE SUM IDENTITY $\qquad \cos(\alpha + \beta) = \cos \alpha \cos \beta - \sin \alpha \sin \beta \qquad$ (10-3)

Similarly,

$$\sin(\alpha + \beta) = \sin(\alpha - (-\beta)) = \sin \alpha \cos(-\beta) - \cos \alpha \sin(-\beta)$$

which leads to the sum identity for sines:

SINE SUM IDENTITY $\qquad \sin(\alpha + \beta) = \sin \alpha \cos \beta + \cos \alpha \sin \beta \qquad$ (10-4)

EXAMPLE 10-5: Find the value of k in

(a) $\sin kx = \sin 3x \cos 4x + \cos 3x \sin 4x$;
(b) $\cos kx = \cos 3x \cos 4x - \sin 3x \sin 4x$;
(c) $\cos kx = \cos 3x \cos 4x + \sin 3x \sin 4x$.

Solution:

(a) Compare the expression on the right-hand side of the equals sign with the expression on the right-hand side of the sum formula for sines. You see that $3x = \alpha$ and $4x = \beta$. Then $\alpha + \beta = 3x + 4x = 7x$. Then, comparing the left-hand side of the given expression and the sum formula for sines, $\alpha + \beta = 7x = kx$. Therefore, $k = 7$.

(b) Compare the expression on the right-hand side of the equals sign with the expression on the right-hand side of the sum formula for cosines. You see that $3x = \alpha$ and $4x = \beta$. Then $\alpha + \beta = 3x + 4x = 7x$. Then, comparing the left-hand side of the given expression and the sum formula for cosines, $\alpha + \beta = 7x = kx$. Therefore, $k = 7$.

(c) The right-hand side of the given expression is like the right-hand side of the difference identity for cosines. Again, $\alpha = 3x$ and $\beta = 4x$. However, since the difference identity has $\cos(\alpha - \beta)$ on the left side, $\alpha - \beta = 3x - 4x = -x = kx$, so $k = -1$.

EXAMPLE 10-6: Use the sum and difference identities for sine and cosine to show that (a) $\sin(\pi + x) = -\sin x$; (b) $\cos(\pi + x) = -\cos x$; (c) $\sin(2\pi + x) = \sin x$; (d) $\cos(2\pi + x) = \cos x$.

Solution:

(a) $\sin(\pi + x) = \sin \pi \cos x + \cos \pi \sin x$
$$= 0 \cos x + (-1 \sin x) = -\sin x$$

(b) $\cos(\pi + x) = \cos \pi \cos x - \sin \pi \sin x$
$$= -1 \cos x + 0 \sin x = -\cos x$$

(c) $\sin(2\pi + x) = \sin 2\pi \cos x + \cos 2\pi \sin x$
$$= 0 \cos x + 1 \sin x = \sin x$$

(d) $\cos(2\pi + x) = \cos 2\pi \cos x - \sin 2\pi \sin x$
$$= 1 \cos x + 0 \sin x = \cos x$$

EXAMPLE 10-7: Without using your calculator, find $\cos 75°$.

Solution: Recall that $\sin 45° = \cos 45° = \frac{\sqrt{2}}{2}$, $\sin 30° = \frac{1}{2}$, and $\cos 30° = \frac{\sqrt{3}}{2}$. Then,

$$\cos 75° = \cos(30° + 45°) = \cos 30° \cos 45° - \sin 30° \sin 45°$$

$$= \frac{\sqrt{3}}{2}\left(\frac{\sqrt{2}}{2}\right) - \frac{1}{2}\left(\frac{\sqrt{2}}{2}\right) = \frac{\sqrt{6} - \sqrt{2}}{4}$$

EXAMPLE 10-8: Without using your calculator, find $\csc \frac{1}{12}\pi$.

Solution: Recall that $\sin \frac{1}{4}\pi = \cos \frac{1}{4}\pi = \frac{\sqrt{2}}{2}$, $\sin \frac{1}{3}\pi = \frac{\sqrt{3}}{2}$, and $\cos \frac{1}{3}\pi = \frac{1}{2}$. Then,

$$\csc \frac{\pi}{12} = \csc\left(\frac{\pi}{3} - \frac{\pi}{4}\right) = \frac{1}{\sin\left(\dfrac{\pi}{3} - \dfrac{\pi}{4}\right)}$$

$$= \frac{1}{\sin \dfrac{\pi}{3} \cos \dfrac{\pi}{4} - \cos \dfrac{\pi}{3} \sin \dfrac{\pi}{4}}$$

$$= \frac{1}{\dfrac{\sqrt{3}}{2}\left(\dfrac{\sqrt{2}}{2}\right) - \dfrac{1}{2}\left(\dfrac{\sqrt{2}}{2}\right)}$$

$$= \frac{4(\sqrt{6} + \sqrt{2})}{6 - 2}$$

$$= \sqrt{6} + \sqrt{2}$$

EXAMPLE 10-9: Show that $\cos(\alpha - \frac{1}{2}\pi) = \sin \alpha$.

Solution: Use the difference identity for cosine with $\beta = \frac{1}{2}\pi$:

$$\cos\left(\alpha - \frac{\pi}{2}\right) = \cos \alpha \cos \frac{\pi}{2} + \sin \alpha \sin \frac{\pi}{2}$$

$$= \cos \alpha (0) + \sin \alpha (1) = \sin \alpha$$

D. Sum and difference identities for tangent

You derive both of these identities by starting with $\tan x = \sin x / \cos x$. For the sum identity $\tan(\alpha + \beta) = [\sin(\alpha + \beta)]/[\cos(\alpha + \beta)]$. Using Equations 10-3 and 10-4,

$$\tan(\alpha + \beta) = \frac{\sin \alpha \cos \beta + \cos \alpha \sin \beta}{\cos \alpha \cos \beta - \sin \alpha \sin \beta}$$

Divide the numerator and denominator of this expression by $\cos \alpha \cos \beta$ to get the sum identity for tangent:

TANGENT SUM IDENTITY
$$\tan(\alpha + \beta) = \frac{\tan \alpha + \tan \beta}{1 - \tan \alpha \tan \beta} \qquad \textbf{(10-5)}$$

Use a similar line of reasoning to determine the difference identity for tangent:

TANGENT DIFFERENCE IDENTITY
$$\tan(\alpha - \beta) = \frac{\tan \alpha - \tan \beta}{1 + \tan \alpha \tan \beta} \qquad \textbf{(10-6)}$$

The proof is left to the reader.

Once you know the tangent for the sum or difference of two angles, you can find the cotangent for this sum or difference by using the reciprocal relationship $\cot \alpha = 1/\tan \alpha$.

EXAMPLE 10-10: Without using your calculator, find the exact value of

$$\frac{\tan 50° - \tan 5°}{1 + \tan 50° \tan 5°}$$

Solution: Compare the given expression with the right-hand side of Equation 10-6. Since $\alpha = 50°$ and $\beta = 5°$, the tangent identity shows that the exact value of the expression must be $\tan(50° - 5°) = \tan 45° = 1$.

10-2. Double-Angle Identities

A. Double-angle identity for cosine

Consider Equation 10-3. If you let $\alpha = \beta$, you can derive the double-angle identity for cosine: $\cos(\alpha + \alpha) = \cos 2\alpha = \cos \alpha \cos \alpha - \sin \alpha \sin \alpha$, or

COSINE
DOUBLE-ANGLE $\cos 2\alpha = \cos^2 \alpha - \sin^2 \alpha$ (10-7)
IDENTITY

There are two useful alternative forms that arise by replacing $\cos^2 \alpha$ by $1 - \sin^2 \alpha$ or by replacing $\sin^2 \alpha$ by $1 - \cos^2 \alpha$:

$$\cos 2\alpha = 1 - 2\sin^2 \alpha = 2\cos^2 \alpha - 1 \qquad (10\text{-}8)$$

B. Double-angle identity for sine

Consider Equation 10-4. If you let $\alpha = \beta$, you can derive the double-angle identity for sine: $\sin(\alpha + \alpha) = \sin 2\alpha = \sin \alpha \cos \alpha + \cos \alpha \sin \alpha$, or

SINE
DOUBLE-ANGLE $\sin 2\alpha = 2\sin \alpha \cos \alpha$ (10-9)
IDENTITY

C. Double-angle identity for tangent

Consider Equation 10-5. If you let $\alpha = \beta$, you can derive the double-angle identity for tangent: $\tan(\alpha + \alpha) = \tan 2\alpha = (\tan \alpha + \tan \alpha)/(1 - \tan \alpha \tan \alpha)$, or

TANGENT
DOUBLE-ANGLE $\tan 2\alpha = \dfrac{2\tan \alpha}{1 - \tan^2 \alpha}$ (10-10)
IDENTITY

EXAMPLE 10-11: For $\cos 2\alpha = \frac{5}{7}$ and $0 < 2\alpha < \pi$, find $\sin \alpha$ and $\cos \alpha$.

Solution: Since $\cos 2\alpha = 1 - 2\sin^2 \alpha$, then $\frac{5}{7} = 1 - 2\sin^2 \alpha$, or $2\sin^2 \alpha = \frac{2}{7}$, so $\sin \alpha = \pm \frac{1}{\sqrt{7}}$. Since $0 < 2\alpha < \pi$, then $0 < \alpha < \frac{1}{2}\pi$. The sine and cosine of any angle between 0 and $\frac{1}{2}\pi$ are positive, so $\sin \alpha = \frac{1}{\sqrt{7}} = \frac{\sqrt{7}}{7}$. Next use $\cos^2 \alpha + \sin^2 \alpha = 1$: $\cos^2 \alpha + \left(\frac{\sqrt{7}}{7}\right)^2 = 1$, so $\cos \alpha = \frac{\sqrt{42}}{7}$.

10-3. Half-Angle Identities

A. Half-angle identity for cosine

In the identity $\cos 2x = 2\cos^2 x - 1$, let x equal $\alpha/2$: $\cos \alpha = 2\cos^2(\alpha/2) - 1$, so $\cos^2(\alpha/2) = (1 + \cos \alpha)/2$. Taking square roots of both sides produces the half-angle identity for cosine:

COSINE
HALF-ANGLE $\cos \dfrac{\alpha}{2} = \pm \sqrt{\dfrac{1 + \cos \alpha}{2}}$ (10-11)
IDENTITY

Use the positive value when $\alpha/2$ lies in quadrant I or IV and the negative value when $\alpha/2$ lies in quadrant II or III.

B. Half-angle identity for sine

In the double-angle identity $\cos 2x = 1 - 2\sin^2 x$, let x equal $\alpha/2$: $\cos \alpha = 1 - 2\sin^2(\alpha/2)$, so $\sin^2(\alpha/2) = (1 - \cos \alpha)/2$. Taking square roots of both sides produces the half-angle identity for sine:

SINE
HALF-ANGLE $\sin \dfrac{\alpha}{2} = \pm \sqrt{\dfrac{1 - \cos \alpha}{2}}$ (10-12)
IDENTITY

Use the positive value when $\alpha/2$ lies in quadrant I or II and the negative value when $\alpha/2$ lies in quadrant III or IV.

C. Half-angle identity for tangent

Begin with the identity $\tan(\alpha/2) = \sin(\alpha/2)/\cos(\alpha/2)$. Replace $\sin(\alpha/2)$ and $\cos(\alpha/2)$ with the half-angle identities just derived. This gives

$$\frac{\sin\dfrac{\alpha}{2}}{\cos\dfrac{\alpha}{2}} = \frac{\sqrt{\dfrac{1-\cos\alpha}{2}}}{\sqrt{\dfrac{1+\cot\alpha}{2}}} = \frac{\sqrt{1-\cos\alpha}}{\sqrt{1+\cos\alpha}}$$

Multiplying numerator and denominator of the last expression by $\sqrt{1+\cos\alpha}$ produces

$$\tan\frac{\alpha}{2} = \frac{\sqrt{1-\cos^2\alpha}}{1+\cos\alpha} = \frac{\sqrt{\sin^2\alpha}}{1+\cos\alpha}$$

Simplifying produces the half-angle identity for tangent:

TANGENT HALF-ANGLE IDENTITY
$$\tan\frac{\alpha}{2} = \frac{\sin\alpha}{1+\cos\alpha} \qquad \text{(10-13)}$$

An alternative form, $\tan\alpha/2 = (1-\cos\alpha)/\sin\alpha$, is also useful.

EXAMPLE 10-12: Use the half-angle identities to find $\sin\frac{1}{12}\pi$.

Solution: From Equation 10-12,

$$\sin\frac{\pi}{12} = \sqrt{\frac{1-\cos\dfrac{\pi}{6}}{2}} = \sqrt{\frac{1-\dfrac{\sqrt{3}}{2}}{2}} = \frac{\sqrt{2-\sqrt{3}}}{2}$$

The positive value is taken because $\sin\frac{1}{12}\pi$ is in quadrant I.

EXAMPLE 10-13: Find $\cos\frac{7}{12}\pi$.

Solution: Notice that $\frac{7}{12}\pi = \frac{1}{2}\pi + \frac{1}{12}\pi$. From Equation 10-3,

$$\cos\frac{7\pi}{12} = \cos\frac{\pi}{2}\cos\frac{\pi}{12} - \sin\frac{\pi}{2}\sin\frac{\pi}{12}$$

$$= (0)\cos\frac{\pi}{12} - (1)\sin\frac{\pi}{12} = -\sin\frac{\pi}{12}$$

From the result of Example 10-12, $\sin\frac{1}{12}\pi = \frac{\sqrt{2-\sqrt{3}}}{2}$, so $\cos\frac{7}{12}\pi = -\frac{\sqrt{2-\sqrt{3}}}{2}$.

SUMMARY

1. Cosines, sines, and tangents of sums and differences of numbers can be expressed as follows:

$$\cos(\alpha-\beta) = \cos\alpha\cos\beta + \sin\alpha\sin\beta$$

$$\cos(\alpha+\beta) = \cos\alpha\cos\beta - \sin\alpha\sin\beta$$

$$\sin(\alpha-\beta) = \sin\alpha\cos\beta - \cos\alpha\sin\beta$$

$$\sin(\alpha+\beta) = \sin\alpha\cos\beta + \cos\alpha\sin\beta$$

$$\tan(\alpha+\beta) = \frac{\tan\alpha+\tan\beta}{1-\tan\alpha\tan\beta}$$

$$\tan(\alpha-\beta) = \frac{\tan\alpha-\tan\beta}{1+\tan\alpha\tan\beta}$$

2. The sine, cosine, or tangent of 2α can be expressed in terms of functions of α as follows:

$$\sin 2\alpha = 2 \sin \alpha \cos \alpha$$

$$\cos 2\alpha = \cos^2 \alpha - \sin^2 \alpha = 1 - 2 \sin^2 \alpha = 2 \cos^2 \alpha - 1$$

$$\tan 2\alpha = \frac{2 \tan \alpha}{1 - \tan^2 \alpha}$$

3. The sine, cosine, or tangent of $\alpha/2$ can be expressed in terms of functions of α as follows (plus or minus, depending on the quadrant):

$$\sin \frac{\alpha}{2} = \pm \sqrt{\frac{1 - \cos \alpha}{2}}$$

$$\cos \frac{\alpha}{2} = \pm \sqrt{\frac{1 + \cos \alpha}{2}}$$

$$\tan \frac{\alpha}{2} = \frac{\sin \alpha}{1 + \cos \alpha} = \frac{1 - \cos \alpha}{\sin \alpha}$$

RAISE YOUR GRADES

Can you . . . ?

☑ evaluate any trigonometric function of a sum or difference of two angles

☑ verify identities involving trigonometric functions of sums or differences of two angles

☑ convert a sum of sines and cosines to the sine of a quantity

☑ evaluate any trigonometric function of a double-angle

☑ verify identities for double-angles

☑ evaluate any trigonometric function of a half-angle

☑ verify identities for half-angles

SOLVED PROBLEMS

PROBLEM 10-1   Find $\cos \frac{3}{4}\pi$ by using the values of the trigonometric functions at $\frac{1}{2}\pi$ and $\frac{1}{4}\pi$.

Solution: Because $\cos \frac{3}{4}\pi = \cos(\frac{1}{2}\pi + \frac{1}{4}\pi)$, you can use Equation 10-3, the sum identity for cosine. Let $\alpha = \frac{1}{2}\pi$ and $\beta = \frac{1}{4}\pi$. Then,

$$\cos\left(\frac{\pi}{2} + \frac{\pi}{4}\right) = \cos \frac{\pi}{2} \cos \frac{\pi}{4} - \sin \frac{\pi}{2} \sin \frac{\pi}{4}$$

$$= (0)\frac{\sqrt{2}}{2} - (1)\frac{\sqrt{2}}{2} = -\frac{\sqrt{2}}{2} \qquad \text{[See Section 10-1.]}$$

PROBLEM 10-2   Find $\tan \frac{3}{4}\pi$ by using the values of the trigonometric functions at π and $\frac{1}{4}\pi$.

Solution: Because $\tan \frac{3}{4}\pi = \tan(\pi - \frac{1}{4}\pi)$, you can use Equation 10-6, the difference identity for tangent. Let $\alpha = \pi$ and $\beta = \frac{1}{4}\pi$. Then,

$$\tan\left(\pi - \frac{\pi}{4}\right) = \frac{\tan \pi - \tan \frac{\pi}{4}}{1 + \tan \pi \tan \frac{\pi}{4}} = \frac{0 - 1}{1 + (0)(1)} = -1$$

You calculate the values of $\tan \pi$ and $\tan \frac{1}{4}\pi$ from the values of sine and cosine at π and $\frac{1}{4}\pi$. If you attempt to use $\tan \frac{3}{4}\pi = \tan(\frac{1}{2}\pi + \frac{1}{4}\pi)$ and the sum identity for tangent, you won't be able to complete the solution because the tangent function is not defined at $\frac{1}{2}\pi$. [See Section 10-1.]

PROBLEM 10-3 Find $\sin \frac{5}{4}\pi$ by using the values of the trigonometric functions at π and $\frac{1}{4}\pi$.

Solution: Because $\sin \frac{5}{4}\pi = \sin(\pi + \frac{1}{4}\pi)$, you can use Equation 10-4, the sum identity for sine. Let $\alpha = \pi$ and $\beta = \frac{1}{4}\pi$. Then,

$$\sin\left(\pi + \frac{\pi}{4}\right) = \sin \pi \cos \frac{\pi}{4} + \cos \pi \sin \frac{\pi}{4}$$

$$= (0)\frac{\sqrt{2}}{2} + (-1)\frac{\sqrt{2}}{2} = -\frac{\sqrt{2}}{2} \qquad \text{[See Section 10-1.]}$$

PROBLEM 10-4 Find $\sec \frac{5}{4}\pi$ by using the values of the trigonometric functions at π and $\frac{1}{4}\pi$.

Solution: First, find $\cos \frac{5}{4}\pi$, and then use the identity $\sec \alpha = 1/\cos \alpha$. Use Equation 10-3, the sum identity for cosine. Let $\alpha = \pi$ and $\beta = \frac{1}{4}\pi$. Then,

$$\cos\left(\pi + \frac{\pi}{4}\right) = \cos \pi \cos \frac{\pi}{4} - \sin \pi \sin \frac{\pi}{4}$$

$$= (-1)\frac{\sqrt{2}}{2} + (0)\frac{\sqrt{2}}{2} = -\frac{\sqrt{2}}{2}$$

Thus, $\sec \frac{5}{4}\pi = \dfrac{1}{-\frac{\sqrt{2}}{2}} = -\sqrt{2}$.

Another way to find $\cos \frac{5}{4}\pi$ is to use the result of Problem 10-3 and the identity $\sin^2 \alpha + \cos^2 \alpha = 1$. Then, $\left(-\frac{\sqrt{2}}{2}\right)^2 + \cos^2(\frac{5}{4}\pi) = 1$, or $\cos^2(\frac{5}{4}\pi) = 1 - \frac{1}{2} = \frac{1}{2}$. Finally, $\cos(\frac{5}{4}\pi) = -\frac{\sqrt{2}}{2}$, since $\frac{5}{4}\pi$ lies in quadrant III. [See Section 10-1.]

PROBLEM 10-5 Find $\cot \frac{5}{4}\pi$ by using the values of the trigonometric functions at π and $\frac{1}{4}\pi$.

Solution: Use the identity $\cot \alpha = \cos \alpha/\sin \alpha$. Then $\cot \frac{5}{4}\pi = \cos \frac{5}{4}\pi/\sin \frac{5}{4}\pi$. In Problem 10-3, you found the value of $\sin \frac{5}{4}\pi$, and in Problem 10-4, you found the value of $\cos \frac{5}{4}\pi$. Using these values,

$$\cot \frac{5\pi}{4} = \frac{\cos \dfrac{5\pi}{4}}{\sin \dfrac{5\pi}{4}} = \frac{-\dfrac{\sqrt{2}}{2}}{-\dfrac{\sqrt{2}}{2}} = 1 \qquad \text{[See Section 10-1.]}$$

PROBLEM 10-6 Explain how to find $\cot \frac{5}{4}\pi$ by using the values of the trigonometric functions at $\frac{3}{2}\pi$ and $\frac{1}{4}\pi$.

Solution: You should first find $\tan \frac{5}{4}\pi$ and then use the identity $\cot \frac{5}{4}\pi = 1/\tan \frac{5}{4}\pi$. Can you find $\tan \frac{5}{4}\pi$ by using $\frac{3}{2}\pi$ and $\frac{1}{4}\pi$? If you attempt to use the fact that $\frac{5}{4}\pi = \frac{3}{2}\pi - \frac{1}{4}\pi$ and the difference identity for tangent, you will encounter the undefined quantity $\tan \frac{3}{2}\pi$. Therefore, you cannot find $\cot \frac{5}{4}\pi$ by using the difference identity for tangent at $\frac{3}{2}\pi$ and $\frac{1}{4}\pi$. To use this approach, you would have to write $\tan \frac{5}{4}\pi$ as $\sin \frac{5}{4}\pi/\cos \frac{5}{4}\pi$ and use difference identities for sine and cosine. [See Section 10-1.]

PROBLEM 10-7 Find $\cos 330°$ using the values of the trigonometric functions at $270°$ and $60°$.

Solution: Use $\cos 330° = \cos(270° + 60°) = \cos 270° \cos 60° - \sin 270° \sin 60° = (0)\frac{1}{2} - (-1) \times \frac{\sqrt{3}}{2} = \frac{\sqrt{3}}{2}$. [See Section 10-1.]

In Problems 10-8 through 10-11, expand the given expression by the appropriate formula and insert all known function values. [See Section 10-1.]

PROBLEM 10-8 $\sin\left(x + \frac{1}{6}\pi\right)$

Solution: Use Equation 10-4. Let $\alpha = x$ and $\beta = \frac{1}{6}\pi$:

$$\sin\left(x + \frac{\pi}{6}\right) = \sin x \cos\frac{\pi}{6} + \cos x \sin\frac{\pi}{6} = \frac{\sqrt{3}}{2}\sin x + \frac{1}{2}\cos x$$

PROBLEM 10-9 $\cos\left(x - \frac{1}{6}\pi\right)$

Solution: Use Equation 10-1. Let $\alpha = x$ and $\beta = \frac{1}{6}\pi$:

$$\cos\left(x - \frac{\pi}{6}\right) = \cos x \cos\frac{\pi}{6} + \sin x \sin\frac{\pi}{6} = \frac{\sqrt{3}}{2}\cos x + \frac{1}{2}\sin x$$

PROBLEM 10-10 $\tan(x + \frac{1}{4}\pi)$

Solution: Use Equation 10-5. Let $\alpha = x$ and $\beta = \frac{1}{4}\pi$:

$$\tan\left(x + \frac{\pi}{4}\right) = \frac{\tan x + \tan\dfrac{\pi}{4}}{1 - \tan x \tan\dfrac{\pi}{4}} = \frac{\tan x + 1}{1 - \tan x}$$

PROBLEM 10-11 $\tan(\frac{1}{4}\pi - x)$

Solution: Use Equation 10-6. Let $\alpha = \frac{1}{4}\pi$ and $\beta = x$:

$$\tan\left(\frac{\pi}{4} - x\right) = \frac{\tan\dfrac{\pi}{4} - \tan x}{1 + \tan\dfrac{\pi}{4}\tan x} = \frac{1 - \tan x}{1 + \tan x}$$

In Problems 10-12 through 10-16, express your answer as a single number or as a trigonometric function of a single number.   [See Section 10-1.]

PROBLEM 10-12 $\sin 25° \cos 20° + \cos 25° \sin 20°$

Solution: Let $\alpha = 25°$ and $\beta = 20°$ in Equation 10-4:

$$\sin\alpha\cos\beta + \cos\alpha\sin\beta = \sin(\alpha + \beta)$$

$$\sin 25° \cos 20° + \cos 25° \sin 20° = \sin(25° + 20°) = \sin 45° = \frac{\sqrt{2}}{2}$$

PROBLEM 10-13 $\cos 25° \cos 20° + \sin 25° \sin 20°$

Solution: Let $\alpha = 25°$ and $\beta = 20°$ in Equation 10-1. You get $\cos(25° - 20°) = \cos 5°$.

PROBLEM 10-14 $\cos 25° \cos 20° - \sin 25° \sin 20°$

Solution: Let $\alpha = 25°$ and $\beta = 20°$ in Equation 10-3. You get $\cos(25° + 20°) = \cos 45° = \frac{\sqrt{2}}{2}$.

PROBLEM 10-15 $\sin 25° \cos 20° - \cos 25° \sin 20°$

Solution: Let $\alpha = 25°$ and $\beta = 20°$ in Equation 10-2. Then $\sin(25° - 20°) = \sin 5°$.

PROBLEM 10-16 $\cos A° \cos 20° - \sin A° \sin 20°$

Solution: Let $\alpha = A$ and $\beta = 20°$ in Equation 10-3. You get $\cos(A° + 20°)$.

PROBLEM 10-17   Given acute angles A and B with $\sin A = \frac{5}{13}$ and $\cos B = \frac{24}{25}$; find $\cos(A + B)$ and $\sin(A + B)$.

Solution: Using the identity $\sin^2\alpha + \cos^2\alpha = 1$, $\left(\frac{5}{13}\right)^2 + \cos^2 A = 1$, or $\cos^2 A = 1 - \frac{25}{169} = \frac{144}{169}$, so $\cos A = \frac{12}{13}$. You choose the positive square root of $\frac{144}{169}$ because A is acute. In the same way,

$\sin^2 B + \left(\frac{24}{25}\right)^2 = 1$, or $\sin^2 B = 1 - \frac{576}{625} = \frac{49}{625}$, so $\sin B = \frac{7}{25}$. Substitute these values into Equations 10-3 and 10-4:

$$\cos(A + B) = \frac{12}{13}\left(\frac{24}{25}\right) - \frac{5}{13}\left(\frac{7}{25}\right) = \frac{253}{325}$$

$$\sin(A + B) = \frac{5}{13}\left(\frac{24}{25}\right) + \frac{12}{13}\left(\frac{7}{25}\right) = \frac{204}{325} \qquad \text{[See Section 10-1.]}$$

PROBLEM 10-18 Show that $\sin(\frac{1}{4}\pi + x) - (\sin\frac{1}{4}\pi - x) = \sqrt{2}\sin x$.

Solution: Use the sum and difference identities for sine:

$$\sin\left(\frac{\pi}{4} + x\right) = \sin\frac{\pi}{4}\cos x + \cos\frac{\pi}{4}\sin x$$

$$\sin\left(\frac{\pi}{4} - x\right) = \sin\frac{\pi}{4}\cos x - \cos\frac{\pi}{4}\sin x$$

Subtract the second equation from the first:

$$\sin\left(\frac{\pi}{4} + x\right) - \sin\left(\frac{\pi}{4} - x\right) = 2\cos\frac{\pi}{4}\sin x$$

$$= \frac{2\sqrt{2}}{2}\sin x = \sqrt{2}\sin x \qquad \text{[See Section 10-1.]}$$

PROBLEM 10-19 Show that $\cos(\pi + x)\cos(\pi - x) + \sin(\pi + x)\sin(\pi - x) = \cos 2x$.

Solution: Compare the left side of the given expression with Equation 10-1: $\cos A \cos B + \sin A \sin B = \cos(A - B)$. You see that $A = \pi + x$ and $B = \pi - x$. The given expression must then equal $\cos[(\pi + x) - (\pi - x)] = \cos 2x$. [See Section 10-1.]

PROBLEM 10-20 Find the sine, cosine, and tangent of $120°$ by using the half- or double-angle formulas and the values of the trigonometric functions of $60°$.

Solution: Let $\alpha = 60°$ and use Equations 10-8, -9, and -10:

$$\cos 120° = \cos 2(60°) = 2\cos^2 60° - 1 = 2\left(\frac{1}{4}\right) - 1 = -\frac{1}{2}$$

$$\sin 120° = \sin 2(60°) = 2\sin 60° \cos 60° = 2\frac{\sqrt{3}}{2}\left(\frac{1}{2}\right) = \frac{\sqrt{3}}{2}$$

$$\tan 120° = \tan 2(60°) = \frac{2\tan 60°}{1 - \tan^2 60°} = \frac{2\sqrt{3}}{1 - 3} = -\sqrt{3}$$

[See Section 10-2.]

PROBLEM 10-21 Find the sine, cosine, and tangent of $60°$ by using the half- or double-angle formulas and the values of the trigonometric functions of $30°$.

Solution: Let $\alpha = 30°$ and use Equations 10-8, -9, and -10:

$$\cos 60° = \cos 2(30°) = 2\cos^2 30° - 1 = 2\left(\frac{\sqrt{3}}{2}\right)^2 - 1 = \frac{1}{2}$$

$$\sin 60° = \sin 2(30°) = 2\sin 30° \cos 30° = 2\left(\frac{1}{2}\right)\left(\frac{\sqrt{3}}{2}\right) = \frac{\sqrt{3}}{2}$$

$$\tan 60° = \tan 2(30°) = \frac{2\tan 30°}{1 - \tan^2 30°} = \frac{2\left(\frac{\sqrt{3}}{3}\right)}{1 - \left(\frac{\sqrt{3}}{3}\right)^2} = \sqrt{3}$$

[See Section 10-2.]

PROBLEM 10-22 Find the sine, cosine, and tangent of $\frac{2}{3}\pi$ by using the half- or double-angle formulas and the values of the trigonometric functions of $\frac{4}{3}\pi$.

Solution: Let $\alpha = \frac{4}{3}\pi$ and use Equations 10-11, -12, and -13:

$$\cos\frac{2\pi}{3} = \cos\frac{\frac{4\pi}{3}}{2} = -\sqrt{\frac{1 + \cos\frac{4\pi}{3}}{2}} = -\sqrt{\frac{1 + \left(-\frac{1}{2}\right)}{2}} = -\sqrt{\frac{1}{4}} = -\frac{1}{2}$$

$$\sin\frac{2\pi}{3} = \sin\frac{\frac{4\pi}{3}}{2} = \sqrt{\frac{1 - \cos\frac{4\pi}{3}}{2}} = \sqrt{\frac{1 - \left(-\frac{1}{2}\right)}{2}} = \sqrt{\frac{3}{4}} = \frac{\sqrt{3}}{2}$$

$$\tan\frac{2\pi}{3} = \tan\frac{\frac{4\pi}{3}}{2} = \frac{\sin\frac{4\pi}{3}}{1 + \cos\frac{4\pi}{3}} = \frac{-\frac{\sqrt{3}}{2}}{1 + \left(-\frac{1}{2}\right)} = -\sqrt{3}$$

You use the negative square root in the cosine identity and the positive square root in the sine identity because $\frac{2}{3}\pi$ lies in quadrant II. [See Section 10-3.]

In Problems 10-23 through 10-27, rewrite the given expression as a trigonometric function of a single number without radicals. [See Section 10-2.]

PROBLEM 10-23 $2\sin 44° \cos 44°$

Solution: Let $\alpha = 44°$ and use $2\sin\alpha\cos\alpha = \sin 2\alpha$:

$$2\sin 44° \cos 44° = \sin 88°$$

PROBLEM 10-24 $2\sin(-10°)\cos(-10°)$

Solution: Let $\alpha = (-10°)$ and use $2\sin\alpha\cos\alpha = \sin 2\alpha$:

$$2\sin(-10°)\cos(-10°) = \sin(-20°) = -\sin 20°$$

PROBLEM 10-25 $\cos^2 30° - \sin^2 30°$

Solution: Let $\alpha = 30°$ and use $\cos^2\alpha - \sin^2\alpha = \cos 2\alpha$:

$$\cos^2 30° - \sin^2 30° = \cos 60°$$

PROBLEM 10-26 $1 - 2\sin^2 70°$

Solution: Let $\alpha = 70°$ and use $1 - 2\sin^2\alpha = \cos 2\alpha$:

$$1 - 2\sin^2 70° = \cos 140°$$

PROBLEM 10-27 $2\cos^2 150° - 1$

Solution: Let $\alpha = 150°$ and use $2\cos^2\alpha - 1 = \cos 2\alpha$:

$$2\cos^2 150° - 1 = \cos 300° = \cos(-60°) = \cos 60°$$

PROBLEM 10-28 Write $2\cos^2 10°$ in terms of a trigonometric function raised to the first power.

Solution: Let $\alpha = 10°$ and use $2\cos^2\alpha = 1 + \cos 2\alpha$:

$$2\cos^2 10° = 1 + \cos 20°$$ [See Section 10-2.]

PROBLEM 10-29 Write $2\sin^2 10°$ in terms of a trigonometric function raised to the first power.

Solution: Let $\alpha = 10°$ and use $2\sin^2\alpha = 1 - \cos 2\alpha$:

$$2\sin^2 10° = 1 - \cos 20°$$ [See Section 10-2.]

PROBLEM 10-30 Write $1 + \cos A$ in terms of a trigonometric function raised to the second power.

Solution: Let $\alpha = A$ and use $1 + \cos\alpha = 2\cos^2(\alpha/2)$:

$$1 + \cos A = 2\cos^2\frac{A}{2}$$

Note: The identity is a form of Equation 10-11. You square both sides of the original identity and multiply by 2. [See Section 10-3.]

PROBLEM 10-31 Write $1 - \cos A$ in terms of a trigonometric function raised to the second power.

Solution: Let $\alpha = A$ and use $1 - \cos\alpha = 2\sin^2(\alpha/2)$:

$$1 - \cos A = 2\sin^2\frac{A}{2}$$ [See Section 10-3 and Problem 10-30.]

PROBLEM 10-32 Show that $\dfrac{1 - \cos\frac{1}{6}\pi}{1 + \cos\frac{1}{6}\pi} = \tan^2\frac{1}{12}\pi$.

Solution: Let $\alpha = \frac{1}{6}\pi$ and use $\dfrac{1 - \cos\alpha}{1 + \cos\alpha} = \tan^2(\alpha/2)$:

$$\frac{1 - \cos\frac{\pi}{6}}{1 + \cos\frac{\pi}{6}} = \tan^2\frac{\pi}{12}$$ [See Section 10-3.]

PROBLEM 10-33 For $\sin\frac{1}{2}\alpha = \frac{3}{5}$ and $\frac{1}{4}\pi < \alpha < \frac{1}{2}\pi$, find $\sin\alpha$, $\sin 2\alpha$, and $\sin 4\alpha$.

Solution: Since α is in quadrant I, $\frac{1}{2}\alpha$ is in quadrant I, so $\cos\frac{1}{2}\alpha$ is positive. Use $\sin^2(\frac{1}{2}\alpha) + \cos^2(\frac{1}{2}\alpha) = 1$:

$$\left(\frac{3}{5}\right)^2 + \cos^2\frac{\alpha}{2} = 1 \quad\text{so}\quad \cos\frac{\alpha}{2} = \frac{4}{5}$$

Now use the double-angle identities for sine and cosine to find $\sin\alpha$, $\sin 2\alpha$, and $\sin 4\alpha$:

$$\sin\alpha = 2\sin\frac{\alpha}{2}\cos\frac{\alpha}{2} = 2\left(\frac{3}{5}\right)\left(\frac{4}{5}\right) = \frac{24}{25}$$

$$\cos\alpha = 2\cos^2\frac{\alpha}{2} - 1 = 2\left(\frac{4}{5}\right)^2 - 1 = \frac{7}{25}$$

$$\sin 2\alpha = 2\sin\alpha\cos\alpha = 2\left(\frac{24}{25}\right)\left(\frac{7}{25}\right) = \frac{336}{625}$$

$$\cos 2\alpha = 2\cos^2\alpha - 1 = 2\left(\frac{7}{25}\right)^2 - 1 = -\frac{527}{625}$$

$$\sin 4\alpha = 2\sin 2\alpha\cos 2\alpha = 2\left(\frac{336}{625}\right)\left(-\frac{527}{625}\right) = -\frac{354\,144}{390\,625}$$ [See Section 10-2.]

PROBLEM 10-34 Show that $\sec(A + B) = \dfrac{\sec A\sec B}{1 - \tan A\tan B}$.

Solution: Use Equation 10-3:

$$\cos(A + B) = \cos A \cos B - \sin A \sin B$$

$$= \frac{1}{\sec A}\left(\frac{1}{\sec B}\right) - \sin A \sin B$$

$$= \frac{1 - \sec A \sec B \sin A \sin B}{\sec A \sec B}$$

$$= \frac{1 - \dfrac{\sin A}{\cos A}\left(\dfrac{\sin B}{\cos B}\right)}{\sec A \sec B}$$

$$= \frac{1 - \tan A \tan B}{\sec A \sec B}$$

$$\sec(A + B) = \frac{1}{\cos(A + B)} = \frac{\sec A \sec B}{1 - \tan A \tan B} \qquad \text{[See Section 10-1.]}$$

PROBLEM 10-35   Find $\cot(A + B)$ in terms of $\cot A$ and $\cot B$.

Solution: Use Equation 10-5:

$$\tan(A + B) = \frac{\tan A + \tan B}{1 - \tan A \tan B}$$

$$\cot(A + B) = \frac{1}{\tan(A + B)} = \frac{1 - \tan A \tan B}{\tan A + \tan B}$$

$$= \frac{1 - \dfrac{1}{\cot A}\left(\dfrac{1}{\cot B}\right)}{\dfrac{1}{\cot A} + \dfrac{1}{\cot B}}$$

$$= \frac{\dfrac{\cot A \cot B - 1}{\cot A \cot B}}{\dfrac{\cot B + \cot A}{\cot A \cot B}}$$

$$= \frac{\cot A \cot B - 1}{\cot B + \cot A} \qquad \text{[See Section 10-1.]}$$

PROBLEM 10-36   Find $\cos \frac{1}{8}\pi$.

Solution: Use the value of $\cos \frac{1}{4}\pi$ to find $\cos \frac{1}{8}\pi$. Let $\alpha = \frac{1}{4}\pi$ and use Equation 10-11:

$$\cos\frac{\alpha}{2} = \pm\sqrt{\frac{1 + \cos\alpha}{2}}$$

$$\cos\frac{\pi}{8} = \sqrt{\frac{1 + \dfrac{\sqrt{2}}{2}}{2}} = \frac{\sqrt{2 + \sqrt{2}}}{2}$$

You choose the positive value because $\frac{1}{8}\pi$ lies in quadrant I.   [See Section 10-3.]

PROBLEM 10-37   Show that $8\sin^4\alpha = 3 - 4\cos 2\alpha + \cos 4\alpha$.

Solution: Use Equations 10-8 and 10-9 in the form $\cos 4\alpha = 1 - 2\sin^2 2\alpha$, $\cos 2\alpha = 1 - 2\sin^2\alpha$, and $\sin 2\alpha = 2\sin\alpha\cos\alpha$:

$$3 - 4\cos 2\alpha + \cos 4\alpha = 3 - 4(1 - 2\sin^2\alpha) + (1 - 2\sin^2 2\alpha)$$
$$= 3 - 4(1 - 2\sin^2\alpha) + (1 - 2(2\sin\alpha\cos\alpha)^2)$$
$$= 3 - 4 + 8\sin^2\alpha + 1 - 8\sin^2\alpha\cos^2\alpha$$
$$= 8\sin^2\alpha(1 - \cos^2\alpha) = 8\sin^2\alpha\sin^2\alpha = 8\sin^4\alpha \qquad \text{[See Section 10-2.]}$$

PROBLEM 10-38 Show that $\sin 3\alpha = 3\sin\alpha - 4\sin^3\alpha$.

Solution:

$$\sin 3\alpha = \sin(2\alpha + \alpha) = \sin 2\alpha\cos\alpha + \cos 2\alpha\sin\alpha$$
$$= 2\sin\alpha\cos\alpha(\cos\alpha) + (1 - 2\sin^2\alpha)\sin\alpha$$
$$= 2\sin\alpha\cos^2\alpha + (1 - 2\sin^2\alpha)\sin\alpha$$
$$= \sin\alpha(2\cos^2\alpha + 1 - 2\sin^2\alpha)$$
$$= \sin\alpha[2(1 - \sin^2\alpha) + 1 - 2\sin^2\alpha]$$
$$= \sin\alpha(2 - 2\sin^2\alpha + 1 - 2\sin^2\alpha)$$
$$= \sin\alpha(3 - 4\sin^2\alpha) = 3\sin\alpha - 4\sin^3\alpha$$

[See Sections 10-1 and 10-2.]

PROBLEM 10-39 Write the left side of the equation $\cos 2x + \sin 2x = -1$ in terms of trigonometric functions of x.

Solution: Use the double-angle identities for cosine and sine on the left side of the given expression to get

$$\cos^2 x - \sin^2 x + 2\sin x\cos x = -1$$

Add 1 to both sides of the equation:

$$\cos^2 x + 1 - \sin^2 x + 2\sin x\cos x = 0$$

Replace $1 - \sin^2 x$ with $\cos^2 x$:

$$\cos^2 x + \cos^2 x + 2\sin x\cos x = 0$$
$$2\cos^2 x + 2\sin x\cos x = 0$$

Factor:

$$2\cos x(\cos x + \sin x) = 0 \qquad \text{[See Section 10-2.]}$$

PROBLEM 10-40 A is a quadrant-II angle with $\sin A = \frac{3}{5}$ and B is a quadrant-III angle with $\cos B = \frac{-\sqrt{2}}{2}$; find $\sin(A + B)$.

Solution: Since A is specified in quadrant II, $\cos A$ is negative. It is found by replacing $\sin A$ by $\frac{3}{5}$ in the identity $\sin^2 A + \cos^2 A = 1$. Then $(\frac{3}{5})^2 + \cos^2 A = 1$, or $\cos A = -\frac{4}{5}$. By a similar argument, $\sin B = \frac{-\sqrt{2}}{2}$. Now use the sum identity for sine to find $\sin(A + B)$:

$$\sin(A + B) = \sin A\cos B + \cos A\sin B$$
$$= \frac{3}{5}\left(\frac{-\sqrt{2}}{2}\right) + \left(-\frac{4}{5}\right)\left(\frac{-\sqrt{2}}{2}\right) = \frac{\sqrt{2}}{10}$$

[See Section 10-1.]

PROBLEM 10-41 Find the sign of $\tan\frac{1}{2}\alpha$ for: **(a)** $0 < \alpha < \frac{1}{2}\pi$; **(b)** $\pi < \alpha < \frac{3}{2}\pi$; **(c)** $-\frac{1}{2}\pi < \alpha < 0$; **(d)** $\frac{3}{2}\pi < \alpha < 2\pi$.

Solution:

(a) Since $0 < \alpha < \frac{1}{2}\pi$, $\frac{1}{2}\alpha$ lies between 0 and $\frac{1}{4}\pi$, so $\tan\frac{1}{2}\alpha$ is positive.
(b) Since $\pi < \alpha < \frac{3}{2}\pi$, $\frac{1}{2}\alpha$ lies between $\frac{1}{2}\pi$ and $\frac{3}{4}\pi$, so $\tan\frac{1}{2}\alpha$ is negative.
(c) Since $-\frac{1}{2}\pi < \alpha < 0$, $\frac{1}{2}\alpha$ lies between $-\frac{1}{4}\pi$ and 0, so $\tan\frac{1}{2}\alpha$ is negative.
(d) Since $\frac{3}{2}\pi < \alpha < 2\pi$, $\frac{1}{2}\alpha$ lies between $\frac{3}{4}\pi$ and π, so $\tan\frac{1}{2}\alpha$ is negative. \qquad [See Section 10-3.]

PROBLEM 10-42   Show that $\tan(45° + \alpha) =$

$$\frac{\cos\alpha + \sin\alpha}{\cos\alpha - \sin\alpha}$$

Solution:

$$\tan(45° + \alpha) = \frac{\sin(45° + \alpha)}{\cos(45° + \alpha)}$$

$$= \frac{\sin 45° \cos\alpha + \cos 45° \sin\alpha}{\cos 45° \cos\alpha - \sin 45° \sin\alpha}$$

$$= \frac{\dfrac{\sqrt{2}}{2}(\cos\alpha + \sin\alpha)}{\dfrac{\sqrt{2}}{2}(\cos\alpha - \sin\alpha)} = \frac{\cos\alpha + \sin\alpha}{\cos\alpha - \sin\alpha}$$

Supplementary Exercises

PROBLEM 10-43   Find $\cot\frac{2}{3}\pi$ by using trigonometric function values at $\frac{1}{3}\pi$.

PROBLEM 10-44   Find $\tan\frac{1}{12}\pi$ by using the values of the trigonometric functions at $\frac{1}{3}\pi$ and $\frac{1}{4}\pi$.

PROBLEM 10-45   Find $\sin\frac{7}{12}\pi$ by using the values of the trigonometric functions at $\frac{1}{3}\pi$ and $\frac{1}{4}\pi$.

PROBLEM 10-46   Find $\sec 45°$ using the values of the trigonometric functions at $180°$ and $135°$.

In Problems 10-47 through 10-50, expand the given expression by using the appropriate formula and inserting all known function values.

PROBLEM 10-47 $\tan\left(\frac{3}{4}\pi + x\right)$

PROBLEM 10-48 $\cos\left(\frac{7}{4}\pi - x\right)$

PROBLEM 10-49 $\sin\left(x + \frac{5}{6}\pi\right)$

PROBLEM 10-50 $\csc\left(x + \frac{1}{3}\pi\right)$

In Problems 10-51 through 10-53, express your answer as a single number or as a trigonometric function of a single number.

PROBLEM 10-51 $\sin 3A \cos A + \cos 3A \sin A$

PROBLEM 10-52 $\dfrac{\tan 15° + \tan 60°}{1 - \tan 15° \tan 60°}$

PROBLEM 10-53 $\dfrac{\tan 60° - \tan 15°}{1 + \tan 15° \tan 60°}$

PROBLEM 10-54   Given that A and B are acute angles with $\sin A = \frac{3}{5}$ and $\cos B = \frac{12}{13}$, find $\cos(A - B)$ and $\sin(A - B)$.

PROBLEM 10-55 Show that $\tan(\frac{1}{4}\pi - x) = \dfrac{\cot x - 1}{\cot x + 1}$.

PROBLEM 10-56 Find the sine, cosine, and tangent of $\frac{1}{12}\pi$ by using the half- or double-angle formulas and the trigonometric functions of $\frac{1}{6}\pi$.

PROBLEM 10-57 Find the sine, cosine, and tangent of $\frac{5}{3}\pi$ by using the half- or double-angle formulas and the trigonometric functions of $\frac{5}{6}\pi$.

PROBLEM 10-58 Find the sine, cosine, and tangent of $-\frac{1}{3}\pi$ by using the half- or double-angle formulas and the trigonometric functions of $-\frac{1}{6}\pi$.

In Problems 10-59 through 10-62, express your answer as a trigonometric function of a single number without radicals.

PROBLEM 10-59 $\cos 2A + 1$

PROBLEM 10-60 $1 - \cos 2A$

PROBLEM 10-61 $\dfrac{2 \tan 40°}{1 - \tan^2 40°}$

PROBLEM 10-62 $\dfrac{1 + \cos \frac{1}{6}\pi}{1 - \cos \frac{1}{6}\pi}$

PROBLEM 10-63 Find $\csc(A + B)$ in terms of $\csc A$, $\csc B$, $\cot A$, and $\cot B$.

PROBLEM 10-64 Find $\cos \frac{11}{8}\pi$.

PROBLEM 10-65 Show that $\cos^4 \alpha - \sin^4 \alpha = 1 - 2 \sin^2 \alpha$.

PROBLEM 10-66 Show that $\sin 2\alpha + \tan 2\alpha = \dfrac{\sin 2\alpha \tan 2\alpha}{\tan \alpha}$.

PROBLEM 10-67 Show that $\dfrac{\cos 5\alpha}{\sin \alpha} - \dfrac{\sin 5\alpha}{\cos \alpha} = \dfrac{2 \cos 6\alpha}{\sin 2\alpha}$.

PROBLEM 10-68 Show that $\tan 50° - \tan 40° = 2 \tan 10°$.

PROBLEM 10-69 Show that $\dfrac{\sin 3x}{\sin x} - \dfrac{\cos 3x}{\cos x} = 2$.

Answers to Supplementary Exercises

10-43: $\cot \frac{2}{3}\pi = \frac{-\sqrt{3}}{3}$

10-44: $\tan \dfrac{\pi}{12} = \dfrac{\sqrt{3} - 1}{\sqrt{3} + 1}$

10-45: $\sin \dfrac{7\pi}{12} = \dfrac{\sqrt{2} + \sqrt{6}}{4}$

10-46: $\sec 45° = \sqrt{2}$. Use $\sec 45° = 1/\cos 45°$ and $\cos 45° = \cos(180° - 135°)$.

10-47: $\dfrac{-1 + \tan x}{1 + \tan x}$

10-48: $\frac{\sqrt{2}}{2}(\cos x - \sin x)$

10-49: $\frac{-\sqrt{3}}{2}\sin x + \frac{1}{2}\cos x$

10-50: $\dfrac{1}{\frac{1}{2}\sin x + \frac{\sqrt{3}}{2}\cos x}$

10-51: $\sin 4A$

10-52: $\tan 75°$

10-53: $\tan 45° = 1$

10-54: $\frac{63}{65}, \frac{16}{65}$. Use $\sin^2\alpha + \cos^2\alpha = 1$ and Equation 10-1.

10-55: Use Equation 10-6 and $\tan x = \dfrac{1}{\cot x}$.

10-56: $\cos\dfrac{\pi}{12} = \dfrac{\sqrt{2 + \sqrt{3}}}{2}$

$\sin\dfrac{\pi}{12} = \dfrac{\sqrt{2 - \sqrt{3}}}{2}$

$\tan\dfrac{\pi}{12} = 2 - \sqrt{3}$

10-57: $\cos\dfrac{5\pi}{3} = \dfrac{1}{2}$

$\sin\dfrac{5\pi}{3} = \dfrac{-\sqrt{3}}{2}$

$\tan\dfrac{5\pi}{3} = -\sqrt{3}$

10-58: $\cos\dfrac{-\pi}{3} = \dfrac{1}{2}$

$\sin\dfrac{-\pi}{3} = \dfrac{-\sqrt{3}}{2}$

$\tan\dfrac{-\pi}{3} = -\sqrt{3}$

10-59: $\cos 2A + 1 = 2\cos^2 A$. Use Equation 10-11.

10-60: $1 - \cos 2A = 2\sin^2 A$. Use Equation 10-12.

10-61: $\tan 80°$. Use Equation 10-10.

10-62: $\cot\dfrac{\pi}{12}$

10-63: $\csc(A + B) = \dfrac{\csc A \csc B}{\cot B - \cot A}$. Use Equation 10-4 and the reciprocal relationships.

10-64: $\cos\dfrac{11\pi}{8} = -\dfrac{\sqrt{2 - \sqrt{2}}}{2}$

10-65: $\cos^4\alpha - \sin^4\alpha = (\cos^2\alpha - \sin^2\alpha)(\cos^2\alpha + \sin^2\alpha) = (\cos^2\alpha - \sin^2\alpha)(1) = 1 - \sin^2\alpha - \sin^2\alpha = 1 - 2\sin^2\alpha.$

10-66: $\sin 2\alpha + \tan 2\alpha = \sin 2\alpha + \dfrac{\sin 2\alpha}{\cos 2\alpha}$

$= \dfrac{\sin 2\alpha \cos 2\alpha + \sin 2\alpha}{\cos 2\alpha}$

$= \dfrac{\sin 2\alpha}{\cos 2\alpha}(\cos 2\alpha + 1)$

$= \tan 2\alpha\, 2\cos^2\alpha$

$= \tan 2\alpha\, 2\sin\alpha\cos\alpha\dfrac{\cos\alpha}{\sin\alpha}$

$= \dfrac{\tan 2\alpha \sin 2\alpha}{\tan\alpha}.$

10-67: $\cos 6\alpha = \cos(5\alpha + \alpha)$

$= \cos 5\alpha\cos\alpha - \sin 5\alpha\sin\alpha$

Then use $\sin 2\alpha = 2\sin\alpha\cos\alpha$:

$$\dfrac{2\cos 6\alpha}{\sin 2\alpha} = \dfrac{2(\cos 5\alpha\cos\alpha - \sin 5\alpha\sin\alpha)}{2\sin\alpha\cos\alpha}$$

$$= \dfrac{\cos 5\alpha}{\sin\alpha} - \dfrac{\sin 5\alpha}{\cos\alpha}$$

10-68: $\tan 50° - \tan 40° = \dfrac{\sin 50°}{\cos 50°} - \dfrac{\sin 40°}{\cos 40°}$

$= \dfrac{\sin 50°\cos 40° - \cos 50°\sin 40°}{\cos 50°\cos 40°}$

$= \dfrac{\sin(50° - 40°)}{\sin 40°\cos 40°}$

$= \dfrac{\sin 10°}{\frac{1}{2}\sin 80°} = \dfrac{\sin 10°}{\frac{1}{2}\cos 10°}$

$= 2\tan 10°$

10-69: $\dfrac{\sin 3x}{\sin x} - \dfrac{\cos 3x}{\cos x} =$

$\dfrac{\sin 3x\cos x - \cos 3x\sin x}{\sin x\cos x}$

$= \dfrac{\sin(3x - x)}{\sin x\cos x} = \dfrac{\sin 2x}{\sin x\cos x}$

$= \dfrac{2\sin x\cos x}{\sin x\cos x} = 2$

11 CONVERSION FORMULAS

11-1. Conversion of Products to Sums and Differences

A. sin A cos B

Recall Equations 10-4 and 10-2:

$$\sin(A + B) = \sin A \cos B + \cos A \sin B$$

$$\sin(A - B) = \sin A \cos B - \cos A \sin B$$

Adding these two equations and multiplying both sides by $\frac{1}{2}$ yields

$$\sin A \cos B = \frac{1}{2}\left[\sin(A + B) + \sin(A - B)\right] \qquad \textbf{(11-1)}$$

B. cos A sin B

Again, begin with Equations 10-4 and 10-2:

$$\sin(A + B) = \sin A \cos B + \cos A \sin B$$

$$\sin(A - B) = \sin A \cos B - \cos A \sin B$$

Subtract the second equation from the first and multiply both sides by $\frac{1}{2}$:

$$\cos A \sin B = \frac{1}{2}\left[\sin(A + B) - \sin(A - B)\right] \qquad \textbf{(11-2)}$$

C. cos A cos B

Recall Equations 10-3 and 10-1:

$$\cos(A + B) = \cos A \cos B - \sin A \sin B$$

$$\cos(A - B) = \cos A \cos B + \sin A \sin B$$

Adding these two equations and multiplying both sides by $\frac{1}{2}$ yields

$$\cos A \cos B = \frac{1}{2}\left[\cos(A + B) + \cos(A - B)\right] \qquad \textbf{(11-3)}$$

D. sin A sin B

Again, begin with Equations 10-3 and 10-1:

$$\cos(A + B) = \cos A \cos B - \sin A \sin B$$

$$\cos(A - B) = \cos A \cos B + \sin A \sin B$$

Subtract the second equation from the first and multiply both sides by $-\frac{1}{2}$:

$$\sin A \sin B = \frac{1}{2}\left[\cos(A - B) - \cos(A + B)\right] \qquad \textbf{(11-4)}$$

EXAMPLE 11-1: Convert the following products to sums and differences: (a) $\cos 10° \sin 20°$; (b) $\sin 10° \sin 20°$; (c) $\cos 10° \cos 20°$; (d) $\sin 10° \cos 20°$.

Solution:

(a) In Equation 11-2 let $A = 10°$ and $B = 20°$. Then substituting into the right-hand side of that identity you get

$$\cos 10° \sin 20° = \frac{1}{2} [\sin(10° + 20°) - \sin(10° - 20°)]$$

$$= \frac{1}{2} [\sin 30° - \sin(-10°)] = \frac{1}{2}(\sin 30° + \sin 10°)$$

The identity $\sin(-\alpha) = -\sin\alpha$ was used in the last step.

(b) In Equation 11-4 let $A = 10°$ and $B = 20°$:

$$\sin 10° \sin 20° = \frac{1}{2} [\cos(10° - 20°) - \cos(10° + 20°)]$$

$$= \frac{1}{2} [\cos(-10°) - \cos(30°)] = \frac{1}{2}(\cos 10° - \cos 30°)$$

The identity $\cos(-\alpha) = \cos\alpha$ was used in the last step.

(c) In Equation 11-3 let $A = 10°$ and $B = 20°$:

$$\cos 10° \cos 20° = \frac{1}{2} [\cos(10° + 20°) + \cos(10° - 20°)]$$

$$= \frac{1}{2} [\cos 30° + \cos(-10°)] = \frac{1}{2}(\cos 30° + \cos 10°)$$

The identity $\cos(-\alpha) = \cos\alpha$ was used in the last step.

(d) In Equation 11-1 let $A = 10°$ and $B = 20°$:

$$\sin 10° \cos 20° = \frac{1}{2} [\sin(10° + 20°) + \sin(10° - 20°)]$$

$$= \frac{1}{2} [\sin 30° + \sin(-10°)] = \frac{1}{2}(\sin 30° - \sin 10°)$$

The identity $\sin(-\alpha) = -\sin\alpha$ was used in the last step.

EXAMPLE 11-2: Convert the following products to sums and differences: (a) $\cos\alpha \cos 4\alpha$; (b) $\sin(-\alpha) \sin 3\alpha$; (c) $\sin 4\alpha \cos(-\alpha)$; (d) $\cos 3\alpha \sin 4\alpha$.

Solution:

(a) Use Equation 11-3 and let $A = \alpha$ and $B = 4\alpha$:

$$\cos\alpha \cos 4\alpha = \frac{1}{2} [\cos(\alpha + 4\alpha) + \cos(\alpha - 4\alpha)]$$

$$= \frac{1}{2} [\cos 5\alpha + \cos(-3\alpha)] = \frac{1}{2}(\cos 5\alpha + \cos 3\alpha)$$

(b) Use Equation 11-4 and let $A = -\alpha$ and $B = 3\alpha$:

$$\sin(-\alpha) \sin 3\alpha = \frac{1}{2} [\cos(-\alpha - 3\alpha) - \cos(-\alpha + 3\alpha)]$$

$$= \frac{1}{2} [\cos(-4\alpha) - \cos(2\alpha)]$$

$$= \frac{1}{2}(\cos 4\alpha - \cos 2\alpha)$$

(c) Use Equation 11-1 and let $A = 4\alpha$ and $B = -\alpha$:

$$\sin 4\alpha \cos(-\alpha) = \frac{1}{2}[\sin(4\alpha + (-\alpha)) + \sin(4\alpha - (-\alpha))]$$

$$= \frac{1}{2}(\sin 3\alpha + \sin 5\alpha)$$

(d) Use Equation 11-2 and let $A = 3\alpha$ and $B = 4\alpha$:

$$\cos 3\alpha \sin 4\alpha = \frac{1}{2}[\sin(3\alpha + 4\alpha) - \sin(3\alpha - 4\alpha)]$$

$$= \frac{1}{2}[\sin 7\alpha + \sin(-\alpha)]$$

$$= \frac{1}{2}(\sin 7\alpha - \sin \alpha)$$

11-2. Conversion of Sums and Differences to Products

A. $\sin A + \sin B$

Recall that

$$\sin(\alpha + \beta) = \sin \alpha \cos \beta + \cos \alpha \sin \beta$$
$$\sin(\alpha - \beta) = \sin \alpha \cos \beta - \cos \alpha \sin \beta$$

so

$$\sin(\alpha + \beta) + \sin(\alpha - \beta) = 2 \sin \alpha \cos \beta$$

Let $A = \alpha + \beta$ and $B = \alpha - \beta$. If you add the equations,

$$A + B = 2\alpha \quad \text{or} \quad \alpha = \frac{1}{2}(A + B)$$

If you subtract,

$$A - B = 2\beta \quad \text{or} \quad \beta = \frac{1}{2}(A - B)$$

Substituting into the first derived relationship,

$$\sin A + \sin B = 2 \sin \frac{1}{2}(A + B) \cos \frac{1}{2}(A - B) \qquad \textbf{(11-5)}$$

B. $\sin A - \sin B$

Following the reasoning of Section A,

$$\sin(\alpha + \beta) - \sin(\alpha - \beta) = 2 \cos \alpha \sin \beta$$

so

$$\sin A - \sin B = 2 \cos \frac{1}{2}(A + B) \sin \frac{1}{2}(A - B) \qquad \textbf{(11-6)}$$

C. $\cos A + \cos B$

Recall that

$$\cos(\alpha + \beta) = \cos \alpha \cos \beta - \sin \alpha \sin \beta$$
$$\cos(\alpha - \beta) = \cos \alpha \cos \beta + \sin \alpha \sin \beta$$

so

$$\cos(\alpha + \beta) + \cos(\alpha - \beta) = 2 \cos \alpha \cos \beta$$

Following the reasoning of Section **A**,

$$\cos A + \cos B = 2\cos\frac{1}{2}(A + B)\cos\frac{1}{2}(A - B) \qquad \textbf{(11-7)}$$

D. cos A − cos B

Following the reasoning of Section **C**,

$$\cos(\alpha + \beta) - \cos(\alpha - \beta) = -2\sin\alpha\sin\beta$$

so

$$\cos A - \cos B = -2\sin\frac{1}{2}(A + B)\sin\frac{1}{2}(A - B) \qquad \textbf{(11-8)}$$

EXAMPLE 11-3: Convert the following sums and differences to products:
(a) $\sin\frac{1}{3}\pi + \sin\frac{1}{4}\pi$; **(b)** $\cos\frac{1}{4}\pi + \cos\frac{1}{3}\pi$; **(c)** $\sin\frac{1}{3}\pi - \sin\frac{1}{4}\pi$; **(d)** $\cos\frac{1}{4}\pi - \cos\frac{1}{3}\pi$.

Solution:

(a) Use Equation 11-5 and let $A = \frac{1}{3}\pi$ and $B = \frac{1}{4}\pi$:

$$\sin\frac{\pi}{3} + \sin\frac{\pi}{4} = 2\sin\frac{1}{2}\left(\frac{\pi}{3} + \frac{\pi}{4}\right)\cos\frac{1}{2}\left(\frac{\pi}{3} - \frac{\pi}{4}\right)$$

$$= 2\sin\frac{7\pi}{24}\cos\frac{\pi}{24}$$

(b) Use Equation 11-7 and let $A = \frac{1}{4}\pi$ and $B = \frac{1}{3}\pi$:

$$\cos\frac{\pi}{4} + \cos\frac{\pi}{3} = 2\cos\frac{1}{2}\left(\frac{\pi}{4} + \frac{\pi}{3}\right)\cos\frac{1}{2}\left(\frac{\pi}{4} - \frac{\pi}{3}\right)$$

$$= 2\cos\frac{7\pi}{24}\cos\left(-\frac{\pi}{24}\right)$$

$$= 2\cos\frac{7\pi}{24}\cos\frac{\pi}{24}$$

(c) Use Equation 11-6 and let $A = \frac{1}{3}\pi$ and $B = \frac{1}{4}\pi$:

$$\sin\frac{\pi}{3} - \sin\frac{\pi}{4} = 2\cos\frac{1}{2}\left(\frac{\pi}{3} + \frac{\pi}{4}\right)\sin\frac{1}{2}\left(\frac{\pi}{3} - \frac{\pi}{4}\right)$$

$$= 2\cos\frac{7\pi}{24}\sin\frac{\pi}{24}$$

(d) Use Equation 11-8 and let $A = \frac{1}{4}\pi$ and $B = \frac{1}{3}\pi$:

$$\cos\frac{\pi}{4} - \cos\frac{\pi}{3} = -2\sin\frac{1}{2}\left(\frac{\pi}{4} + \frac{\pi}{3}\right)\sin\frac{1}{2}\left(\frac{\pi}{4} - \frac{\pi}{3}\right)$$

$$= -2\sin\frac{7\pi}{24}\sin\left(\frac{-\pi}{24}\right)$$

$$= 2\sin\frac{7\pi}{24}\sin\frac{\pi}{24}$$

EXAMPLE 11-4: Convert the expressions in the numerator and denominator
of $\dfrac{\cos 4A + \cos 2A}{\sin 4A + \sin 2A}$ to product form and simplify. What identity have you
verified?

Solution: By the sum identities,

$$\cos 4A + \cos 2A = 2\cos \frac{1}{2}(4A + 2A)\cos \frac{1}{2}(4A - 2A)$$

$$= 2\cos 3A \cos A$$

and

$$\sin 4A + \sin 2A = 2\sin \frac{1}{2}(4A + 2A)\cos \frac{1}{2}(4A - 2A)$$

$$= 2\sin 3A \cos A$$

Then

$$\frac{\cos 4A + \cos 2A}{\sin 4A + \sin 2A} = \frac{2\cos 3A \cos A}{2\sin 3A \cos A}$$

$$= \frac{\cos 3A}{\sin 3A} = \cot 3A$$

You have verified the identity

$$\frac{\cos 4A + \cos 2A}{\sin 4A + \sin 2A} = \cot 3A$$

SUMMARY

1. Products of sines or cosines can be converted to sums of sines and cosines.
2. The formulas for converting products of sines and cosines to sums and products are

$$\sin A \cos B = \frac{1}{2}\left[\sin(A + B) + \sin(A - B)\right]$$

$$\cos A \sin B = \frac{1}{2}\left[\sin(A + B) - \sin(A - B)\right]$$

$$\cos A \cos B = \frac{1}{2}\left[\cos(A + B) + \cos(A - B)\right]$$

$$\sin A \sin B = \frac{1}{2}\left[\cos(A - B) - \cos(A + B)\right]$$

3. Sums of sines or cosines can be converted to products of sines or cosines.
4. The formulas for converting sums of sines or cosines to products are

$$\sin A + \sin B = 2\sin \frac{1}{2}(A + B)\cos \frac{1}{2}(A - B)$$

$$\sin A - \sin B = 2\cos \frac{1}{2}(A + B)\sin \frac{1}{2}(A - B)$$

$$\cos A + \cos B = 2\cos \frac{1}{2}(A + B)\cos \frac{1}{2}(A - B)$$

$$\cos A - \cos B = -2\sin \frac{1}{2}(A + B)\sin \frac{1}{2}(A - B)$$

5. Conversions from sums to products or products to sums are carried out in order to prove trigonometric identities or to solve trigonometric equations.

RAISE YOUR GRADES
Can you...?

☑ convert a product of sines or cosines to a sum or difference
   of sines or cosines

☑ convert a sum or difference of sines or cosines to a product
   of sines or cosines

☑ verify trigonometric identities with the sum and difference identities

SOLVED PROBLEMS

In Problems 11-1 through 11-10, convert each expression to a sum or difference.

[See Section 11-1.]

PROBLEM 11-1 $\sin \alpha \cos 3\alpha$

Solution: Use Equation 11-1 and let $A = \alpha$ and $B = 3\alpha$:

$$\sin \alpha \cos 3\alpha = \frac{1}{2} \sin(\alpha + 3\alpha) + \sin(\alpha - 3\alpha)$$

$$= \frac{1}{2}[\sin 4\alpha + \sin(-2\alpha)] = \frac{1}{2}(\sin 4\alpha - \sin 2\alpha)$$

PROBLEM 11-2 $\cos \alpha \sin \alpha$

Solution: Use Equation 11-2 and let both A and B equal α:

$$\cos \alpha \sin \alpha = \frac{1}{2}[\sin(\alpha + \alpha) - \sin(\alpha - \alpha)]$$

$$= \frac{1}{2}(\sin 2\alpha - \sin 0) = \frac{1}{2}\sin 2\alpha$$

PROBLEM 11-3 $\cos(-\alpha)\cos \alpha$

Solution: Use Equation 11-3 and let $A = -\alpha$ and $B = \alpha$:

$$\cos(-\alpha)\cos \alpha = \frac{1}{2}[\cos((-\alpha) + \alpha) + \cos((-\alpha) - \alpha)]$$

$$= \frac{1}{2}[\cos 0 + \cos(-2\alpha)] = \frac{1}{2}[1 + \cos 2\alpha]$$

PROBLEM 11-4 $\sin 3\alpha \sin 5\alpha$

Solution: Use Equation 11-4 and let $A = 3\alpha$ and $B = 5\alpha$:

$$\sin 3\alpha \sin 5\alpha = \frac{1}{2}[\cos(3\alpha - 5\alpha) - \cos(3\alpha + 5\alpha)]$$

$$= \frac{1}{2}[\cos(-2\alpha) - \cos 8\alpha] = \frac{1}{2}(\cos 2\alpha - \cos 8\alpha)$$

PROBLEM 11-5 $\cos 135° \sin 25°$

Solution: Use Equation 11-2 and let $A = 135°$ and $B = 25°$:

$$\cos 135° \sin 25° = \frac{1}{2} \left[\sin(135° + 25°) - \sin(135° - 25°) \right]$$

$$= \frac{1}{2} \left(\sin 160° - \sin 110° \right)$$

PROBLEM 11-6 $\sin 45° \sin 15°$

Solution: Use Equation 11-4 and let $A = 45°$ and $B = 15°$:

$$\sin 45° \sin 15° = \frac{1}{2} \left[\cos(45° - 15°) - \cos(45° + 15°) \right]$$

$$= \frac{1}{2} \left(\cos 30° - \cos 60° \right)$$

PROBLEM 11-7 $\sin(-45°) \cos(-15°)$

Solution: Use Equation 11-1 and let $A = -45°$ and $B = -15°$:

$$\sin(-45°) \cos(-15°) = \frac{1}{2} \left[\sin\left(-45° + (-15°)\right) + \sin\left(-45° - (-15°)\right) \right]$$

$$= \frac{1}{2} \left[\sin(-60°) + \sin(-30°) \right]$$

$$= -\frac{1}{2} \left(\sin 60° + \sin 30° \right)$$

PROBLEM 11-8 $\sin 2 \cos 3$

Solution: Use Equation 11-1 and let $A = 2$ and $B = 3$:

$$\sin 2 \cos 3 = \frac{1}{2} \left[\sin(2 + 3) + \sin(2 - 3) \right] = \frac{1}{2} \left(\sin 5 - \sin 1 \right)$$

PROBLEM 11-9 $\cos 1 \cos 2$

Solution: Use Equation 11-3 and let $A = 1$ and $B = 2$:

$$\cos 1 \cos 2 = \frac{1}{2} \left[\cos(1 + 2) + \cos(1 - 2) \right]$$

$$= \frac{1}{2} \left[\cos 3 + \cos(-1) \right] = \frac{1}{2} \left(\cos 3 + \cos 1 \right)$$

PROBLEM 11-10 $\sin x \cos \frac{1}{2} x$

Solution: Use Equation 11-1 and let $A = x$ and $B = \frac{1}{2} x$:

$$\sin x \cos \frac{x}{2} = \frac{1}{2} \left[\sin\left(x + \frac{x}{2}\right) + \sin\left(x - \frac{x}{2}\right) \right]$$

$$= \frac{1}{2} \left(\sin \frac{3x}{2} + \sin \frac{x}{2} \right)$$

In Problems 11-11 through 11-17, convert each expression to a product.   [See Section 11-2.]

PROBLEM 11-11 $\sin 50° - \sin 20°$

Solution: Use Equation 11-6 and let $A = 50°$ and $B = 20°$:

$$\sin 50° - \sin 20° = 2\cos\frac{1}{2}(50° + 20°)\sin\frac{1}{2}(50° - 20°)$$

$$= 2\cos\frac{1}{2}(70°)\sin\frac{1}{2}(30°) = 2\cos 35°\sin 15°$$

PROBLEM 11-12 $\sin(-50°) + \sin 20°$

Solution: Use Equation 11-5 and let $A = -50°$ and $B = 20°$:

$$\sin(-50°) + \sin 20° = 2\sin\frac{1}{2}((-50°) + 20°)\cos\frac{1}{2}((-50°) - 20°)$$

$$= 2\sin\frac{1}{2}(-30°)\cos\frac{1}{2}(-70°)$$

$$= 2\sin(-15°)\cos(-35°) = -2\sin 15°\cos 35°$$

PROBLEM 11-13 $\cos 4x - \cos x$

Solution: Use Equation 11-8 and let $A = 4x$ and $B = x$:

$$\cos 4x - \cos x = -2\sin\frac{1}{2}(4x + x)\sin\frac{1}{2}(4x - x)$$

$$= -2\sin\frac{1}{2}(5x)\sin\frac{1}{2}(3x) = -2\sin\frac{5x}{2}\sin\frac{3x}{2}$$

PROBLEM 11-14 $\cos(-x) - \cos 4x$

Solution: Use Equation 11-8 and let $A = -x$ and $B = 4x$:

$$\cos(-x) - \cos 4x = -2\sin\frac{1}{2}((-x) + 4x)\sin\frac{1}{2}((-x) - 4x)$$

$$= -2\sin\frac{1}{2}(3x)\sin\frac{1}{2}(-5x)$$

$$= -2\sin\frac{3x}{2}\sin\left(-\frac{5x}{2}\right) = 2\sin\frac{3x}{2}\sin\frac{5x}{2}$$

PROBLEM 11-15 $\cos(-x) + \cos 4x$

Solution: Use Equation 11-7 and let $A = -x$ and $B = 4x$:

$$\cos(-x) + \cos 4x = 2\cos\frac{1}{2}[(-x) + 4x]\cos\frac{1}{2}[(-x) - 4x]$$

$$= 2\cos\frac{1}{2}(3x)\cos\frac{1}{2}(-5x)$$

$$= 2\cos\frac{3x}{2}\cos\left(-\frac{5x}{2}\right) = 2\cos\frac{3x}{2}\cos\frac{5x}{2}$$

PROBLEM 11-16 $\sin\frac{1}{10}\pi + \sin\frac{1}{5}\pi$

Solution: Use Equation 11-5 and let $A = \frac{1}{10}\pi$ and $B = \frac{1}{5}\pi$:

$$\sin\frac{\pi}{10} + \sin\frac{\pi}{5} = 2\sin\frac{1}{2}\left(\frac{\pi}{10} + \frac{\pi}{5}\right)\cos\frac{1}{2}\left(\frac{\pi}{10} - \frac{\pi}{5}\right)$$

$$= 2\sin\frac{1}{2}\left(\frac{3\pi}{10}\right)\cos\frac{1}{2}\left(-\frac{\pi}{10}\right) = 2\sin\frac{3\pi}{20}\cos\frac{\pi}{20}$$

PROBLEM 11-17 $\sin\frac{3}{4} - \sin\frac{1}{4}$

Solution: Use Equation 11-6 and let $A = \frac{3}{4}$ and $B = \frac{1}{4}$:

$$\sin\frac{3}{4} - \sin\frac{1}{4} = 2\cos\frac{1}{2}\left(\frac{3}{4} + \frac{1}{4}\right)\sin\frac{1}{2}\left(\frac{3}{4} - \frac{1}{4}\right)$$

$$= 2\cos\frac{1}{2}(1)\sin\frac{1}{2}\left(\frac{1}{2}\right) = 2\cos\frac{1}{2}\sin\frac{1}{4}$$

Verify the identities in Problems 11-18 through 11-24. Use the conversion identities to convert numerator and denominator of the left-hand side from sums and differences to products.

[See Section 11-2.]

PROBLEM 11-18 $\dfrac{\sin 5x - \sin 3x}{\sin 5x + \sin 3x} = \dfrac{\tan x}{\tan 4x}$

Solution:

$$\sin 5x - \sin 3x = 2\cos\frac{1}{2}(5x + 3x)\sin\frac{1}{2}(5x - 3x)$$

$$= 2\cos 4x \sin x$$

and

$$\sin 5x + \sin 3x = 2\sin\frac{1}{2}(5x + 3x)\cos\frac{1}{2}(5x - 3x)$$

$$= 2\sin 4x \cos x$$

Then

$$\frac{2\cos 4x \sin x}{2\sin 4x \cos x} = \frac{2}{2}\left(\frac{\cos 4x}{\sin 4x}\right)\left(\frac{\sin x}{\cos x}\right) = \cot 4x \tan x = \frac{\tan x}{\tan 4x}$$

PROBLEM 11-19 $\dfrac{\sin 4x + \sin 2x}{\sin 4x - \sin 2x} = \dfrac{\tan 3x}{\tan x}$

Solution:

$$\sin 4x + \sin 2x = 2\sin\frac{1}{2}(4x + 2x)\cos\frac{1}{2}(4x - 2x)$$

$$= 2\sin 3x \cos x$$

and

$$\sin 4x - \sin 2x = 2\cos\frac{1}{2}(4x + 2x)\sin\frac{1}{2}(4x - 2x)$$

$$= 2\cos 3x \sin x$$

Then

$$\frac{2\sin 3x \cos x}{2\cos 3x \sin x} = \frac{\tan 3x}{\tan x}$$

PROBLEM 11-20 $\dfrac{\cos 3x + \cos(-x)}{\cos 3x - \cos(-x)} = \dfrac{-\cot 2x}{\tan x}$

Solution:

$$\cos 3x + \cos(-x) = 2\cos\frac{1}{2}[3x + (-x)]\cos\frac{1}{2}[3x - (-x)]$$

$$= 2\cos x \cos 2x$$

and

$$\cos 3x - \cos(-x) = -2\sin\frac{1}{2}[3x + (-x)]\sin\frac{1}{2}[3x - (-x)]$$

$$= -2\sin x \sin 2x$$

Then

$$\frac{2\cos x \cos 2x}{-2\sin x \sin 2x} = \frac{-\cot 2x}{\tan x}$$

PROBLEM 11-21 $\dfrac{\sin 5x + \sin 2x}{\cos 5x + \cos 2x} = \tan\dfrac{7x}{2}$

Solution:

$$\sin 5x + \sin 2x = 2\sin\frac{1}{2}(5x + 2x)\cos\frac{1}{2}(5x - 2x)$$

$$= 2\sin\frac{7x}{2}\cos\frac{3x}{2}$$

and

$$\cos 5x + \cos 2x = 2\cos\frac{1}{2}(5x + 2x)\cos\frac{1}{2}(5x - 2x)$$

$$= 2\cos\frac{7x}{2}\cos\frac{3x}{2}$$

Then

$$\frac{2\sin\dfrac{7x}{2}\cos\dfrac{3x}{2}}{2\cos\dfrac{7x}{2}\cos\dfrac{3x}{2}} = \tan\frac{7x}{2}$$

PROBLEM 11-22 $\dfrac{\sin A + \sin B}{\cos A - \cos B} = -\cot\dfrac{A - B}{2}$

Solution:

$$\sin A + \sin B = 2\sin\frac{1}{2}(A + B)\cos\frac{1}{2}(A - B)$$

$$= 2\sin\frac{A + B}{2}\cos\frac{A - B}{2}$$

and

$$\cos A - \cos B = -2\sin\frac{1}{2}(A + B)\sin\frac{1}{2}(A - B)$$

Then

$$\frac{2\sin\dfrac{A + B}{2}\cos\dfrac{A - B}{2}}{-2\sin\dfrac{A + B}{2}\sin\dfrac{A - B}{2}} = -\cot\frac{A - B}{2}$$

PROBLEM 11-23 $\dfrac{\cos 3A + \cos 5A}{\sin 3A - \sin 5A} = -\cot A$

Solution:

$$\cos 3A + \cos 5A = 2\cos\frac{1}{2}(3A + 5A)\cos\frac{1}{2}(3A - 5A)$$

$$= 2\cos 4A\cos(-A) = 2\cos 4A\cos A$$

and

$$\sin 3A - \sin 5A = 2\cos\frac{1}{2}(3A + 5A)\sin\frac{1}{2}(3A - 5A)$$

$$= 2\cos 4A\sin(-A) = -2\cos 4A\sin A$$

Then

$$\frac{2\cos 4A\cos A}{-2\cos 4A\sin A} = -\cot A$$

PROBLEM 11-24 $\dfrac{\sin(x + h) - \sin x}{h} = \dfrac{\cos\left(x + \dfrac{h}{2}\right)\sin\dfrac{h}{2}}{\dfrac{h}{2}}$

Solution: Using the difference identity for sines, convert $\sin(x + h) - \sin x$ to

$$2\cos\frac{1}{2}(x + h + x)\sin\frac{1}{2}(x + h - x) = 2\cos\left(x + \frac{h}{2}\right)\sin\frac{h}{2}$$

Then

$$\frac{\sin(x + h) - \sin x}{h} = \frac{\cos\left(x + \dfrac{h}{2}\right)\sin\dfrac{h}{2}}{\dfrac{h}{2}}$$

PROBLEM 11-25 Show that $\cos\alpha + \cos 3\alpha + \cos 5\alpha + \cos 7\alpha = 4\cos\alpha\cos 2\alpha\cos 4\alpha$.

Solution: Write $\cos\alpha + \cos 3\alpha + \cos 5\alpha + \cos 7\alpha$ as $(\cos\alpha + \cos 7\alpha) + (\cos 3\alpha + \cos 5\alpha)$. Now apply the conversion formula for sums of cosines to each of the quantities inside parentheses:

$$\cos 3\alpha + \cos 5\alpha = 2\cos\frac{1}{2}(3\alpha + 5\alpha)\cos\frac{1}{2}(3\alpha - 5\alpha)$$

$$= 2\cos 4\alpha\cos(-\alpha) = 2\cos 4\alpha\cos\alpha$$

$$\cos\alpha + \cos 7\alpha = 2\cos\frac{1}{2}(\alpha + 7\alpha)\cos\frac{1}{2}(\alpha - 7\alpha)$$

$$= 2\cos 4\alpha\cos(-3\alpha) = 2\cos 4\alpha\cos 3\alpha$$

Then

$$(\cos\alpha + \cos 7\alpha) + (\cos 3\alpha + \cos 5\alpha) = 2\cos 4\alpha\cos\alpha + 2\cos 4\alpha\cos 3\alpha$$

$$= 2\cos 4\alpha(\cos\alpha + \cos 3\alpha)$$

Apply the conversion formula to $\cos\alpha + \cos 3\alpha$:

$$\cos\alpha + \cos 3\alpha = 2\cos\frac{1}{2}(\alpha + 3\alpha)\cos\frac{1}{2}(\alpha - 3\alpha)$$

$$= 2\cos 2\alpha\cos(-\alpha) = 2\cos 2\alpha\cos\alpha$$

Then

$$2\cos 4\alpha(\cos\alpha + \cos 3\alpha) = 2\cos 4\alpha(2\cos 2\alpha\cos\alpha)$$

$$= 4\cos\alpha\cos 2\alpha\cos 4\alpha \qquad \text{[See Section 11-2.]}$$

PROBLEM 11-26 Without using your calculator, find the value of **(a)** $\sin 75° - \sin 15°$; **(b)** $\cos 105° + \cos 15°$; **(c)** $\sin 105° + \sin 15°$.

Solution: For each of these expressions, convert the sum or difference to a product and evaluate the resulting trigonometric functions.

(a) $\sin 75° - \sin 15° = 2\cos\dfrac{1}{2}(75° + 15°)\sin\dfrac{1}{2}(75° - 15°)$

$$= 2\cos 45°\sin 30° = 2\left(\frac{\sqrt{2}}{2}\right)\left(\frac{1}{2}\right) = \frac{\sqrt{2}}{2}$$

(b) $\cos 105° + \cos 15° = 2\cos\dfrac{1}{2}(105° + 15°)\cos\dfrac{1}{2}(105° - 15°)$

$$= 2\cos 60°\cos 45° = 2\left(\frac{1}{2}\right)\left(\frac{\sqrt{2}}{2}\right) = \frac{\sqrt{2}}{2}$$

(c) $\sin 105° + \sin 15° = 2\sin\dfrac{1}{2}(105° + 15°)\cos\dfrac{1}{2}(105° - 15°)$

$$= 2\sin 60°\cos 45° = 2\left(\frac{\sqrt{3}}{2}\right)\left(\frac{\sqrt{2}}{2}\right) = \frac{\sqrt{6}}{2} \qquad \text{[See Section 11-2.]}$$

Supplementary Exercises

Express each of the following products as a sum or difference.

PROBLEM 11-27 $-2\sin(-\alpha)\sin 9\alpha$

PROBLEM 11-28 $\cos\alpha\cos(-9\alpha)$

PROBLEM 11-29 $\cos\alpha\sin(-\alpha)$

PROBLEM 11-30 $\cos 35°\sin(-15°)$

PROBLEM 11-31 $\cos 30°\sin 330°$

PROBLEM 11-32 $\sin\frac{1}{12}\pi\sin\frac{1}{2}\pi$

PROBLEM 11-33 $\cos\frac{1}{7}\pi\sin\frac{7}{5}\pi$

PROBLEM 11-34 $\sin\frac{1}{5}\pi\cos\frac{1}{5}\pi$

PROBLEM 11-35 $\sin(-x)\sin\frac{1}{2}x$

Express each of the following sums or differences as a product.

PROBLEM 11-36 $\cos 120° + \cos 20°$

PROBLEM 11-37 $\cos(-120°) - \cos(-20°)$

PROBLEM 11-38 $\sin 3x + \sin 7x$

PROBLEM 11-39 $\sin 9x - \sin x$

PROBLEM 11-40 $\cos 3y + \cos 8y$

PROBLEM 11-41 $\cos \frac{1}{6}\pi - \cos \frac{1}{18}\pi$

PROBLEM 11-42 $\cos 1 + \cos 2$

Prove the following identities.

PROBLEM 11-43 $\dfrac{\sin 5A - \sin 3A}{\sin 5A + \sin 3A} = \dfrac{\cot 4A}{\cot A}$

PROBLEM 11-44 $\dfrac{2 \sin 3\alpha}{\sin 4\alpha + \sin 2\alpha} = \sec \alpha$

PROBLEM 11-45 $\dfrac{\cos 3x + \cos x}{\sin 3x + \sin x} = \cot 2x$

PROBLEM 11-46 $\dfrac{\sin A - \sin 3A}{\sin^2 A - \cos^2 A} = 2 \sin A$

PROBLEM 11-47 $\dfrac{\sin 5\alpha + \sin 3\alpha}{2 \sin \alpha \cos \alpha (\cos^2\alpha - \sin^2\alpha)} = 4 \cos \alpha$

PROBLEM 11-48 Without using your calculator, find the value for: **(a)** $\sin 52.5° \cos 7.5°$; **(b)** $\cos 37.5° \sin 7.5°$; **(c)** $\cos 37.5° \cos 7.5°$.

Answers to Supplementary Exercises

11-27: $\cos 8\alpha - \cos 10\alpha$

11-28: $\dfrac{1}{2}(\cos 8\alpha + \cos 10\alpha)$

11-29: $-\dfrac{1}{2}\sin 2\alpha$

11-30: $\dfrac{1}{2}(\sin 20° - \sin 50°)$

11-31: $\dfrac{1}{2}\sin 300°$

11-32: $\dfrac{1}{2}\left(\cos \dfrac{5\pi}{12} - \cos \dfrac{7\pi}{12}\right)$

11-33: $\dfrac{1}{2}\left(\sin \dfrac{54\pi}{35} + \sin \dfrac{44\pi}{35}\right)$

11-34: $\dfrac{1}{2}\sin \dfrac{2\pi}{5}$

11-35: $\dfrac{1}{2}\left(\cos \dfrac{3x}{2} - \cos \dfrac{x}{2}\right)$

11-36: $2 \cos 70° \cos 50°$

11-37: $-2 \sin 70° \sin 50°$

11-38: $2 \sin 5x \cos 2x$

11-39: $2 \cos 5x \sin 4x$

11-40: $2 \cos \dfrac{11y}{2} \cos \dfrac{5y}{2}$

11-41: $-2 \sin \dfrac{\pi}{9} \sin \dfrac{\pi}{18}$

11-42: $2 \cos \dfrac{3}{2} \cos \dfrac{1}{2}$

11-43:

$\sin 5A - \sin 3A$

$\quad = 2\cos\frac{1}{2}(5A + 3A)\sin\frac{1}{2}(5A - 3A)$

$\quad = 2\cos 4A \sin A$

$\sin 5A + \sin 3A$

$\quad = 2\sin\frac{1}{2}(5A + 3A)\cos\frac{1}{2}(5A - 3A)$

$\quad = 2\sin 4A \cos A$

$\dfrac{2\cos 4A \sin A}{2\sin 4A \cos A} = \dfrac{\cot 4A}{\cot A}$

11-44:

$\sin 4\alpha + \sin 2\alpha$

$\quad = 2\sin\frac{1}{2}(4\alpha + 2\alpha)\cos\frac{1}{2}(4\alpha - 2\alpha)$

$\quad = 2\sin 3\alpha \cos\alpha$

$\dfrac{2\sin 3\alpha}{2\sin 3\alpha \cos\alpha} = \dfrac{1}{\cos\alpha} = \sec\alpha$

11-45:

$\cos 3x + \cos x$

$\quad = 2\cos\frac{1}{2}(3x + x)\cos\frac{1}{2}(3x - x)$

$\quad = 2\cos 2x \cos x$

$\sin 3x + \sin x$

$\quad = 2\sin\frac{1}{2}(3x + x)\cos\frac{1}{2}(3x - x)$

$\quad = 2\sin 2x \cos x$

$\dfrac{2\cos 2x \cos x}{2\sin 2x \cos x} = \cot 2x$

11-46:

$\sin A - \sin 3A$

$\quad = 2\cos\frac{1}{2}(A + 3A)\sin\frac{1}{2}(A - 3A)$

$\quad = 2\cos 2A \sin(-A)$

$\quad = -2\cos 2A \sin A$

$\sin^2 A - \cos^2 A = -\cos 2A$

$\dfrac{-2\cos 2A \sin A}{-\cos 2A} = 2\sin A$

11-47:

$\sin 5\alpha + \sin 3\alpha$

$\quad = 2\sin\frac{1}{2}(5\alpha + 3\alpha)\cos\frac{1}{2}(5\alpha - 3\alpha)$

$\quad = 2\sin 4\alpha \cos\alpha$

$\cos^2\alpha - \sin^2\alpha = \cos 2\alpha$

$2\sin\alpha \cos\alpha = \sin 2\alpha$

$\dfrac{\sin 5\alpha + \sin 3\alpha}{2\sin\alpha \cos\alpha(\cos^2\alpha - \sin^2\alpha)} = \dfrac{2\sin 4\alpha \cos\alpha}{\sin 2\alpha \cos 2\alpha}$

$\qquad\qquad = \dfrac{2\sin 4\alpha \cos\alpha}{\frac{1}{2}\sin 4\alpha}$

$\qquad\qquad = 4\cos\alpha$

11-48: (a) $\dfrac{\sqrt{3} + \sqrt{2}}{4}$; (b) $\dfrac{\sqrt{2} - 1}{4}$; (c) $\dfrac{\sqrt{2} + \sqrt{3}}{4}$

12 INVERSE FUNCTIONS

THIS CHAPTER IS ABOUT

☑ **General Inverse Functions**
☑ **Inverse Trigonometric Functions**
☑ **Graphing Inverse Trigonometric Functions**

12-1. General Inverse Functions

A. Review of definitions

Recall that a *function*, F, is a relation with the property that the first component of each ordered pair in the relation is paired with exactly one second component. Thus if (x, y_1) and (x, y_2) are ordered pairs of a function, then y_1 must equal y_2. If F is a function, there corresponds a single y in the range to each x in the domain.

EXAMPLE 12-1: Tell which of the following relations are functions:

(a) $\{(4, -2), (1, 4), (3, 5)\}$; (b) $\{(2, 8), (4, 5), (2, 6)\}$;
(c) $\{(1, -2), (2, 2), (3, -2), (4, 6)\}$.

Solution:

(a) Relation $\{(4, -2), (1, 4), (3, 5)\}$ is a function because each of the first components, 1, 3, and 4, is paired with exactly one second component; 1 is paired with 4, 3 is paired with 5, and 4 is paired with -2.
(b) Relation $\{(2, 8), (4, 5), (2, 6)\}$ is not a function because first component 2 is paired with both second component 8 and second component 6.
(c) Relation $\{(1, -2), (2, 2), (3, -2), (4, 6)\}$ is a function even though the first components 1 and 3 are paired with the same second component, -2.

Function-F is said to be *one-to-one* if whenever $F(x_1) = F(x_2)$, then $x_1 = x_2$ (see Chapter 4). This also means that in the set of ordered pairs that constitutes the function, there cannot be an ordered pair with the same second component and different first components. For example, $(2, 3)$ and $(8, 3)$ couldn't appear in the set of ordered pairs of a one-to-one function. Example 12-1(c) is an example of a function that is not one-to-one.

EXAMPLE 12-2: Tell whether the following functions are one-to-one:
(a) $F(x) = 2x$; (b) $F(x) = x^2$.

Solution:

(a) If $F(x_1) = 2x_1 = F(x_2) = 2x_2$, then $x_1 = x_2$, so $F(x) = 2x$ is one-to-one.
(b) $F(-2) = 4$ and $F(2) = 4$ so $F(x) = x^2$ is not one-to-one.

B. Inverse Functions

The function $y = F(x) = x + 1$ has the set of all real numbers as its domain. Using set notation to express the ordered pairs defined by this function, you can write $F = \{(x, y) \mid y = x + 1\}$; three ordered pairs in this set are $(0, 1)$,

$(-1, 0)$, and $(2, 3)$. Function F increases each number by 1. The **inverse function** of F, F^{-1}, must decrease each number by 1. You can write $F^{-1} = \{(x, y) \,|\, y = x - 1\}$; three ordered pairs in this set are $(1, 0)$, $(0, -1)$, and $(3, 2)$. Each ordered pair in F^{-1} has first and second components in reverse order of an ordered pair in F. Note that $F(F^{-1}(x)) = (x - 1) + 1 = x$ and $F^{-1}(F(x)) = (x + 1) - 1 = x$. The relationship, $F(F^{-1}(x)) = F^{-1}(F(x)) = x$, is true for any function and its inverse.

EXAMPLE 12-3: Find the inverse of $F(x) = x^3$.

Solution: Function F consists of the set $\{(x, y) \,|\, x \text{ is real and } y = x^3\}$; applying F to a number produces the third power of that number. The inverse function must take the cube root of each number (raise the number to the $\frac{1}{3}$ power). Therefore, F^{-1} is the set $\{(x, y) \,|\, x \text{ is real and } y = x^{1/3}\}$. Again note that $F(F^{-1}(x)) = (x^{1/3})^3 = x$ and $F^{-1}(F(x)) = (x^3)^{1/3} = x$ so $F(F^{-1}(x)) = F^{-1}(F(x)) = x$.

EXAMPLE 12-4: Find the inverse function for $F(x) = \sin x$.

Solution: For any real number x, $\sin x$ is a number between -1 and 1. Thus, $\sin^{-1} x$ for x between -1 and 1 must equal t, where t is a number such that $\sin t = x$. Since there are an infinite number of t's satisfying $\sin t = x$, a convention that designates only one of these numbers as the principal value of the inverse sine function at x must be established. The convention is to let the number t between $-\frac{1}{2}\pi$ and $\frac{1}{2}\pi$ that satisfies $\sin t = x$ be the value of the inverse sine function at x; that is, $\sin^{-1} x = t$, or $\arcsin x = t$. For any x between -1 and 1, $F(F^{-1}(x)) = F^{-1}(F(x)) = x$.

Note: Don't equate $\sin^{-1} x$ with $\dfrac{1}{\sin x}$: They are very different expressions. The expression $\dfrac{1}{\sin x}$ is equivalent to $(\sin x)^{-1}$.

EXAMPLE 12-5: Find the inverse function for $G(x) = x^2$.

Solution: Function $G = \{(x, y) \,|\, x \text{ is real and } y = x^2\}$; G produces the square of any real number. To reverse the action of G, G^{-1} must take the square root of a number. If you start with -2, $G(-2) = (-2)^2 = 4$, so $G^{-1}(4)$ must equal -2. Suppose that you define $G^{-1}(x) = -\sqrt{x}$ for all positive x. Then $G(2) = 2^2 = 4$ but $G^{-1}(4) = -\sqrt{4} = -2$, so $G^{-1}(G(2)) \neq 2$. You could define $G^{-1}(x) = \sqrt{x}$, but then $G^{-1}(G(-2)) \neq -2$. No matter which definition you use for the inverse, you cannot define an inverse function for all positive numbers. If, however, the domain of x^2 is restricted to the set of all positive numbers, then you can define the inverse as the positive square root of any positive number. Likewise, the inverse for the function x^2 with domain restricted to the set of all negative numbers can be defined as the set of negative square roots of positive numbers.

C. Only one-to-one functions have inverses.

What is the difference between $F(x) = x^3$, which has an inverse, and $G(x) = x^2$, which doesn't have an inverse? The graphs in Figure 12-1 show that F is a one-to-one function but G is not. When a function is not one-to-one, more than one value of y is associated with a single value of x. For example, the ordered pairs $(2, 4)$ and $(-2, 4)$ are among the ordered pairs belonging to G. In attempting to define an inverse function for G, should 2 or -2 be associated with 4 under your definition of the inverse function? Since this question has no single answer, it is *impossible* to define a complete inverse function for G.

- An inverse function for all values in the domain of a function exists if and only if the function is one-to-one.

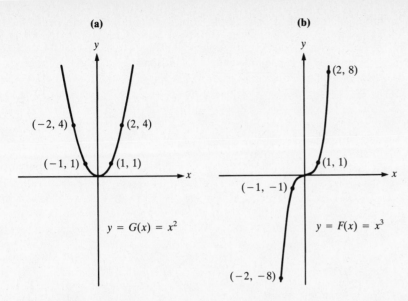

Figure 12-1
(a) $G(x) = x^2$; **(b)** $F(x) = x^3$

If the domain of $G(x) = x^2$ is restricted to positive values of x, an inverse can be defined for this modified one-to-one function. The inverse is the positive square root of any nonnegative number.

EXAMPLE 12-6: Can you find inverses for the following functions: **(a)** $f(x) = \tan x$; **(b)** $g(x) = 3/x$; **(c)** $h(x) = x^5$? The graphs of the functions are shown in Figure 12-2.

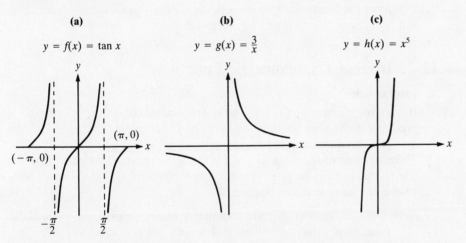

Figure 12-2

Solution: Inverses exist for **(b)** and **(c)** because they are one-to-one functions, but $\tan x$ is not a one-to-one function. If the domain of $\tan x$ is restricted to all x between $-\frac{1}{2}\pi$ and $\frac{1}{2}\pi$ radians, an inverse can be defined for this restricted function.

If $y = F(x)$ is a one-to-one function that can be expressed algebraically, you find the inverse function by solving the equation $y = F(x)$ for x in terms of y.

EXAMPLE 12-7: Find the inverse of $F(x) = 2x$.

Solution: Solve $y = 2x$ for x; you get $x = y/2$. To find the inverse of $F(x)$, as a function of x, replace y by x and x by y, so $F^{-1}(x) = x/2$.

If a function is one-to-one on some domain and its values are given in the form of a table, then you find values of the inverse function by reversing the columns of the table. For example, consider the table for tan x restricted to $-\frac{1}{3}\pi$ to $\frac{1}{3}\pi$:

x in radians	$\tan x$
$-\dfrac{\pi}{3}$	$-\sqrt{3}$
$-\dfrac{\pi}{4}$	-1
$-\dfrac{\pi}{6}$	$-\dfrac{\sqrt{3}}{3}$
0	0
$\dfrac{\pi}{6}$	$\dfrac{\sqrt{3}}{3}$
$\dfrac{\pi}{4}$	1
$\dfrac{\pi}{3}$	$\sqrt{3}$

An inverse exists for this restricted tangent function, and the values of the inverse function for a number in the right-hand column can be determined by finding the corresponding number in the left column. For example, $\tan^{-1}(1) = \frac{1}{4}\pi$, $\tan^{-1}(-\sqrt{3}) = -\frac{1}{3}\pi$, and $\tan^{-1}0 = 0$.

Your calculator provides another means of finding values of the inverse trigonometric functions. This was discussed briefly in Chapters 6 and 7 and will be discussed again later in this chapter.

12-2. Inverse Trigonometric Functions

A. Inverse sine

If x is a real number on $[-1, 1]$, the **inverse sine function**, $\sin^{-1}x$, or arc sin x produces the number between $-\frac{1}{2}\pi$ and $\frac{1}{2}\pi$ whose sine is equal to x. You can interpret this number as the angle measured in radians whose sine equals x. The restriction of the range of $\sin^{-1}$ is made because the sine function is not one-to-one and has no general inverse; however, on $[-\frac{1}{2}\pi, \frac{1}{2}\pi]$, sine is one-to-one and an inverse can be defined.

Note: Arc sin, when written with a capital A, means the principal value of the function; principal values of Arc sin x are restricted to $[-\frac{1}{2}\pi, \frac{1}{2}\pi]$. This notation also applies to the other trigonometric functions.

The graph of the sine function shows that sin x varies from -1 to $+1$ as x varies from $-\frac{1}{2}\pi$ to $\frac{1}{2}\pi$, inclusive (see Chapter 8). Other restrictions on the range can be made, but the convention is to make $[-\frac{1}{2}\pi, \frac{1}{2}\pi]$ the range of $\sin^{-1}x$. The domain of $\sin^{-1}x$ is $[-1, 1]$. The graph of the inverse sine function is shown in Figure 12-3.

To evaluate $\sin^{-1}0.5$ on your calculator, use the key strokes

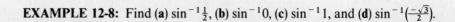

depending on the configuration of your calculator. To evaluate $\sin^{-1}0.5$ without a calculator, consult a table of the sine function to find the number between $-\frac{1}{2}\pi$ and $\frac{1}{2}\pi$ whose sine is equal to 0.5. You can find the inverse sine of any other number between -1 and 1 in a similar manner.

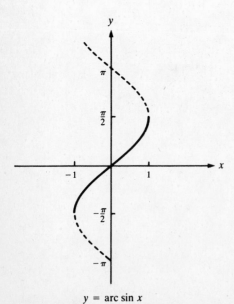

$y = \text{arc sin } x$

Figure 12-3
The inverse sine function.

EXAMPLE 12-8: Find (a) $\sin^{-1}\frac{1}{2}$, (b) $\sin^{-1}0$, (c) $\sin^{-1}1$, and (d) $\sin^{-1}\left(\frac{-\sqrt{3}}{2}\right)$.

Solution:

(a) $\sin^{-1}\frac{1}{2}$ is the number between $-\frac{1}{2}\pi$ and $\frac{1}{2}\pi$ whose sine is $\frac{1}{2}$. From Table 6-1, you see that $\sin\frac{1}{6}\pi = \frac{1}{2}$, so $\sin^{-1}\frac{1}{2} = \frac{1}{6}\pi$.

(b) $\sin^{-1}0$ is the number between $-\frac{1}{2}\pi$ and $\frac{1}{2}\pi$ whose sine is 0. Table 6-1 shows that $\sin 0 = 0$; therefore, $\sin^{-1}0 = 0$.

(c) $\sin^{-1}1$ is the number between $-\frac{1}{2}\pi$ and $\frac{1}{2}\pi$ whose sine is 1. Table 6-1 shows $\sin\frac{1}{2}\pi = 1$; therefore, $\sin^{-1}1 = \frac{1}{2}\pi$.

(d) $\sin^{-1}\left(\frac{-\sqrt{3}}{2}\right)$ is the number between $-\frac{1}{2}\pi$ and $\frac{1}{2}\pi$ whose sine is $\frac{-\sqrt{3}}{2}$. Table 6-1 shows that $\sin\frac{1}{3}\pi = \frac{\sqrt{3}}{2}$. You know that $\sin(-\alpha) = -\sin\alpha$. This means that $\sin(-\frac{1}{3}\pi) = \frac{-\sqrt{3}}{2}$, or $\sin^{-1}\left(\frac{-\sqrt{3}}{2}\right) = -\frac{1}{3}\pi$.

B. Inverse cosine

The **inverse cosine function**, $\cos^{-1}x$, or arc cos x, produces the number between 0 and π, angle in radians, whose cosine is equal to x. Because cosine is not one-to-one, its range must be restricted. By convention, $[0, \pi]$ is the range of inverse cosine. The domain of $\cos^{-1}x$ is the interval $[-1, 1]$. The graph is shown in Figure 12-4.

Follow the instructions for inverse sine to find the value of $\cos^{-1}x$ with your calculator or from a table, substituting cosine for sine.

EXAMPLE 12-9: Find **(a)** $\cos^{-1}\frac{1}{2}$, **(b)** $\cos^{-1}0$, **(c)** $\cos^{-1}1$, and **(d)** $\cos^{-1}\left(\frac{-\sqrt{3}}{2}\right)$.

Solution:

(a) $\cos^{-1}\frac{1}{2}$ is the number between 0 and π whose cosine is $\frac{1}{2}$. Table 6-1 reveals that $\cos\frac{1}{3}\pi = \frac{1}{2}$; therefore, $\cos^{-1}\frac{1}{2} = \frac{1}{3}\pi$.

(b) $\cos^{-1}0$ is the number between 0 and π whose cosine is 0. Table 6-1 shows that $\cos\frac{1}{2}\pi = 0$; therefore, $\cos^{-1}0 = \frac{1}{2}\pi$.

(c) $\cos^{-1}1$ is the number between 0 and π whose cosine is 1. Table 6-1 shows that $\cos 0 = 1$; therefore, $\cos^{-1}1 = 0$.

(d) $\cos^{-1}\left(\frac{-\sqrt{3}}{2}\right)$ is the number between 0 and π whose cosine is $\frac{-\sqrt{3}}{2}$. Table 6-1 shows that $\cos\frac{1}{6}\pi = \frac{\sqrt{3}}{2}$. Recall the identity $\cos(\pi - \alpha) = -\cos\alpha$. Then $\cos(\pi - \frac{1}{6}\pi) = -\cos\frac{1}{6}\pi = \frac{-\sqrt{3}}{2}$, so $\cos^{-1}\left(\frac{-\sqrt{3}}{2}\right) = \frac{5}{6}\pi$.

C. Inverse tangent

The **inverse tangent function**, $\tan^{-1}x$, or arc tan x, produces the angle between $-\frac{1}{2}\pi$ and $\frac{1}{2}\pi$, whose tangent is equal to x. The domain of $\tan^{-1}x$ is the interval $[-\infty, +\infty]$. The graph of inverse tangent appears in Figure 12-5.

Follow the instructions for inverse sine to find the value of $\tan^{-1}x$ with your calculator or from a table, substituting tangent for sine.

EXAMPLE 12-10: Find **(a)** $\tan^{-1}(-1)$, **(b)** $\tan^{-1}\sqrt{3}$, and **(c)** $\tan^{-1}\left(\frac{\sqrt{3}}{3}\right)$.

Solution:

(a) $\tan^{-1}(-1)$ is the number whose tangent is -1. From Table 6-1 you see that $\tan\frac{1}{4}\pi = 1$. Then, from the identity $\tan(-\alpha) = -\tan\alpha$, $\tan^{-1}(-1) = -\frac{1}{4}\pi$.

(b) $\tan^{-1}\sqrt{3}$ is the number whose tangent is $\sqrt{3}$. From Table 6-1, $\tan\frac{1}{3}\pi = \sqrt{3}$, so $\tan^{-1}\sqrt{3} = \frac{1}{3}\pi$.

(c) $\tan^{-1}\left(\frac{\sqrt{3}}{3}\right)$ is the number whose tangent is $\frac{\sqrt{3}}{3}$. From Table 6-1, $\tan\frac{1}{6}\pi = \frac{\sqrt{3}}{3}$, so $\tan^{-1}\left(\frac{\sqrt{3}}{3}\right) = \frac{1}{6}\pi$.

Values of the inverse cosecant, inverse secant, and inverse tangent functions are found by using the reciprocal relationships $\csc x = 1/\sin x$, $\sec x = 1/\cos x$, and $\cot x = 1/\tan x$.

D. Inverse cosecant

The **inverse cosecant function**, $\csc^{-1}x$, or arc csc x, produces the number between $-\frac{1}{2}\pi$ and $\frac{1}{2}\pi$, excluding 0, whose cosecant is equal to x. The domain

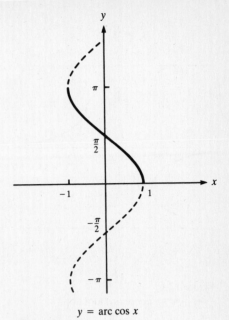

$y = $ arc cos x

Figure 12-4
The inverse cosine function.

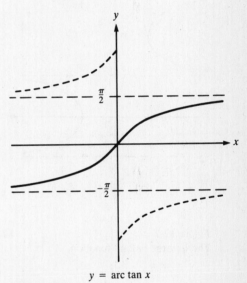

$y = $ arc tan x

Figure 12-5
The inverse tangent function.

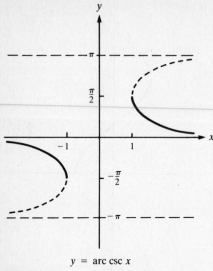

$y = \text{arc csc } x$

Figure 12-6
The inverse cosecant function.

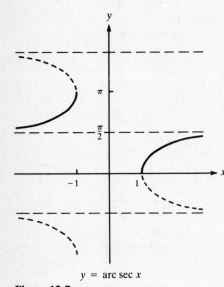

$y = \text{arc sec } x$

Figure 12-7
The inverse secant function.

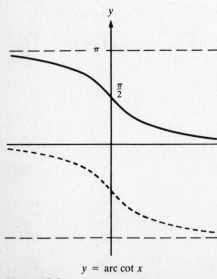

$y = \text{arc cot } x$

Figure 12-8
The inverse cotangent function.

of $\csc^{-1}x$ consists of intervals $(-\infty, -1]$ and $[1, +\infty)$. You can use the reciprocal relationship $\csc x = 1/\sin x$ to evaluate the inverse cosecant function. The graph of inverse cosecant is shown in Figure 12-6.

To evaluate $\csc^{-1}1.5$ on your calculator, use the key strokes

$\boxed{1.5}\ \boxed{1/x}\ \boxed{\text{INV}}\ \boxed{\sin}$, $\boxed{1.5}\ \boxed{1/x}\ \boxed{\text{arc sin}}$,   or $\boxed{1.5}\ \boxed{1/x}\ \boxed{\sin^{-1}}$

depending on the configuration of your calculator. To evaluate $\csc^{-1}1.5$ without a calculator, consult a table of the sine function to find the number between $-\frac{1}{2}\pi$ and $\frac{1}{2}\pi$, excluding 0, whose sine is equal to $\frac{1}{1.5}$. You can find the inverse cosecant of any other number on the intervals $(-\infty, -1]$ or $[1, +\infty)$ in the same way. If the result is not in the range for the inverse cosecant function, find a number in the range by using related angles.

EXAMPLE 12-11: Find (**a**) $\csc^{-1}2$, (**b**) $\csc^{-1}(-1)$, and (**c**) $\csc^{-1}\frac{1}{2}$.

Solution:

(**a**) $\csc^{-1}2$ is the number whose cosecant is 2, or whose sine is $\frac{1}{2}$. From Table 6-1, $\sin\frac{1}{6}\pi = \frac{1}{2}$, so $\csc^{-1}2 = \frac{1}{6}\pi$.
(**b**) $\csc^{-1}(-1)$ is the number whose cosecant is -1, or whose sine is -1. From Table 6-1, $\sin\frac{1}{2}\pi = 1$.   Since $\sin(-\alpha) = -\sin\alpha$, $\sin(-\frac{1}{2}\pi) = -1$ and $\csc^{-1}(-1) = -\frac{1}{2}\pi$.
(**c**) $\csc^{-1}\frac{1}{2}$ is the number whose cosecant is $\frac{1}{2}$, or whose sine is 2. Since sine can never equal 2, $\csc^{-1}\frac{1}{2}$ is undefined.

E. Inverse secant

The **inverse secant function**, $\sec^{-1}x$, or arc sec x, produces the number between 0 and π, excluding $\frac{1}{2}\pi$, whose secant is equal to x. The domain of $\sec^{-1}x$ consists of the intervals $(-\infty, -1]$ and $[1, +\infty)$. The graph of the inverse secant function is shown in Figure 12-7.

Follow the instructions for inverse cosecant to find the value of $\sec^{-1}x$ with your calculator or from a table, substituting secant for cosecant.

EXAMPLE 12-12: Find (**a**) $\sec^{-1}2$, (**b**) $\sec^{-1}(-1)$, and (**c**) $\sec^{-1}\frac{1}{2}$.

Solution:

(**a**) $\sec^{-1}2$ is the number whose secant is 2, or whose cosine is $\frac{1}{2}$. From Table 6-1 $\cos\frac{1}{3}\pi = \frac{1}{2}$, so $\sec^{-1}2 = \frac{1}{3}\pi$.
(**b**) $\sec^{-1}(-1)$ is the number whose secant is -1, or whose cosine is -1. From Table 6-1, $\cos\pi = -1$, so, taking into account the restrictions on the range of the inverse secant function given above, $\sec^{-1}(-1) = \pi$.
(**c**) $\sec^{-1}\frac{1}{2}$ is the number whose secant is $\frac{1}{2}$, or whose cosine is 2. Because cosine can never equal 2, $\sec^{-1}\frac{1}{2}$ is undefined.

F. Inverse cotangent

The **inverse cotangent function**, $\cot^{-1}x$, or arc cot x, produces the number between 0 and π whose cotangent is equal to x. The domain of $\cot^{-1}x$ is the interval $(-\infty, +\infty)$. The graph of the inverse cotangent function is shown in Figure 12-8.

Follow the instructions for inverse cosecant to find the value of $\cot^{-1}x$ with your calculator or from a table, substituting cotangent for cosecant.

EXAMPLE 12-13: Find (**a**) $\cot^{-1}(-1)$, (**b**) $\cot^{-1}(-\sqrt{3})$, and (**c**) $\cot^{-1}\left(\frac{\sqrt{3}}{3}\right)$.

Solution:

(**a**) $\cot^{-1}(-1)$ is the number between 0 and π whose cotangent is -1 or whose tangent is -1. From Table 6-1, $\tan\frac{1}{4}\pi = 1$, so $\cot\frac{1}{4}\pi = 1$. Thus, from related angles, $\cot(\pi - \frac{1}{4}\pi) = \cot\frac{3}{4}\pi = -1$, or $\cot^{-1}(-1) = \frac{3}{4}\pi$.

(b) $\cot^{-1}(-\sqrt{3})$ is the number between 0 and π whose cotangent is $-\sqrt{3}$. From Table 6-1, $\tan\frac{1}{6}\pi = \frac{\sqrt{3}}{3}$, so $\cot\frac{1}{6}\pi = \sqrt{3}$. As in **(a)**, $\cot^{-1}(-\sqrt{3}) = \pi - \frac{1}{6}\pi = \frac{5}{6}\pi$.

(c) $\cot^{-1}(\frac{\sqrt{3}}{3})$ is the number whose cotangent is $\frac{\sqrt{3}}{3}$, or whose tangent is $1/(\sqrt{3}/3) = \sqrt{3}$. From Table 6-1, $\tan\frac{1}{3}\pi = \sqrt{3}$, so $\cot^{-1}(\frac{\sqrt{3}}{3}) = \frac{1}{3}\pi$.

12-3. Graphing Inverse Trigonometric Functions

We'll illustrate the procedures for graphing inverse trigonometric functions with the inverse sine function, $y = K\sin^{-1}(Lx + M) + C$. The procedures also apply, with appropriate changes, to the other inverse trigonometric functions.

A. For $y = K\sin^{-1}(Lx + M) + C$, the domain depends on L and M.

Since the domain of $y = \sin^{-1}x$ is $[-1, 1]$, the domain of $y = \sin^{-1}Lx$ is $-1 \leqslant Lx \leqslant 1$ or $\frac{-1}{L} \leqslant x \leqslant \frac{1}{L}$. Parameter L expands or contracts the domain of the inverse function by the factor $\frac{1}{L}$. The domain of $y = \sin^{-1}(Lx + M)$ is

$$-1 \leqslant Lx + M \leqslant 1 \qquad \text{or} \qquad \frac{-1}{L} - \frac{M}{L} \leqslant x \leqslant \frac{1}{L} - \frac{M}{L}$$

EXAMPLE 12-14: Find the domains and sketch graphs of the following inverse functions: **(a)** $y = \sin^{-1}2x$; **(b)** $y = \cos^{-1}0.5x$; **(c)** $y = \tan^{-1}3x$.

Solution:

(a) The domain of $\sin^{-1}x$ is $[-1, 1]$. The domain of $\sin^{-1}2x$ is $-1 \leqslant 2x \leqslant 1$, or $[-\frac{1}{2}, \frac{1}{2}]$ (see Figure 12-9).

(b) The domain of $\cos^{-1}x$ is also $[-1, 1]$. The domain of $\cos^{-1}0.5x$ is $[-\frac{1}{0.5}, \frac{1}{0.5}] = [-2, 2]$ (see Figure 12-9).

(c) The domain of $\tan^{-1}x$ is the set of all real numbers. Then the domain of $\tan^{-1}3x$ is also the set of all real numbers (see Figure 12-9).

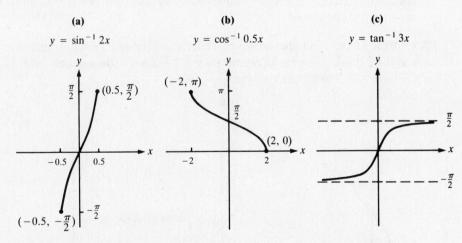

Figure 12-9

B. For $y = K\sin^{-1}(Lx + M) + C$, the range depends on K and C.

The range of $y = \sin^{-1}x$ is $[-\frac{1}{2}\pi, \frac{1}{2}\pi]$. Then the range of $y = K\sin^{-1}x$ is $[K(-\frac{1}{2}\pi), K(\frac{1}{2}\pi)]$. The range of the inverse function is expanded or contracted by the factor K. The range of $K\sin^{-1}x + C$ is $[K(-\frac{1}{2}\pi) + C, K(\frac{1}{2}\pi) + C]$.

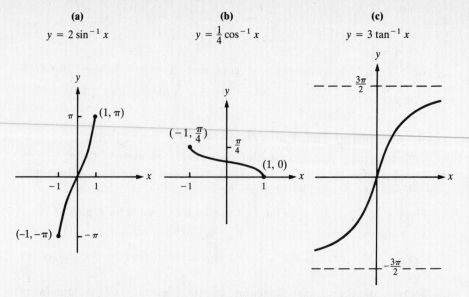

(a)

$$y = 2\sin^{-1}x$$

(b)

$$y = \tfrac{1}{4}\cos^{-1}x$$

(c)

$$y = 3\tan^{-1}x$$

Figure 12-10

EXAMPLE 12-15: Find the ranges and sketch graphs of the following inverse functions: **(a)** $y = 2\sin^{-1}x$; **(b)** $y = \tfrac{1}{4}\cos^{-1}x$; **(c)** $y = 3\tan^{-1}x$ (see Figure 12-10).

Solution:

(a) The range of $\sin^{-1}x$ is $\left[-\tfrac{1}{2}\pi,\tfrac{1}{2}\pi\right]$. The range of $2\sin^{-1}x$ is then $\left[2(-\tfrac{1}{2}\pi), 2(\tfrac{1}{2}\pi)\right]$, or $[-\pi,\pi]$.

(b) The range of $\cos^{-1}x$ is $[0,\pi]$. Then the range of $\tfrac{1}{4}\cos^{-1}x$ is $\left[(\tfrac{1}{4})0, (\tfrac{1}{4})\pi\right]$, or $\left[0,\tfrac{1}{4}\pi\right]$.

(c) The range of $\tan^{-1}x$ is $\left[-\tfrac{1}{2}\pi,\tfrac{1}{2}\pi\right]$. Then the range of $3\tan^{-1}x$ is $\left[3(-\tfrac{1}{2}\pi), 3(\tfrac{1}{2}\pi)\right] = \left[-\tfrac{3}{2}\pi,\tfrac{3}{2}\pi\right]$.

C. For $y = K\sin^{-1}(Lx + M) + C$, vertical placement depends on C.

From Figures 12-3 through 12-8, you see that the graphs of the inverse trigonometric functions can be graphed with respect to the line $y = 0$. Call $y = 0$ the horizontal reference line. The graph of $y = \sin^{-1}x + C$ is just like the graph of $y = \sin^{-1}x$ except that it is translated upward by C units if C is positive and downward by C units if C is negative. This means that the horizontal reference line is also translated upward or downward by C units, according to the sign of C.

EXAMPLE 12-16: Find the horizontal reference line and sketch graphs for each of the following inverse functions: **(a)** $y = 2 + \sin^{-1}x$; **(b)** $y = \cos^{-1}x - \tfrac{1}{4}$; **(c)** $y = \tan^{-1}x - 3$ (see Figure 12-11).

(a)

$$y = 2 + \sin^{-1}x$$

(b)

$$y = \cos^{-1}x - \tfrac{1}{4}$$

(c)

$$y = \tan^{-1}x - 3$$

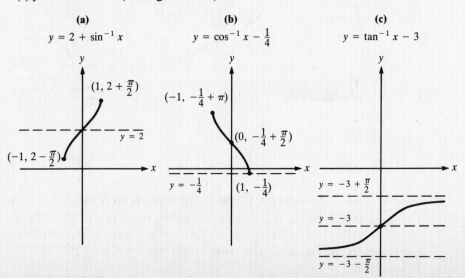

Figure 12-11

Solution:

(a) The horizontal reference line is the line $y = 2$.
(b) The horizontal reference line is the line $y = -\frac{1}{4}$.
(c) The horizontal reference line is the line $y = -3$.

D. For $y = K\sin^{-1}(Lx + M) + C$, the phase shift depends on $\frac{M}{L}$.

The domain of $y = \sin^{-1}x$ is $[-1, 1]$. Then the domain of $y = \sin^{-1}(Lx + M)$ is $-1 \leqslant Lx + M \leqslant 1$, or $\dfrac{-1}{L} - \dfrac{M}{L} \leqslant x \leqslant \dfrac{1}{L} - \dfrac{M}{L}$.

The expression $\dfrac{M}{L}$ shifts the graph of the inverse function to the left by $\dfrac{M}{L}$ units if $\dfrac{M}{L}$ is positive and to the right by $\dfrac{-M}{L}$ units if $\dfrac{M}{L}$ is negative.

EXAMPLE 12-17: Find the phase shifts and sketch graphs of the following inverse functions: **(a)** $y = \sin^{-1}(x - 2)$; **(b)** $y = \cos^{-1}(x + 1)$; **(c)** $y = \tan^{-1}(2x - 1)$ (see Figure 12-12).

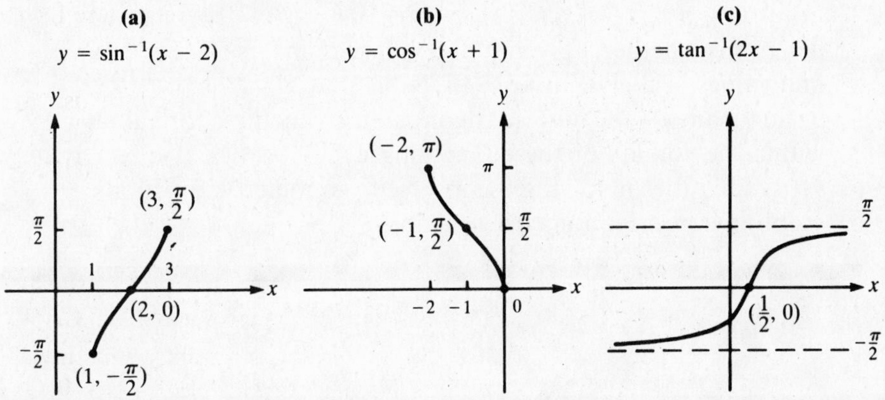

(a) $y = \sin^{-1}(x - 2)$ **(b)** $y = \cos^{-1}(x + 1)$ **(c)** $y = \tan^{-1}(2x - 1)$

Figure 12-12

Solution:

(a) The domain of $\sin^{-1}x$ is $[-1, 1]$. The domain of $\sin^{-1}(x - 2) = \sin^{-1}(x + (-2))$ is then $[-1 - (-2), 1 - (-2)]$, or $[1, 3]$. This means that the phase shift is 2 units to the right.
(b) The domain of $\cos^{-1}x$ is $[-1, 1]$. The domain of $\cos^{-1}(x + 1)$ is $[-1 - 1, 1 - 1]$, or $[-2, 0]$. This is a phase shift of 1 unit to the left.
(c) The domain of $\tan^{-1}x$ is the set of all real numbers. The domain of $\tan^{-1}(2x - 1) = \tan^{-1}[2(x + (-\frac{1}{2}))]$ is also the set of all real numbers but the phase has been shifted to the right $\frac{1}{2}$ unit.

SUMMARY

1. Only one-to-one functions have complete inverses.
2. You can find the values of inverse functions by using your calculator, function tables, or by solving algebraic equations.
3. Inverse sine, $\sin^{-1}x$ or $\arcsin x$, is the number whose sine equals x. The domain of inverse sine is $-1 \leqslant x \leqslant 1$ and the range is $\left[-\frac{1}{2}\pi, \frac{1}{2}\pi\right]$.
4. Inverse cosine, $\cos^{-1}x$ or $\arccos x$, is the number whose cosine equals x. The domain of inverse cosine is $-1 \leqslant x \leqslant 1$ and the range is $[0, \pi]$.
5. Inverse tangent, $\tan^{-1}x$ or $\arctan x$, is the number whose tangent equals x. The domain of inverse tangent is the set of all real numbers and the range is $\left(-\frac{1}{2}\pi, \frac{1}{2}\pi\right)$.

6. Inverse cosecant, $\csc^{-1}x$ or arc csc x, is the number whose cosecant equals x. The domain of inverse cosecant is the set of all $x \geqslant 1$ or $x \leqslant -1$ and the range is $\left[-\frac{1}{2}\pi, 0\right)$ or $\left(0, \frac{1}{2}\pi\right]$.

7. Inverse secant, $\sec^{-1}x$ or arc sec x, is the number whose secant equals x. The domain of inverse secant is the set of all $x \geqslant 1$ or $x \leqslant -1$ and the range is $\left[0, \frac{1}{2}\pi\right)$ or $\left(\frac{1}{2}\pi, \pi\right]$.

8. Inverse cotangent, $\cot^{-1}x$ or arc cot x, is the number whose cotangent equals x. The domain of inverse cotangent is the set of all real numbers and the range is $(0, \pi)$.

9. In $y = K \sin^{-1}(Lx + M) + C$, the domain depends on L, the range depends on K, the horizontal reference line depends on C, and the phase shift depends on $\dfrac{M}{L}$. These conditions also apply to the other inverse trigonometric functions.

RAISE YOUR GRADES

Can you ...?

☑ determine whether a given function has an inverse
☑ find values of a given inverse function
☑ evaluate any of the inverse trigonometric functions for values within the domain of these functions
☑ verify identities involving the inverse trigonometric functions
☑ graph inverse trigonometric functions

SOLVED PROBLEMS

PROBLEM 12-1   Show that $\sin^{-1}x + \cos^{-1}x = \frac{1}{2}\pi$.

Solution: Let $\sin^{-1}x = y$ and $\cos^{-1}x = z$, so $\sin y = x$ and $\cos z = x$. Then, since $\sin y = \cos z$ and $0 \leqslant z \leqslant \pi$, $y + z = \frac{1}{2}\pi$, $\sin y = \cos(\frac{1}{2}\pi - y)$ by complementary angles (see Chapter 6), so $\cos(\frac{1}{2}\pi - y) = \cos z$. By replacing z and y by their equivalent expressions in terms of inverse trigonometric functions, you get $\sin^{-1}x + \cos^{-1}x = \frac{1}{2}\pi$.   [See Section 12-2.]

PROBLEM 12-2   Find $\cos(\text{arc } \sin\frac{4}{5})$.

Solution: If you let $z = \text{arc } \sin\frac{4}{5}$, then $\sin z = \frac{4}{5}$. Using $\sin^2 z + \cos^2 z = 1$, you get $\cos(\text{arc } \sin\frac{4}{5}) = \cos z = \sqrt{1 - \left(\frac{4}{5}\right)^2} = \frac{3}{5}$.   [See Section 12-2.]

PROBLEM 12-3   Find (a) $\sin^{-1}\frac{\sqrt{2}}{2}$, (b) $\sin^{-1}\left(\frac{-\sqrt{2}}{2}\right)$.

Solution:

(a) If $\sin^{-1}\frac{\sqrt{2}}{2} = y$, then $\sin y = \frac{\sqrt{2}}{2}$, so $y = \frac{1}{4}\pi$.
(b) If $\sin^{-1}\left(\frac{-\sqrt{2}}{2}\right) = y$, then $\sin y = \frac{-\sqrt{2}}{2}$, so $y = -\frac{1}{4}\pi$.   [See Section 12-2.]

PROBLEM 12-4   Solve each of the following expressions for x: (a) $\sin x = a$, where $-1 \leqslant a \leqslant 1$ and x is on $\left[-\frac{1}{2}\pi, \frac{1}{2}\pi\right]$; (b) $\cos x = b$, where $-1 \leqslant b \leqslant 1$ and x is on $[0, \pi]$; (c) $\tan x = c$, where $-1 \leqslant c \leqslant 1$ and x lies on $\left[-\frac{1}{4}\pi, \frac{1}{4}\pi\right]$.

Solution:

(a) Take the inverse sine of both sides of the given equation to get $x = \sin^{-1} a$.
(b) Take the inverse cosine of both sides of the given equation to get $x = \cos^{-1} b$.
(c) Take the inverse tangent of both sides of the given equation to get $x = \tan^{-1} c$.

[See Section 12-2.]

PROBLEM 12-5 Evaluate $\sin(\text{arc}\cos\frac{5}{13} - \text{arc}\sin\frac{8}{17})$.

Solution: Use the identity for the sine of a sum:

$$\sin\left(\text{arc}\cos\frac{5}{13} - \text{arc}\sin\frac{8}{17}\right) = \sin\left(\text{arc}\cos\frac{5}{13}\right)\cos\left(\text{arc}\sin\frac{8}{17}\right)$$

$$- \cos\left(\text{arc}\cos\frac{5}{13}\right)\sin\left(\text{arc}\sin\frac{8}{17}\right)$$

Then,

$$\sin\left(\text{arc}\cos\frac{5}{13}\right) = \sqrt{1 - \left(\frac{5}{13}\right)^2} = \frac{12}{13}$$

$$\cos\left(\text{arc}\sin\frac{8}{17}\right) = \sqrt{1 - \left(\frac{8}{17}\right)^2} = \frac{15}{17}$$

$$\cos\left(\text{arc}\cos\frac{5}{13}\right) = \frac{5}{13}$$

$$\sin\left(\text{arc}\sin\frac{8}{17}\right) = \frac{8}{17}$$

Therefore,

$$\sin\left(\text{arc}\cos\frac{5}{13}\right)\cos\left(\text{arc}\sin\frac{8}{17}\right) - \cos\left(\text{arc}\cos\frac{5}{13}\right)\sin\left(\text{arc}\sin\frac{8}{17}\right)$$

$$= \left(\frac{12}{13}\right)\left(\frac{15}{17}\right) - \left(\frac{5}{13}\right)\left(\frac{8}{17}\right) = \frac{140}{221}$$ [See Section 12-2.]

PROBLEM 12-6 For each of the inverse trigonometric functions, with restricted ranges as defined in Section 12-3, determine if the function is even, odd, or neither even nor odd.

Solution: Recall that if $f(-x) = f(x)$, then the function is even; if $f(-x) = -f(x)$, then the function is odd. Look at Figures 12-3 through 12-8. You can readily determine that arc sin x, arc tan x, and arc csc x are odd; arc cos x, arc cot x, and arc sec x are neither even nor odd.

[See Section 12-1.]

PROBLEM 12-7 For each of the inverse trigonometric functions, with restricted ranges as defined in Section 12-3, determine if $f(x)$ steadily increases, decreases, or neither as x increases from the lowest value in the range to the highest.

Solution: From simple inspection of Figures 12-3 through 12-8 you see that the values of $\sin^{-1} x$ and $\tan^{-1} x$ steadily increase, $\cos^{-1} x$ and $\cot^{-1} x$ decrease, and $\sec^{-1} x$ and $\csc^{-1} x$ do neither.

[See Section 12-1.]

For Problems 12-8 through 12-13, find the value of the given expression without using your calculator.   [See Section 12-2.]

PROBLEM 12-8 $\sin(\text{arc}\tan 2)$

Solution: The expression arc tan 2 can be stated as the angle whose tangent is 2. Construct a triangle with $\tan\theta = 2$ and solve for the hypotenuse (see Figure 12-13). From the triangle, you see that $\sin(\text{arc}\tan 2) = \sin\theta = \frac{2}{\sqrt{5}} = \frac{2\sqrt{5}}{5}$.

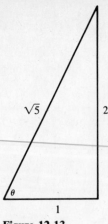

Figure 12-13

PROBLEM 12-9 $\cos(\arccos(-\tfrac{1}{3}))$.

Solution: Using the fact that cosine and arc cosine are inverse functions, you get $\cos(\arccos(-\tfrac{1}{3})) = -\tfrac{1}{3}$.

PROBLEM 12-10 $\tan(\arcsin\tfrac{1}{4})$

Solution: Proceed as in Problem 12-8 (see Figure 12-14). From the triangle, you see that $\tan(\arcsin\tfrac{1}{4}) = \frac{1}{\sqrt{15}} = \frac{\sqrt{15}}{15}$.

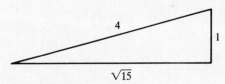

Figure 12-14

PROBLEM 12-11 $\tan(\cos^{-1}\tfrac{2}{5})$

Solution: Proceed as in Problem 12-8 (see Figure 12-15). From the triangle, you see that $\tan(\cos^{-1}\tfrac{2}{5}) = \frac{\sqrt{21}}{2}$.

Figure 12-15

PROBLEM 12-12 $\sin(\arcsin a + \arcsin b)$

Solution: From the identity for the sine of a sum,

$$\sin(\arcsin a + \arcsin b) = \sin(\arcsin a)\cos(\arcsin b) + \cos(\arcsin a)\sin(\arcsin b)$$

$$= a\cos(\arcsin b) + b\cos(\arcsin a)$$

Construct triangles to represent both cases (see Figure 12-16). You see that $\cos(\arcsin a) = \sqrt{1-a^2}$ and $\cos(\arcsin b) = \sqrt{1-b^2}$.
Then,

$$\sin(\arcsin a + \arcsin b) = a\cos(\arcsin b) + b\cos(\arcsin a)$$
$$= a\sqrt{1-b^2} + b\sqrt{1-a^2}$$

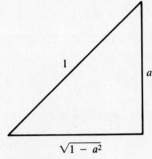
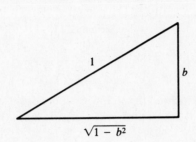

Figure 12-16

PROBLEM 12-13 $\tan(\sin^{-1}x + \tfrac{1}{4}\pi)$

Solution: From the identity for the tangent of a sum,

$$\tan(\sin^{-1}x + \tfrac{1}{4}\pi) = \frac{\tan(\sin^{-1}x) + \tan\tfrac{1}{4}\pi}{1 - (\tan(\sin^{-1}x)\tan\tfrac{1}{4}\pi)}$$
$$= \frac{\tan(\sin^{-1}x) + 1}{1 - \tan(\sin^{-1}x)}$$

Construct an appropriate triangle (see Figure 12-17). You see that $\tan(\sin^{-1}x) = x/\sqrt{1-x^2}$.
Then,

$$\frac{\tan(\sin^{-1}x)+1}{1-\tan(\sin^{-1}x)} = \frac{\dfrac{x}{\sqrt{1-x^2}}+1}{1-\dfrac{x}{\sqrt{1-x^2}}} = \frac{x+\sqrt{1-x^2}}{\sqrt{1-x^2}-x} = \frac{\sqrt{1-x^2}+x}{\sqrt{1-x^2}-x}\left(\frac{\sqrt{1-x^2}+x}{\sqrt{1-x^2}+x}\right) = \frac{(x+\sqrt{1-x^2})^2}{1-2x^2}$$

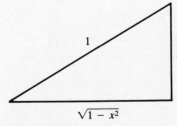
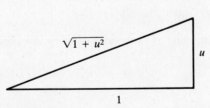

Figure 12-17   **Figure 12-18**

PROBLEM 12-14 Show that $\sin^{-1}(2u/(1+u^2)) = 2\tan^{-1}u$.

Solution: Consider the triangle of Figure 12-18. From the triangle, you see that $\sin(\tan^{-1}u) = u/\sqrt{1+u^2}$ and $\cos(\tan^{-1}u) = 1/\sqrt{1+u^2}$. From the double-angle identities,

$$\sin(2\tan^{-1}u) = 2\sin(\tan^{-1}u)\cos(\tan^{-1}u) = \frac{2u}{\sqrt{1+u^2}}\left(\frac{1}{\sqrt{1+u^2}}\right) = \frac{2u}{1+u^2}$$

By taking the inverse sine of both sides of $\sin(2\tan^{-1}u) = 2u/(1+u^2)$, the identity follows.

[See Section 12-2.]

PROBLEM 12-15 Show that $2\arccos x = \arccos(2x^2 - 1)$.

Solution: From the double-angle identity,

$$\cos(2\arccos x) = \cos^2(\arccos x) - \sin^2(\arccos x)$$

Construct an appropriate triangle (see Figure 12-19). You get $\sin^2(\arccos x) = 1 - x^2$ and $\cos^2(\arccos x) = x^2$. Then,

$$\cos(2\arccos x) = x^2 - (1 - x^2) = 2x^2 - 1$$

Applying the arc cos function to both sides of this expression, you get the identity.

[See Section 12-2.]

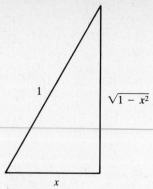

Figure 12-19

PROBLEM 12-16 Which of the following functions are inverses of each other? (a) $f(x) = \frac{1}{2}x$ and $g(x) = x$; (b) $f(x) = 3 + x$ and $g(x) = x - 3$; (c) $f(x) = x^{1/4}$ and $g(x) = x^4$; (d) $f(x) = x^{1/7}$ and $g(x) = x^7$.

Solution:

(a) Because $f(g(x)) = \frac{1}{2}x$, rather than x, these two functions are not inverses.
(b) Because $f(g(x)) = 3 + x - 3 = x$ and $g(f(x)) = 3 + x - 3 = x$, these two functions are inverses.
(c) Recall that the nth root of a number raised to the nth power, where n is even, is the absolute value of the number. Because $f(g(x)) = (x^4)^{1/4} = |x|$, rather than x, the two functions are not inverses.
(d) Recall that the nth root of a number raised to the nth power, where n is odd, is the number. Because $f(g(x)) = (x^7)^{1/7} = x$ and $g(f(x)) = (x^{1/7})^7 = x$, these two functions are inverses.

[See Section 12-1.]

In Problems 12-17 through 12-20, solve for x.

PROBLEM 12-17 $\arcsin x = 0.241$

Solution: Taking the sine of both sides of the given expression,

$$\sin(\arcsin x) = \sin 0.241$$

$$x = 0.24 \text{ rad.}$$

[See Section 12-2.]

PROBLEM 12-18 $\sin^{-1} x = -\frac{1}{2}$

Solution: Taking the sine of both sides of the given expression,

$$\sin(\sin^{-1} x) = \sin\left(-\frac{1}{2}\right) = -\sin\frac{1}{2}$$

$$x = -\sin\frac{1}{2} = -0.48 \text{ rad.}$$

[See Section 12-2.]

PROBLEM 12-19 $\arccos x = 0.796$

Solution: Taking the cosine of both sides of the given expression,

$$\cos(\arccos x) = \cos 0.796$$

$$x = 0.70 \text{ rad}$$

[See Section 12-2.]

PROBLEM 12-20 $\tan^{-1} x = -2$

Solution: This function is undefined, since the inverse tangent function takes values between $-\frac{1}{2}\pi$ and $\frac{1}{2}\pi$.

[See Section 12-2.]

In Problems 12-21 through 12-25, evaluate the given expression.

PROBLEM 12-21 $\tan^{-1}(-3)$

Solution: The desired value is the number between $-\frac{1}{2}\pi$ and $\frac{1}{2}\pi$ whose tangent is -3. Place your calculator in radian mode and enter $\boxed{3}$ $\boxed{+/-}$ $\boxed{\text{INV}}$ $\boxed{\tan}$ to get -1.25 rad.

[See Section 12-2.]

PROBLEM 12-22 $\cos^{-1}(-\frac{1}{2})$

Solution: The desired value is the number between 0 and π whose cosine is $-\frac{1}{2}$. You know that $\cos\frac{1}{3}\pi = \frac{1}{2}$ and, by related angles, that $\cos(\pi - \frac{1}{3}\pi) = \cos\frac{2}{3}\pi = -\frac{1}{2}$. Thus $\cos^{-1}(-\frac{1}{2}) = \frac{2}{3}\pi$.

[See Section 12-2.]

PROBLEM 12-23 $\tan^{-1} 6$

Solution: You must find the number between $-\frac{1}{2}\pi$ and $\frac{1}{2}\pi$ whose tangent is 6. Place your calculator in radian mode and enter $\boxed{6}$ $\boxed{\text{INV}}$ $\boxed{\tan}$ to get 1.41 rad.

[See Section 12-2.]

PROBLEM 12-24 $\sec^{-1}(-\sqrt{2})$

Solution: You must find the number between 0 and π whose secant is $-\sqrt{2}$, or whose cosine is $\frac{1}{-\sqrt{2}} = \frac{-\sqrt{2}}{2}$. You know that $\cos\frac{1}{4}\pi = \frac{\sqrt{2}}{2}$ and, by related angles, that $\cos(\pi - \frac{1}{4}\pi) = \cos\frac{3}{4}\pi = -\frac{\sqrt{2}}{2}$. Thus $\sec^{-1}(-\sqrt{2}) = \frac{3}{4}\pi$.

[See Section 12-2.]

PROBLEM 12-25 $\csc^{-1}(\frac{-2}{\sqrt{3}})$

Solution: You must find the number between $-\frac{1}{2}\pi$ and $\frac{1}{2}\pi$, excluding zero, whose cosecant is $\frac{-2}{\sqrt{3}}$, or whose sine is $\frac{-\sqrt{3}}{2}$. You know that $\sin\frac{1}{3}\pi = \frac{\sqrt{3}}{2}$. Because sine is an odd function, $\sin -\frac{1}{3}\pi = \frac{-\sqrt{3}}{2}$. Thus $\csc^{-1}(\frac{-2}{\sqrt{3}}) = -\frac{1}{3}\pi$.

[See Section 12-2.]

PROBLEM 12-26 Sketch the graph of $y = \frac{1}{4}\arcsin 2x + \frac{1}{2}\pi$.

Solution: Recall that the domain of $y = K\sin^{-1}(Lx + M) + C$ depends on L and M, the range depends on K, the placement of the axis depends on C, and phase shift depends on $\dfrac{M}{L}$. Using these rules, you see that the domain of y is $-\frac{1}{2}$ to $\frac{1}{2}$, the range is $\frac{3}{8}\pi$ to $\frac{5}{8}\pi$, the horizontal reference line is $y = \frac{1}{2}\pi$, and there is no phase shift. The graph is shown in Figure 12-20. [See Section 12-3.]

PROBLEM 12-27 Sketch the graph of $y = \arcsin(x + 2)$.

Solution: Because $-1 \leqslant x + 2 \leqslant 1$ is equivalent to $-3 \leqslant x \leqslant -1$, the graph of $\arcsin(x + 2)$ looks like the graph of $\arcsin x$ shifted to the left by 2 units. The graph is shown in Figure 12-21.

[See Section 12-3.]

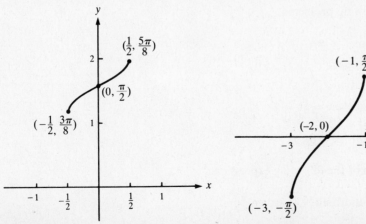

Figure 12-20   **Figure 12-21**

PROBLEM 12-28 Sketch the graph of $y = \text{arc sec}(x - 3)$.

Solution: The graph of $y = \text{arc sec}(x - 3)$ looks like the graph of arc sec x shifted to the right by 3 units. The graph is shown in Figure 12-22. [See Section 12-3.]

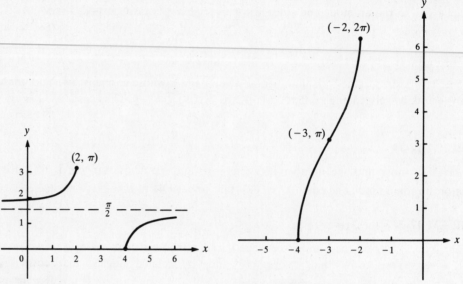

Figure 12-22   **Figure 12-23**

PROBLEM 12-29 Sketch the graph of $y = \pi + 2 \arcsin(x + 3)$.

Solution: The graph of $y = \pi + 2 \arcsin(x + 3)$ has domain $[-4, -2]$ and range $[0, 2\pi]$ and the horizontal reference line is $y = \pi$. The graph is shown in Figure 12-23. [See Section 12-3.]

Supplementary Exercises

PROBLEM 12-30 Use your calculator to evaluate the following, giving answers in radians correct to four decimal places. **(a)** $\cos^{-1} 0.357$; **(b)** $\sin^{-1}\left(-\frac{34}{57}\right)$; **(c)** $\cos^{-1}\left(\frac{5}{\sqrt{29}}\right)$; **(d)** $\cos^{-1}\left(\frac{7}{2\sqrt{3}}\right)$; **(e)** $\cot^{-1}(-2.631)$; **(f)** $\tan^{-1} 5.78$; **(g)** $\sec^{-1} 2.362$; **(h)** $\tan^{-1}(\csc^{-1} 4.385)$

PROBLEM 12-31 Use your calculator to evaluate the following, giving answers in degrees correct to two decimal places. **(a)** $\cos^{-1} 0.527$; **(b)** $\arcsin(\sec 1.78)$; **(c)** $\cos^{-1}(\sin 39°)$; **(d)** $\sin^{-1}(\tan 54°15')$; **(e)** $\tan^{-1}(-0.893)$; **(f)** $\cot^{-1} 0.5678$; **(g)** $\sec^{-1}\left(\frac{5}{\sqrt{3}}\right)$; **(h)** $\csc^{-1}(\cot 138°)$

Solve for x in the following problems.

PROBLEM 12-32 $\cos^{-1} x = \frac{1}{3}$

PROBLEM 12-33 $\sec^{-1} x = 3$

PROBLEM 12-34 $\text{arc csc } x = \frac{1}{2}$

PROBLEM 12-35 $\cot^{-1} x = 3$

Find the value of each of the following expressions.

PROBLEM 12-36 $\cos\left(\text{arc sin} \frac{15}{17} - \text{arc sin} \frac{8}{17}\right)$

PROBLEM 12-37 $\tan^{-1} 0.7683$

PROBLEM 12-38 $\cos^{-1}(-0.3)$

PROBLEM 12-39 $\tan^{-1}(-1)$

PROBLEM 12-40 $\sec^{-1}(-8)$

PROBLEM 12-41 $\csc^{-1}4$

PROBLEM 12-42 $\tan^{-1}(-3)$

PROBLEM 12-43 $\sin(\arc\tan 4)$

PROBLEM 12-44 $\tan(\arc\cos\frac{2}{3})$

PROBLEM 12-45 $\arc\tan(\tan\frac{3}{4}\pi)$

PROBLEM 12-46 $\arc\sin(\tan\frac{17}{4}\pi)$

PROBLEM 12-47 $\sin[\arc\cos(-\frac{3}{5})]$

PROBLEM 12-48 $\tan[\arc\sin(-\frac{5}{13})]$

PROBLEM 12-49 $\arc\sin(\sin\frac{3}{4}\pi)$

PROBLEM 12-50 $\csc(\arc\sec 3)$

PROBLEM 12-51 $\tan(\arc\cot 5)$

PROBLEM 12-52 $\arc\sin\frac{1}{3} + \arc\cos\frac{1}{3}$

PROBLEM 12-53 $\arc\csc 2$

PROBLEM 12-54 $\arc\cos(\sin\frac{8}{5}\pi)$

Sketch the graph of each of the following.

PROBLEM 12-55 $y = \arc\sin 3x$

PROBLEM 12-56 $y = 3\arc\sin 3x$

PROBLEM 12-57 $y = 2\arc\tan 4x$

Answers to Supplementary Exercises

12-30: (a) 1.206; (b) −0.6391; (c) 0.3805;
(d) no angle; (e) −0.3632; (f) 1.3995;
(g) 1.1336 (h) 0.2261

12-31: (a) 58.20°; (b) no angle; (c) 51°;
(d) no angle; (e) 41.76°; (f) 60.41°; (g) 69.73°;
(h) −41.76°

12-32: $x = 0.9450$

12-33: $x = -1.0101$

12-34: no x

12-35: −7.0153

12-36: $\frac{240}{289}$

12-37: 0.6551 rad

12-38: 1.8755 rad

12-39: $-\frac{1}{4}\pi = -0.7854$ rad

12-40: 1.6961 rad

12-41: 0.2527 rad

12-42: −1.2490 rad

12-43: $\frac{4}{\sqrt{17}} = 0.9701$ rad

12-44: $\frac{\sqrt{5}}{2}$

12-45: $-\frac{1}{4}\pi$

12-46: $\frac{1}{2}\pi$

12-47: $\frac{4}{5}$

12-48: $-\frac{5}{12}$

12-49: $\frac{1}{4}\pi$

12-50: $\frac{3}{4}\sqrt{2}$

12-51: $\frac{1}{5}$

12-52: $\frac{1}{2}\pi$

12-53: $\frac{1}{6}\pi$

12-54: $\frac{9}{10}\pi$

12-55:

12-56:

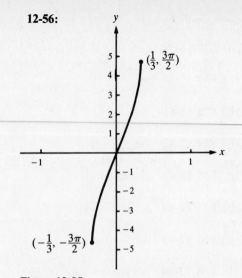

Figure 12-25

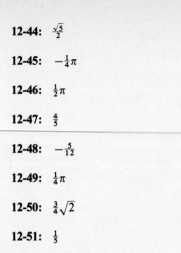

Figure 12-24

12-57:

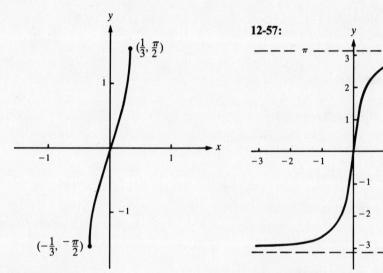

Figure 12-26

13 TRIGONOMETRIC EQUATIONS

THIS CHAPTER IS ABOUT

☑ **Equations and Identities**

☑ **How to Solve sin $x = c$**

☑ **How to Solve General Trigonometric Equations**

13-1. Equations and Identities

A mathematical expression in one or more variables containing an equals sign is called an equation, rather than an identity, if its solution set does not consist of all allowable values of the variables. Two examples of equations are $2x - 7 = 3$, whose only solution is $x = 5$; and $x^2 - x - 2 = 0$, whose solution set is $\{2, -1\}$. An example of an identity is $(x + 1)(x - 1) = x^2 - 1$, whose solution set contains all real numbers.

Trigonometric equations contain at least one of the trigonometric functions. Examples of trigonometric equations are $\sin x = 1$, $\sec x = \tan x$, $\sin^2 x + \cot x + 1 = \frac{3}{4}$, $(1/\cos x) + (1/\tan x) - 3 = 0$, and $\tan x = 1$. The solution sets of these equations do not include all real numbers. For example, the solution set of $\tan x = 1$ is $\{x \mid x = \frac{1}{4}\pi + n\pi, n \text{ any integer}\}$ if x is measured in radians, or $\{x \mid x = 45 + n180, n \text{ any integer}\}$ if x is in degrees. Unless indicated, we'll use radians in the trigonometric equations in this chapter.

13-2. How to Solve sin $x = c$

The sine function is used as a model for solving trigonometric equations. Everything said about the sine function applies, with appropriate modifications, to the other trigonometric functions.

A. For $c > 0$

To solve $\sin x = c$ for $c > 0$, find a number between 0 and $\frac{1}{2}\pi$ that satisfies the equation. Use your calculator, a table of trigonometric functions, or your knowledge of the values of the trigonometric ratios for special acute angles to find this solution.

EXAMPLE 13-1: Solve $\sin x = \frac{1}{2}$ for $0 \leqslant x \leqslant \frac{1}{2}\pi$.

Solution: A value of x between 0 and $\frac{1}{2}\pi$ with $\sin x = \frac{1}{2}$ is sought. Table 6-1 shows that $x = \frac{1}{6}\pi$ is a solution. Since solutions are restricted to $[0, \frac{1}{2}\pi]$, it is the only solution.

EXAMPLE 13-2: Solve $\tan x = 1$ for $0 \leqslant x \leqslant \frac{1}{2}\pi$.

Solution: Table 6-1 shows that $\frac{1}{4}\pi$ is a solution of $\tan x = 1$. Since solutions are again restricted to $[0, \frac{1}{2}\pi]$, it is the only solution.

Recall that $\sin(\pi - \alpha) = \sin \alpha$, $\sin(\pi + \alpha) = -\sin \alpha$, and $\sin(2\pi - \alpha) = -\sin \alpha$ (see Chapter 10). If you know the value of $\sin \alpha$, you can use these relationships to find other solutions of $\sin x = c$.

EXAMPLE 13-3: Solve $\sin x = \frac{1}{2}$ for $0 \leqslant x \leqslant 2\pi$.

Solution: From Example 13-1, $\frac{1}{6}\pi$ is a solution of $\sin x = \frac{1}{2}$. Recall that sine is positive in quadrants I and II; therefore, $\sin \frac{5}{6}\pi = \sin(\pi - \frac{1}{6}\pi) = \sin \frac{1}{6}\pi = \frac{1}{2}$, so the other solution of $\sin x = \frac{1}{2}$ between 0 and 2π is $\frac{5}{6}\pi$. The solution set is $\{\frac{1}{6}\pi, \frac{5}{6}\pi\}$.

EXAMPLE 13-4: Solve $\tan x = 1$ for $0 \leqslant x \leqslant 2\pi$.

Solution: From Example 13-2, $\frac{1}{4}\pi$ is a solution of $\tan x = 1$. Recall that $\tan x$ is positive in quadrants I and III; therefore $\tan \frac{5}{4}\pi = \tan(\pi + \frac{1}{4}\pi) = \tan \frac{1}{4}\pi = 1$, so the other solution of $\tan x = 1$ is $\frac{5}{4}\pi$. The solution set is $\{\frac{1}{4}\pi, \frac{5}{4}\pi\}$.

B. For $c < 0$

To solve $\sin x = c$ for $c < 0$, find a solution of $\sin x = -c$ for x between 0 and $\frac{1}{2}\pi$, then use related angles and periodicity to solve $\sin x = c$.

EXAMPLE 13-5: Solve $\sin x = -\frac{1}{2}$ for $0 \leqslant x \leqslant 2\pi$.

Solution: From Example 13-1, $\frac{1}{6}\pi$ is the solution of $\sin x = \frac{1}{2}$ between 0 and $\frac{1}{2}\pi$. Recall that sine is negative in quadrants III and IV. Then the solutions of $\sin x = -\frac{1}{2}$ are the angles in quadrants III and IV whose related angle is $\frac{1}{6}\pi$. These are $\pi + \frac{1}{6}\pi = \frac{7}{6}\pi$ and $2\pi - \frac{1}{6}\pi = \frac{11}{6}\pi$.

EXAMPLE 13-6: Solve $\tan x = -1$ for $0 \leqslant x \leqslant 2\pi$.

Solution: From Example 13-2, a number between 0 and $\frac{1}{2}\pi$ for which $\tan x = 1$ is $\frac{1}{4}\pi$. Now find the angles in quadrants II and IV whose related angle is $\frac{1}{4}\pi$. These numbers are $\pi - \frac{1}{4}\pi = \frac{3}{4}\pi$ and $2\pi - \frac{1}{4}\pi = \frac{7}{4}\pi$. The solution set is $\{\frac{3}{4}\pi, \frac{7}{4}\pi\}$.

C. Unrestricted solutions

You'll use the periodicity of the trigonometric functions to find solutions of trigonometric equations whose solution set is not restricted to $[0, 2\pi]$.

EXAMPLE 13-7: Solve $\sin x = \frac{1}{2}$ for x any real number.

Solution: In Example 13-3 you found all numbers between 0 and 2π that satisfy $\sin x = \frac{1}{2}$. Using the periodicity of the sine function (see Chapter 8), any number that differs from $\frac{1}{6}\pi$ or $\frac{5}{6}\pi$ by a multiple of 2π is also a solution of $\sin x = \frac{1}{2}$. Therefore, the solution set is $\{x \mid x = \frac{1}{6}\pi + 2n\pi$ or $x = \frac{5}{6}\pi + 2n\pi, n$ any integer$\}$. Some of the solutions are $x = -\frac{11}{6}\pi, \frac{1}{6}\pi, \frac{13}{6}\pi, \frac{25}{6}\pi, -\frac{7}{6}\pi, \frac{5}{6}\pi, \frac{17}{6}\pi$, and $\frac{29}{6}\pi$.

EXAMPLE 13-8: Solve $\cos x = \frac{1}{2}$.

Solution: First find a value of x between 0 and $\frac{1}{2}\pi$ for which $\cos x = \frac{1}{2}$; from Table 6-1, $\cos \frac{1}{3}\pi = \frac{1}{2}$. Since $\cos x$ is positive in quadrants I and IV, find the angle in quadrant IV whose related angle is $\frac{1}{3}\pi$. It is $2\pi - \frac{1}{3}\pi = \frac{5}{3}\pi$. Finally, since cosine is periodic with period 2π, the solution set is $\{x \mid x = \frac{1}{3}\pi + 2n\pi$ or $x = \frac{5}{3}\pi + 2n\pi$, n any integer$\}$.

EXAMPLE 13-9: Solve $\cos x = -\frac{1}{2}$.

Solution: From Example 13-8, $\cos \frac{1}{3}\pi = \frac{1}{2}$. Cosine is negative in quadrants II and III, so you must find the angles in those quadrants whose related angle is $\frac{1}{3}\pi$. These angles are $\pi - \frac{1}{3}\pi = \frac{2}{3}\pi$ and $\pi + \frac{1}{3}\pi = \frac{4}{3}\pi$. Then, by periodicity of the cosine function, the solution set is $\{x \mid x = \frac{2}{3}\pi + 2n\pi$ or $x = \frac{4}{3}\pi + 2n\pi, n$ any integer$\}$.

EXAMPLE 13-10: Solve $\tan x = 1$.

Solution: You solved $\tan x = 1$ for x in $[0, 2\pi]$ in Example 13-4. Since tangent has period π, the solution set of the equation is $\{x \mid x = \frac{1}{4}\pi + n\pi$ or $x = \frac{5}{4}\pi + n\pi$, n any integer$\}$.

EXAMPLE 13-11: Solve $\sec x = 1.45$.

Solution: You need to find a number whose secant is 1.45 or whose cosine is $\frac{1}{1.45}$. Set your calculator in radian mode; enter $\boxed{1.45}$ $\boxed{1/x}$ $\boxed{\cos^{-1}}$. You get $x = 0.8098$ radians. Next, find any other number between 0 and 2π for which $\sec x = 1.45$. Secant is positive in quadrants I and IV. Therefore, the angle in quadrant IV whose related angle is 0.8098 radians, $2\pi - 0.8098 = 5.4734$, is also a solution. The secant function has period 2π, so the solution set is $\{x \mid x = 0.8098 + 2n\pi$ or $x = 5.4734 + 2n\pi$, n any integer$\}$.

13-3. How to Solve General Trigonometric Equations

If a trigonometric equation is not of the form $f(x) = c$ where f is one of the trigonometric functions, use identities or algebraic manipulation to reduce it to this form, and then apply the techniques discussed in Section 13-2 to complete the solution. There are no general rules for bringing trigonometric equations into the form $f(x) = c$. The examples that follow demonstrate several types of solutions.

Whenever a solution is reached by raising both sides of an equation to a power, or by multiplication or division of both sides of an equation by an expression involving the variable, check each potential solution by substitution into the equation. Potential solutions that do not check are discarded; they are called **extraneous solutions**.

EXAMPLE 13-12: Solve $\sin x = \cos x$ for $0 < x < \pi$.

Solution: Dividing both sides by $\cos x$ leads to $\dfrac{\sin x}{\cos x} = \dfrac{\cos x}{\cos x}$ or $\tan x = 1$. The only solution of this equation for $0 < x < \pi$ is $x = \frac{1}{4}\pi$. Checking, $\sin \frac{1}{4}\pi = \cos \frac{1}{4}\pi = \frac{\sqrt{2}}{2}$ so $x = \frac{1}{4}\pi$ is the solution.

EXAMPLE 13-13: Find the solution in degrees of $\sin x \cos x = \frac{\sqrt{3}}{4}$ for x between $0°$ and $360°$.

Solution: Convert this to $f(x) = c$ by using the double-angle identity:

$$\sin 2x = 2 \sin x \cos x$$

Divide both sides by 2:

$$\frac{1}{2}\sin 2x = \sin x \cos x$$

Replace $\sin x \cos x$ by $\frac{\sqrt{3}}{4}$:

$$\frac{1}{2}\sin 2x = \frac{\sqrt{3}}{4}$$

Multiply both sides by 2:

$$\sin 2x = \frac{\sqrt{3}}{2}$$

Let $y = 2x$. Then solve $\sin y = \frac{\sqrt{3}}{2}$. From Table 6-1, $y = 60°$ is a solution, and the angle in quadrant II whose related angle is $60°$, or $180 - 60 = 120°$, is also a solution. Then $2x = 60°$, $x = 30°$, and $2x = 120°$, $x = 60°$. There are other potential solutions for x: $\sin y = \frac{\sqrt{3}}{2}$ also has $60 + 360 = 420°$ and $120 + 360 = 480°$

as solutions. Thus $2x = 420$, $x = 210°$, and $2x = 480$, $x = 240°$ are also possible solutions. Check your results in the original equation:

$$\sin 60° \cos 60° = \frac{\sqrt{3}}{2}\left(\frac{1}{2}\right) = \frac{\sqrt{3}}{4}$$

$$\sin 120° \cos 120° = \frac{\sqrt{3}}{2}\left(-\frac{1}{2}\right) = \frac{-\sqrt{3}}{4} \neq \frac{\sqrt{3}}{4} \text{ (extraneous)}$$

$$\sin 210° \cos 210° = -\frac{1}{2}\left(\frac{-\sqrt{3}}{2}\right) = \frac{\sqrt{3}}{4}$$

$$\sin 240° \cos 240° = -\frac{\sqrt{3}}{2}\left(\frac{1}{2}\right) = -\frac{\sqrt{3}}{4} \neq \frac{\sqrt{3}}{4} \text{ (extraneous)}$$

From this you see that two of the possible solutions are extraneous. The solution set is $\{60°, 210°\}$.

Note: Whenever you are solving equations of the type $\sin kx = c$, $\cos kx = c$, etc., for x between 0 and 2π or 0 and $360°$, you must find all solutions of $\sin y = c$ between 0 and $2k\pi$ to be assured of finding all solutions of $\sin kx = c$ between 0 and 2π.

EXAMPLE 13-14: Find all x such that $0 \leqslant x \leqslant 2\pi$ and $2\cos^2 x - \cos x - 1 = 0$.

Solution: First notice that the left side of the equation is a factorable quadratic. The result is

$$(2\cos x + 1)(\cos x - 1) = 0$$

Setting the right factor equal to zero, $\cos x - 1 = 0$, or $\cos x = 1$, leads to the solutions $x = 0$ and $x = 2\pi$. Setting the left factor equal to zero, $2\cos x + 1 = 0$, or $\cos x = -\frac{1}{2}$, and since $\cos\frac{1}{3}\pi = \frac{1}{2}$, this leads to the solutions $x = \pi - \frac{1}{3}\pi = \frac{2}{3}\pi$ and $x = \pi + \frac{1}{3}\pi = \frac{4}{3}\pi$. The solution set is $\{0, \frac{2}{3}\pi, \frac{4}{3}\pi, 2\pi\}$. Note—no squaring or multiplication or division by a variable was performed, so checking is not required.

EXAMPLE 13-15: Find all x such that $0 \leqslant x \leqslant 2\pi$ and $\sin^2 x + \cos x - 1 = 0$.

Solution: Use the identity $\sin^2 x + \cos^2 x = 1$:　　$1 - \cos^2 x + \cos x - 1 = 0$

Simplify and factor:　　　　　　　　　　　　　　$\cos x(1 - \cos x) = 0$

Thus $\cos x = 0$, producing the solutions $x = \frac{1}{2}\pi$ and $\frac{3}{2}\pi$, or $1 - \cos x = 0$, producing the solutions $x = 0$ and $x = 2\pi$. The solution set is $\{0, \frac{1}{2}\pi, \frac{3}{2}\pi, 2\pi\}$.

EXAMPLE 13-16: Find all x such that $\tan x + \sin x = 0$ and give the answer in degrees.

Solution: Use the identity $\tan x = \frac{\sin x}{\cos x}$:　　　　$\frac{\sin x}{\cos x} + \sin x = 0$

Factor:　　　　　　　　　　　　　　　　$\sin x\left[\left(\frac{1}{\cos x}\right) + 1\right] = 0$

Thus $\sin x = 0$, producing the solutions $x = k(180°)$, k an integer, or $\frac{1}{\cos x} + 1 = 0$, which is equivalent to $\cos x = -1$. From this you get the solutions $x = (2k - 1)180°$, k an integer. Combining these two solutions, the solution set is $\{x \mid x = k(180°) \text{ for } k \text{ any integer}\}$.

EXAMPLE 13-17: Find all t between 0 and 2π for which $2\sec t - \cos t = 0$.

Solution: Use $\sec t = \frac{1}{\cos t}$:　　　$\frac{2}{\cos t} - \cos t = 0$

Multiply by $\cos t$:
$$2 - \cos^2 t = 0$$

Solving this last equation produces $\cos t = \pm\sqrt{2}$; however, $\cos t$ is always less than 1. Therefore, there are no solutions.

EXAMPLE 13-18: Find all x such that $8 \sin x - \csc x = 2$.

Solution: Use $\csc x = \dfrac{1}{\sin x}$: $\qquad 8 \sin x - \dfrac{1}{\sin x} - 2 = 0$

Multiply by $\sin x$: $\qquad\qquad 8 \sin^2 x - 2 \sin x - 1 = 0$

Factor: $\qquad\qquad\qquad (4 \sin x + 1)(2 \sin x - 1) = 0$

Then $4 \sin x + 1 = 0$ or $\sin x = -\frac{1}{4}$; first, use your calculator to find x between 0 and $\frac{1}{2}\pi$ for which $\sin x = \frac{1}{4}$; this leads to $x = 0.253$ radians; then the x's for which $\sin x = -\frac{1}{4}$ are $\pi + 0.253 = 3.395$ and $2\pi - 0.253 = 6.031$. Next, $2 \sin x - 1 = 0$ or $\sin x = \frac{1}{2}$. This expression has $\frac{1}{6}\pi = 0.524$ and $\pi - \frac{1}{6}\pi = 2.618$ as solutions.

Check your answers: $\qquad 8 \sin 0.524 - \csc 0.524 = 2$
$$8 \sin 2.618 - \csc 2.618 = 2$$
$$8 \sin 3.395 - \csc 3.395 = 2$$
$$8 \sin 6.031 - \csc 6.031 = 2$$

The solution set is $\{x \mid x = 0.524 + 2n\pi \text{ or } x = 2.618 + 2n\pi \text{ or } x = 3.395 + 2n\pi$ or $x = 6.031 + 2n\pi, n$ any integer$\}$.

Use the following techniques to bring an equation into the form $f(x) = c$, where f is one of the six trigonometric functions: (1) algebraic manipulation; (2) basic trigonometric identities; (3) properties of the trigonometric functions, specifically periodicity and boundedness.

SUMMARY

1. Trigonometric equations contain at least one of the trigonometric functions and are only true for certain values of the variable.
2. You solve $f(x) = c$, where f is any of the trigonometric functions and $c > 0$, by finding a solution between 0 and $\frac{1}{2}\pi$. Use related angles and periodicity to find other solutions.
3. You solve $f(x) = c$, where f is any of the trigonometric functions, and $c < 0$, by finding a solution of $f(x) = -c$ between 0 and $\frac{1}{2}\pi$. Use related angles and periodicity to find other solutions.
4. If the trigonometric equation is not of the form $f(x) = c$, where f is one of the trigonometric functions, use trigonometric identities and algebraic principles to get factors of the form $f(x) = c$. Solve each of these equations. The solution set is the union of the solution sets of each of these equations.

RAISE YOUR GRADES

Can you...?

☑ solve $\sin x = c$ for $c > 0$
☑ solve $\sin x = c$ for $c < 0$
☑ solve $f(x) = c$ where f is any trigonometric function
☑ solve a general trigonometric equation

SOLVED PROBLEMS

Unless otherwise indicated, find solutions only on the interval from 0 to 2π, including 0 but not 2π, for each of the following problems.

PROBLEM 13-1 Solve $\sin 3x \cos 2x + \sin 2x \cos 3x = -\frac{1}{2}$ for x in degrees and $0° \leqslant x < 180°$.

Solution: Use the sum identity for sine to write

$$\sin 3x \cos 2x + \sin 2x \cos 3x = \sin(3x + 2x) = \sin 5x$$

You know that $\sin \theta = -\frac{1}{2}$ when $\theta = 180 + 30 = 210°$, $\theta = 210 + 360 = 570°$, $\theta = 360 - 30 = 330°$, and $\theta = 330 + 360 = 690°$. Let $5x = \theta$, so $5x = 210°, 330°, 570°,$ and $690°$. Then $x = 42°$, $66°$, $114°$, and $138°$. [See Section 13-3.]

PROBLEM 13-2 Solve $\sec x = 2$ for all x.

Solution: From the identity $\cos x = 1/\sec x$, the given equation is equivalent to $\cos x = \frac{1}{2}$. The solutions of this equation are $\frac{1}{3}\pi + 2n\pi$ and $\frac{5}{3}\pi + 2n\pi$ for n any integer. [See Section 13-2.]

PROBLEM 13-3 Solve $\csc x = -3$ for all x.

Solution: From the identity $\sin x = 1/\csc x$, the given equation is equivalent to $\sin x = -\frac{1}{3}$. Using your calculator, you find that $\sin x = \frac{1}{3}$ when $x = 0.34$ rad.

The solutions of $\sin x = -\frac{1}{3}$ between 0 and 2π are the angles in quadrants III and IV whose related angle is 0.34 rad, or $3.14 + 0.34 = 3.48$ rad or $6.28 - 0.34 = 5.94$ rad. Any other solution is obtained by adding a multiple of 2π to one of these solutions. The solution set is $\{3.48 + 2n\pi, 5.94 + 2n\pi, n$ any integer$\}$. [See Section 13-2.]

PROBLEM 13-4 Solve $\cot x = \frac{-\sqrt{3}}{3}$ for all x.

Solution: From the identity $\cot x = 1/\tan x$, the given equation is equivalent to $\tan x = -\sqrt{3}$. You know that $\tan x = \sqrt{3}$ when $x = \frac{1}{3}\pi$ rad. The solutions of $\tan x = -\sqrt{3}$ between 0 and 2π are the angles in quadrants II and IV whose related angle is $\frac{1}{3}\pi$ rad, or $\pi - \frac{1}{3}\pi = \frac{2}{3}\pi$ rad or $2\pi - \frac{1}{3}\pi = \frac{5}{3}\pi$ rad. Any other solution is obtained by adding a multiple of π to these solutions. The solution set is $\{\frac{2}{3}\pi + n\pi, \frac{5}{3}\pi + n\pi, n$ any integer$\}$, or more simply, $\{\frac{2}{3}\pi + n\pi, n$ any integer$\}$. [See Section 13.2.]

PROBLEM 13-5 Solve $\sin 2x - \tan x = 0$.

Solution: Using the double angle identity for $\sin 2x$ and the ratio identity for $\tan x$,

$$2 \sin x \cos x - \frac{\sin x}{\cos x} = 0$$

$$\sin x \left(2 \cos x - \frac{1}{\cos x} \right) = 0$$

Either $\sin x = 0$ and $x = 0, \pi$, or $2 \cos x - (1/\cos x) = 0$, $\cos^2 x = \frac{1}{2}$, and $x = \frac{1}{4}\pi, \frac{3}{4}\pi, \frac{5}{4}\pi, \frac{7}{4}\pi$. The solution set is $\{0, \frac{1}{4}\pi, \frac{3}{4}\pi, \frac{5}{4}\pi, \frac{7}{4}\pi\}$. [See Section 13-3.]

PROBLEM 13-6 Solve $1 + \cos x = \sin x$.

Solution: From the identity $\sin^2 x + \cos^2 x = 1$, $\sin x = \sqrt{1 - \cos^2 x}$.

Substitute: $1 + \cos x = \sqrt{1 - \cos^2 x}$

Square both sides: $(1 + \cos x)^2 = 1 - \cos^2 x$

Factor: $= (1 - \cos x)(1 + \cos x)$

Rearrange terms: $(1 + \cos x)^2 - (1 - \cos x)(1 + \cos x) = 0$

$$(1 + \cos x)[(1 + \cos x) - (1 - \cos x)] = 0$$

$$(1 + \cos x)(2 \cos x) = 0$$

Either $1 + \cos x = 0$, $\cos x = -1$, and $x = \pi$, or $2 \cos x = 0$, $\cos x = 0$, and $x = \frac{1}{2}\pi$. The solution set is $\{\frac{1}{2}\pi, \pi\}$. [See Section 13-3.]

PROBLEM 13-7 Solve $\sec x - 2 \cos x = \tan x$.

Solution:

Use the reciprocal identities: $$\frac{1}{\cos x} - 2 \cos x = \frac{\sin x}{\cos x}$$

Multiply by $\cos x$: $$1 - 2 \cos^2 x = \sin x$$

Use $\cos^2 x = 1 - \sin^2 x$: $$1 - 2(1 - \sin^2 x) = \sin x$$

$$1 - 2 + 2 \sin^2 x = \sin x$$

Regroup: $$2 \sin^2 x - \sin x - 1 = 0$$

Factor: $$(2 \sin x + 1)(\sin x - 1) = 0$$

Either $2 \sin x + 1 = 0$, $\sin x = -\frac{1}{2}$, and $x = \frac{7}{6}\pi, \frac{11}{6}\pi$, or $\sin x - 1 = 0$, $\sin x = 1$, and $x = \frac{1}{2}\pi$. Since both sides of the equation were multiplied by $\cos x$, you should check whether $x = \frac{1}{2}\pi$ is a solution: $\sec \frac{1}{2}\pi$ is undefined, $\cos \frac{1}{2}\pi = 0$, and $\tan \frac{1}{2}\pi$ is undefined, so $\frac{1}{2}\pi$ is not a solution. The solution set is $\{\frac{7}{6}\pi, \frac{11}{6}\pi\}$. [See Section 13-3.]

PROBLEM 13-8 Solve $\sin(x + \frac{1}{6}\pi) = \frac{1}{2}$.

Solution: You know that $\sin \theta = \frac{1}{2}$ when $\theta = \frac{1}{6}\pi$ or $\theta = \pi - \frac{1}{6}\pi = \frac{5}{6}\pi$. Setting $\theta = x + \frac{1}{6}\pi$ you get $\frac{1}{6}\pi = x + \frac{1}{6}\pi$ or $x = 0$, and $\frac{5}{6}\pi = x + \frac{1}{6}\pi$ or $x = \frac{2}{3}\pi$. [See Section 13-3.]

PROBLEM 13-9 Solve $\sin[2(x - \frac{1}{3}\pi)] + 1 = 0$, for all x.

Solution: The equation is equivalent to $\sin[2(x - \frac{1}{3}\pi)] = -1$. If you let $\theta = 2(x - \frac{1}{3}\pi)$, you get $\sin \theta = -1$. The solutions of this equation are $\theta = \frac{3}{2}\pi + 2n\pi$ for n any integer. Then,

$$2(x - \tfrac{1}{3}\pi) = \tfrac{3}{2}\pi + 2n\pi$$

$$x - \tfrac{1}{3}\pi = \tfrac{3}{4}\pi + n\pi$$

$$x = \tfrac{3}{4}\pi + \tfrac{1}{3}\pi + n\pi = \tfrac{13}{12}\pi + n\pi = \tfrac{1}{12}\pi + (n + 1)\pi$$

for n any integer. [See Section 13-2.]

PROBLEM 13-10 Solve $\cos(x - \frac{1}{2}\pi) = \sin(x + \frac{1}{2}\pi)$, for all x.

Solution: Expanding $\cos(x - \frac{1}{2}\pi)$,

$$\cos x \cos \tfrac{1}{2}\pi + \sin x \sin \tfrac{1}{2}\pi = \sin x$$

and $\sin(x + \frac{1}{2}\pi)$,

$$\sin x \cos \tfrac{1}{2}\pi + \cos x \sin \tfrac{1}{2}\pi = \cos x$$

the original equation becomes $\sin x = \cos x$. Dividing both sides of this equation by $\cos x$, you get $\tan x = 1$. The solutions of this equation, from Example 13-10, are $\frac{1}{4}\pi + n\pi$ and $\frac{5}{4}\pi + n\pi$, for n any integer. You've divided both sides of $\sin x = \cos x$ by $\cos x$, but since $\cos x \neq 0$ at the solutions, you don't need to check them. [See Section 13-3.]

PROBLEM 13-11 Solve $\sin x = -\sin x \tan x$.

Solution: Write the equation as $\sin x + \sin x \tan x = 0$, or by factoring the left side, as $\sin x(1 + \tan x) = 0$. Then $\sin x = 0$ and $x = 0, \pi$, or $1 + \tan x = 0$, $\tan x = -1$, and $x = \frac{3}{4}\pi, \frac{7}{4}\pi$. The solution set is $\{0, \frac{3}{4}\pi, \pi, \frac{7}{4}\pi\}$. [See Section 13-3.]

PROBLEM 13-12 Solve $\cot x - \tan x = 0$.

Solution: Write the equation as $(1/\tan x) - \tan x = 0$. Multiply by $\tan x$ to get $1 - \tan^2 x = 0$ or $\tan^2 x = 1$. Then $\tan x = 1$ and $x = \frac{1}{4}\pi, \frac{5}{4}\pi$, or $\tan x = -1$ and $x = \frac{3}{4}\pi, \frac{7}{4}\pi$. Since you have multiplied by $\tan x$, you should check the solutions. You'll see that the solution set is $\{\frac{1}{4}\pi, \frac{3}{4}\pi, \frac{5}{4}\pi, \frac{7}{4}\pi\}$. [See Section 13-3.]

PROBLEM 13-13 Solve $2\sin^3 x - \sin x = 0$.

Solution: Factor the left-hand side of the equation and set each factor equal to zero:

$$2\sin^3 x - \sin x = \sin x(2\sin^2 x - 1) = 0$$

$$\sin x = 0 \qquad 2\sin^2 x - 1 = 0$$

$$x = 0, \pi \qquad 2\sin^2 x = 1$$

$$\sin^2 x = \frac{1}{2}$$

$$\sin x = \pm\frac{\sqrt{2}}{2}$$

$$x = \frac{\pi}{4}, \frac{3\pi}{4}, \frac{5\pi}{4}, \frac{7\pi}{4}$$

The solution set is $\{0, \frac{1}{4}\pi, \frac{3}{4}\pi, \pi, \frac{5}{4}\pi, \frac{7}{4}\pi\}$. [See Section 13-3.]

PROBLEM 13-14 Solve $2\sin x + \cot x = \csc x$.

Solution:

Use $\cot x = \cos x/\sin x$ and
$\csc x = 1/\sin x$: $\qquad 2\sin x + \dfrac{\cos x}{\sin x} = \dfrac{1}{\sin x}$

Multiply by $\sin x$: $\qquad 2\sin^2 x + \cos x = 1$

Use $\sin^2 x = 1 - \cos^2 x$: $\qquad 2 - 2\cos^2 x + \cos x = 1$

$$2\cos^2 x - \cos x - 1 = 0$$

Factor: $\qquad (2\cos x + 1)(\cos x - 1) = 0$

Either $2\cos x + 1 = 0$, $\cos x = -\frac{1}{2}$, and $x = \frac{2}{3}\pi, \frac{4}{3}\pi$, or $\cos x - 1 = 0$, $\cos x = 1$, and $x = 0$. You've multiplied by $\sin x$, but if $\sin x = 0$, $\csc x$ is undefined so the original equation has no solution for these x values. The solution set is $\{\frac{2}{3}\pi, \frac{4}{3}\pi\}$. [See Section 13-3.]

PROBLEM 13-15 Solve $\cos x = \cot x$.

Solution:

Use $\cot x = \cos x/\sin x$: $\qquad \cos x = \dfrac{\cos x}{\sin x}$

$$\cos x - \dfrac{\cos x}{\sin x} = 0$$

Factor: $\qquad \cos x\left(1 - \dfrac{1}{\sin x}\right) = 0$

Either $\cos x = 0$, and $x = \frac{1}{2}\pi, \frac{3}{2}\pi$, or $1 - 1/\sin x = 0$, $\sin x = 1$, and $x = \frac{1}{2}\pi$. The solution set is $\{\frac{1}{2}\pi, \frac{3}{2}\pi\}$. [See Section 13-3.]

PROBLEM 13-16 Solve $\sin x - \sin 2x = 0$.

Solution: The double angle identity for $\sin 2x$ is the key to the solution of this problem:

$$\sin x - \sin 2x = \sin x - 2\sin x \cos x = 0$$

$$\sin x(1 - 2\cos x) = 0$$

Either $\sin x = 0$ and $x = 0$, π, or $1 - 2\cos x = 0$, $\cos x = \frac{1}{2}$ and $x = \frac{1}{3}\pi$, $\frac{5}{3}\pi$. The solution set is $\{0, \frac{1}{3}\pi, \frac{5}{3}\pi, \pi\}$. [See Section 13-3.]

PROBLEM 13-17   Solve $3\cos x - 2\cos^2(\frac{1}{2}x) = 0$.

Solution: Use the half-angle identity $\cos^2(\frac{1}{2}x) = (1 + \cos x)/2$:

$$3\cos x - \frac{2(1 + \cos x)}{2} = 0$$

$$3\cos x - 1 - \cos x = 0$$

$$2\cos x = 1$$

$$\cos x = \frac{1}{2}$$

$$x = \frac{\pi}{3}, \frac{5\pi}{3}$$

The solution set is $\{\frac{1}{3}\pi, \frac{5}{3}\pi\}$. [See Section 13-3.]

PROBLEM 13-18   Solve $\sin 3x + \sin x = \sin 2x$.

Solution: Use the identity

$$\sin \theta + \sin \alpha = 2\sin \frac{\theta + \alpha}{2} \cos \frac{\theta - \alpha}{2}$$

to convert the sum of sines to a product:

$$\sin 3x + \sin x = 2\sin \frac{3x + x}{2} \cos \frac{3x - x}{2} = 2\sin 2x \cos x$$

Then you can rewrite the original equation:

$$2\sin 2x \cos x = \sin 2x$$

$$2\sin 2x \cos x - \sin 2x = 0$$

Factoring this last expression, you get $\sin 2x(2\cos x - 1) = 0$. Either $\sin 2x = 0$, $2x = 0$, π, 2π, 3π, and $x = 0, \frac{1}{2}\pi, \pi, \frac{3}{2}\pi$, or $2\cos x - 1 = 0$, $\cos x = \frac{1}{2}$, and $x = \frac{1}{3}\pi, \frac{5}{3}\pi$. (Since you want all values of x between 0 and 2π, you must consider values of $2x$ between 0 and 4π.) The solution set is $\{0, \frac{1}{3}\pi, \frac{1}{2}\pi, \pi, \frac{3}{2}\pi, \frac{5}{3}\pi\}$. [See Section 13-3.]

PROBLEM 13-19   Solve $\sin 3x - \sin x = \sin x$.

Solution: There are two approaches to the solution of this problem. We could reduce $\sin 3x$ to an expression involving sines and cosines of x, but we'll use the identity that expresses the left-hand side as a product:

$$\sin 3x - \sin x = 2\cos \frac{3x + x}{2} \sin \frac{3x - x}{2}$$

$$= 2\cos 2x \sin x$$

Substituting and moving all terms to the left, you get

$$2\cos 2x \sin x - \sin x = 0$$

$$\sin x(2\cos 2x - 1) = 0$$

Either $\sin x = 0$, and $x = 0$, π, or $2\cos 2x - 1 = 0$, $\cos 2x = \frac{1}{2}$, $2x = \frac{1}{3}\pi, \frac{5}{3}\pi, \frac{7}{3}\pi, \frac{11}{3}\pi$, and $x = \frac{1}{6}\pi$, $\frac{5}{6}\pi, \frac{7}{6}\pi, \frac{11}{6}\pi$. The solution set is $\{0, \frac{1}{6}\pi, \frac{5}{6}\pi, \pi, \frac{7}{6}\pi, \frac{11}{6}\pi\}$. [See Section 13-3.]

PROBLEM 13-20 Solve $\cos x - (2/\cos x) = 1$.

Solution: Multiply by $\cos x$ and group terms:

$$\cos^2 x - \cos x - 2 = 0$$

Factor: $(\cos x - 2)(\cos x + 1) = 0$

Either $\cos x - 2 = 0$, and $\cos x = 2$, which has no solutions because $\cos x \leqslant 1$, or $\cos x + 1 = 0$, $\cos x = -1$, and $x = \pi$. The solution set is $\{\pi\}$. [See Section 13-3.]

PROBLEM 13-21 Solve $\sin 4x = 1$.

Solution: Since $0 \leqslant x \leqslant 2\pi$, you must find solutions of $\sin 4x = 1$ for $4x$ on $[0, 8\pi]$. So $\sin 4x = 1$ for $4x = \frac{1}{2}\pi, \frac{1}{2}\pi + 2\pi, \frac{1}{2}\pi + 4\pi, \frac{1}{2}\pi + 6\pi$, or $x = \frac{1}{8}\pi, \frac{1}{8}\pi + \frac{1}{2}\pi, \frac{1}{8}\pi + \pi$, and $\frac{1}{8}\pi + \frac{3}{2}\pi$. The solution set is $\{\frac{1}{8}\pi, \frac{5}{8}\pi, \frac{9}{8}\pi, \frac{13}{8}\pi\}$. [See Section 13-2.]

PROBLEM 13-22 Solve $\tan^2 x = 1 + \sec x$.

Solution: You could use identities to change everything to sines and cosines, or you could observe that the identity $\tan^2 x = \sec^2 x - 1$ will change everything to secants:

$$\sec^2 x - 1 = 1 + \sec x$$

$$\sec^2 x - \sec x - 2 = 0$$

$$(\sec x - 2)(\sec x + 1) = 0$$

Either $\sec x - 2 = 0$, $\sec x = 2$, and $x = \frac{1}{3}\pi, \frac{5}{3}\pi$, or $\sec x + 1 = 0$, $\sec x = -1$, and $x = \pi$. The solution set is $\{\frac{1}{3}\pi, \pi, \frac{5}{3}\pi\}$. [See Section 13-3.]

PROBLEM 13-23 Solve $\sin x - \sqrt{3}\cos x = \sqrt{2}$, for all x.

Solution:

$$\sin x - \sqrt{3}\cos x = \sqrt{2}$$

Factor:

$$2\left(\frac{1}{2}\sin x - \frac{\sqrt{3}}{2}\cos x\right) = \sqrt{2}$$

Substitute $\cos\frac{\pi}{3}$ for $\frac{1}{2}$ and $\sin\frac{\pi}{3}$ for $\frac{\sqrt{3}}{2}$: $2\left(\cos\frac{\pi}{3}\sin x - \sin\frac{\pi}{3}\cos x\right) = \sqrt{2}$

Use the difference formula for sine: $2\sin\left(x - \frac{\pi}{3}\right) = \sqrt{2}$

Simplify: $\sin\left(x - \frac{\pi}{3}\right) = \frac{\sqrt{2}}{2}$

The solutions of $\sin\theta = \frac{\sqrt{2}}{2}$ are $\theta = \frac{1}{4}\pi + 2n\pi$ or $\theta = \frac{3}{4}\pi + 2n\pi$. For $\theta = x - \frac{1}{3}\pi$, the solutions are $x = \frac{1}{4}\pi + \frac{1}{3}\pi + 2n\pi = \frac{7}{12}\pi + 2n\pi$ or $x = \frac{3}{4}\pi + \frac{1}{3}\pi + 2n\pi = \frac{13}{12}\pi + 2n\pi$. [See Section 13-3.]

PROBLEM 13-24 Solve **(a)** $\sin x + \cos x = 1$; **(b)** $\tan x + \sec x = 1$; **(c)** $\cot x + \csc x = 1$.

Solution:

(a) You can write $\sin x + \cos x = 1$ as

$$\frac{\sqrt{2}}{2}\sin x + \frac{\sqrt{2}}{2}\cos x = \frac{\sqrt{2}}{2}$$

or

$$\sin\left(x + \frac{\pi}{4}\right) = \frac{\sqrt{2}}{2}$$

Now $\sin\theta = \frac{\sqrt{2}}{2}$ when $\theta = \frac{1}{4}\pi$ or $\frac{3}{4}\pi$. Setting $x + \frac{1}{4}\pi = \frac{1}{4}\pi$, you get $x = 0$; setting $x + \frac{1}{4}\pi = \frac{3}{4}\pi$, you get $x = \frac{1}{2}\pi$. The solution set is $\{0, \frac{1}{2}\pi\}$.

(b)

$$\tan x + \sec x = 1$$

Use $\tan x = \dfrac{\sin x}{\cos x}$ and $\sec x = \dfrac{1}{\cos x}$:

$$\frac{\sin x}{\cos x} + \frac{1}{\cos x} = 1$$

Multiply by $\cos x$:

$$\sin x + 1 = \cos x$$

Regroup:

$$\sin x - \cos x = -1$$

Use $\cos x = \sin\left(\dfrac{\pi}{2} - x\right)$:

$$\sin x - \sin\left(\frac{\pi}{2} - x\right) = -1$$

Use the sine difference identity:

$$2\sin\left(x - \frac{\pi}{4}\right)\cos\frac{\pi}{4} = -1$$

Use $\cos\dfrac{\pi}{4} = \dfrac{\sqrt{2}}{2}$:

$$\sqrt{2}\sin\left(x - \frac{\pi}{4}\right) = -1$$

$$\sin\left(x - \frac{\pi}{4}\right) = \frac{-\sqrt{2}}{2}$$

The solutions of $\sin\theta = \frac{-\sqrt{2}}{2}$ are $\theta = \frac{5}{4}\pi + 2n\pi$ or $\theta = \frac{7}{4}\pi + 2n\pi$. If $\theta = x - \frac{1}{4}\pi$, the solutions are $x = \frac{5}{4}\pi + \frac{1}{4}\pi + 2n\pi = \frac{3}{2}\pi + 2n\pi$ or $x = \frac{7}{4}\pi + \frac{1}{4}\pi + 2n\pi = 2\pi + 2n\pi$. You've multiplied by $\cos x$, so you must check the solutions: $\tan\frac{3}{2}\pi$ is not defined, so discard these solutions; $\tan 2\pi + \sec 2\pi = 1$, which checks. The solution set is $\{2n\pi$, for n any integer$\}$.

(c) Follow the procedure of part **b**. The solution set is $\{\frac{1}{2}\pi + 2n\pi$, for n any integer$\}$.

[See Section 13-3.]

PROBLEM 13-25 Solve $\cos x - 2\cos x \sin x = 0$.

Solution:

$$\cos x - 2\cos x \sin x = 0$$

Factor:

$$\cos x(1 - 2\sin x) = 0$$

Either $\cos x = 0$ and $x = \frac{1}{2}\pi, \frac{3}{2}\pi$, or $1 - 2\sin x = 0$, $\sin x = \frac{1}{2}$ and $x = \frac{1}{6}\pi, \frac{5}{6}\pi$. The solution set is $\{\frac{1}{6}\pi, \frac{1}{2}\pi, \frac{5}{6}\pi, \frac{3}{2}\pi\}$.

[See Section 13-3.]

PROBLEM 13-26 Solve $\sin x + \sqrt{3}\cos x = 1$, for all x.

Solution:

$$\sin x + \sqrt{3}\cos x = 1$$

Factor:

$$2\left(\frac{1}{2}\sin x + \frac{\sqrt{3}}{2}\cos x\right) = 1$$

From Example 13-23,

$$2\sin\left(x + \frac{\pi}{3}\right) = 1$$

$$\sin\left(x + \frac{\pi}{3}\right) = \frac{1}{2}$$

The solutions of $\sin\theta = \frac{1}{2}$ are $\theta = \frac{1}{6}\pi + 2n\pi$ or $\theta = \frac{5}{6}\pi + 2n\pi$. If $\theta = x + \frac{1}{3}\pi$, $x = \frac{1}{6}\pi - \frac{1}{3}\pi + 2n\pi = -\frac{1}{6}\pi + 2n\pi = \frac{11}{6}\pi + 2n\pi$ or $x = \frac{5}{6}\pi - \frac{1}{3}\pi + 2n\pi = \frac{1}{2}\pi + 2n\pi$ n any integer.

[See Section 13-3.]

PROBLEM 13-27 Solve $2\sin x - \cos x - 1 = 0$.

Solution:

$$2\sin x - \cos x - 1 = 0$$

Regroup:

$$2\sin x - \cos x = 1$$

Factor:
$$\sqrt{5}\left(\frac{2}{\sqrt{5}}\sin x - \frac{1}{\sqrt{5}}\cos x\right) = 1$$

From Example 13-23,
$$\sqrt{5}\sin(x - 0.46) = 1$$

$$\sin(x - 0.46) = \frac{1}{\sqrt{5}}$$

$$x - 0.46 = \text{arc}\sin\frac{1}{\sqrt{5}} = 0.46 \text{ or } 2.68$$

$$x = 0.92 \text{ or } 3.14$$

The solution set is $\{0.92, \pi\}$. [See Section 13-3.]

Supplementary Exercises

PROBLEM 13-28   Solve $\sin x \cos^2 x = \sin x$.

PROBLEM 13-29   Solve $\sin x \cos^2 x = \cos x$.

PROBLEM 13-30   Solve $2\sin^3 x = \sin x$.

PROBLEM 13-31   Solve $\cot^2 x + 2\cot x + 1 = 0$.

PROBLEM 13-32   Solve $\cos 2x - \sin x = 0$.

PROBLEM 13-33   Solve $\sin^2 x + \cos 2x = 1$.

PROBLEM 13-34   Solve $\sec^2 x = 2\sin^2 x + \tan^2 x$.

PROBLEM 13-35   Solve $2\sin^4 x - \cos^2 x = 0$.

PROBLEM 13-36   Solve $\sin 2x + \sin x = 0$.

PROBLEM 13-37   Solve **(a)** $\sec^2 x + \tan^2 x = 1$ and **(b)** $\csc^2 x + \cot^2 x = 1$, for all x.

PROBLEM 13-38   Solve **(a)** $\sin x + \cos x = 0$ and **(b)** $\sin x - \cos x = 0$.

PROBLEM 13-39   Solve $2\cos^2 x + 3\cos x + 1 = 0$.

PROBLEM 13-40   Solve $\tan^2 x - 1 = 0$.

PROBLEM 13-41   Solve $\cos^2 x - 4\cos x + 4 = 0$.

PROBLEM 13-42   Solve $\tan x + \sec x = 3$.

PROBLEM 13-43   Solve $\cos 2\theta - 2\sin^2\theta = 1$.

PROBLEM 13-44   Solve $\sin 2\theta + 2\sin\theta\cos\theta = 0$.

PROBLEM 13-45   Solve $2\sin^2(\frac{1}{2}x) - \cos x = 1$.

PROBLEM 13-46   Solve $\tan 2x + 2\tan x = \tan 2x \tan^2 x$.

PROBLEM 13-47 Solve $4\sin^2 x - 3 = 0$.

PROBLEM 13-48 Solve $\sin x + \csc x + 2 = 0$.

PROBLEM 13-49 Solve $\sqrt{3}(\sec x + 1) = \tan x$.

PROBLEM 13-50 Solve $\tan^2 x - \cos^2 x = \sin^2 x$.

PROBLEM 13-51 Solve $2\sin^3 x - \sin x = 0$.

PROBLEM 13-52 Solve $2\cos^2 x \sin x - \sin x - 2\cos^2 x + 1 = 0$.

Answers to Supplementary Exercises

13-28: $0, \pi$

13-29: $\frac{1}{2}\pi, \frac{3}{2}\pi$

13-30: $0, \frac{1}{4}\pi, \frac{3}{4}\pi, \pi, \frac{5}{4}\pi, \frac{7}{4}\pi$

13-31: $\frac{3}{4}\pi, \frac{7}{4}\pi$

13-32: $\frac{1}{6}\pi, \frac{5}{6}\pi, \frac{3}{2}\pi$

13-33: $0, \pi$

13-34: $\frac{1}{4}\pi, \frac{3}{4}\pi, \frac{5}{4}\pi, \frac{7}{4}\pi$

13-35: $\frac{1}{4}\pi, \frac{3}{4}\pi, \frac{5}{4}\pi, \frac{7}{4}\pi$

13-36: $0, \frac{2}{3}\pi, \pi, \frac{4}{3}\pi$

13-37: (a) $x = n\pi$; (b) $x = \frac{1}{2}\pi + n\pi$

13-38: (a) $\frac{3}{4}\pi, \frac{7}{4}\pi$; (b) $\frac{1}{4}\pi, \frac{5}{4}\pi$

13-39: $\frac{2}{3}\pi, \pi, \frac{4}{3}\pi$

13-40: $\frac{1}{4}\pi, \frac{3}{4}\pi, \frac{5}{4}\pi, \frac{7}{4}\pi$

13-41: no solutions

13-42: 0.93

13-43: $0, \pi$

13-44: $0, \frac{1}{2}\pi, \pi, \frac{3}{2}\pi$

13-45: $\frac{1}{2}\pi, \frac{3}{2}\pi$

13-46: $0, \pi$

13-47: $\frac{1}{3}\pi, \frac{2}{3}\pi, \frac{4}{3}\pi, \frac{5}{3}\pi$

13-48: $\frac{3}{2}\pi$

13-49: $\frac{2}{3}\pi, \pi$

13-50: $\frac{1}{4}\pi, \frac{3}{4}\pi, \frac{5}{4}\pi, \frac{7}{4}\pi$

13-51: $0, \frac{1}{4}\pi, \frac{3}{4}\pi, \pi, \frac{5}{4}\pi, \frac{7}{4}\pi$

13-52: $\frac{1}{4}\pi, \frac{1}{2}\pi, \frac{3}{4}\pi, \frac{5}{4}\pi, \frac{7}{4}\pi$

14 OBLIQUE TRIANGLES

THIS CHAPTER IS ABOUT

☑ **Law of Sines**
☑ **Law of Cosines**
☑ **Applications of Oblique Triangles**

Recall that an oblique triangle is any triangle without a right angle. Some examples are shown in Figure 14-1. (All problems and examples in this chapter are based on the nomenclature of the model oblique triangle shown in Figure 14-2.) You'll need at least three pieces of information about an oblique triangle to solve it. The possibilities are as follows:

Case 1. Given one side and two angles.
Case 2. Given an angle, the side opposite, and one other side.
Case 3. Given two sides and the angle between them.
Case 4. Given the three sides.

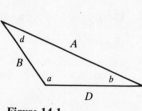
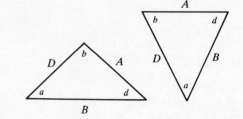

Figure 14-1
Oblique triangles.

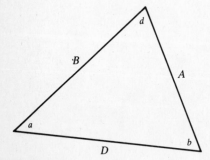

Figure 14-2
Model oblique triangle.

Note: The length of at least one side *must* be among the given information.

14-1. Law of Sines

Use the **law of sines** to solve cases 1 and 2:

LAW OF SINES
$$\frac{A}{\sin a} = \frac{B}{\sin b} = \frac{D}{\sin d} \tag{14-1}$$

where A, B, and D are the lengths of the sides; a, b, and d are the measures of the angles opposite A, B, and D, respectively.

A. Given one side and two angles (Case 1)

The third angle can be found, since $a + b + d = 180°$. If A, B, or D is known, then the ratio in the law of sines can be found and used to calculate the lengths of the other two sides.

EXAMPLE 14-1: Given that $a = 30°$, $b = 80°$, and $A = 5$ as shown in Figure 14-3, solve the triangle.

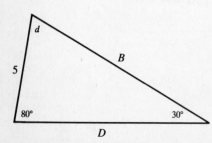

Figure 14-3

Solution: Because $a + b + d = 180°$, $d = 70°$. You know that

$$\frac{A}{\sin a} = \frac{5}{\sin 30°} = \frac{5}{1/2} = 10$$

From the law of sines,

$$\frac{B}{\sin b} = \frac{B}{\sin 80°} = \frac{A}{\sin a} = 10$$

$$B = 10 \sin 80° = 9.8480$$

Also,

$$\frac{D}{\sin d} = \frac{D}{\sin 70°} = \frac{A}{\sin a} = 10$$

$$D = 10 \sin 70° = 9.3969$$

EXAMPLE 14-2: Given that $a = 40°$, $d = 60°$, and $B = 6$ as shown in Figure 14-4, solve the triangle.

Solution: Because $a + b + d = 180°$, $b = 80°$. From the law of sines,

$$\frac{A}{\sin 40°} = \frac{B}{\sin b} = \frac{6}{\sin 80°}$$

$$A = \sin 40° \left(\frac{6}{\sin 80°}\right) = 3.9162$$

Also,

$$\frac{D}{\sin 60°} = \frac{B}{\sin b} = \frac{6}{\sin 80°}$$

$$D = \sin 60° \left(\frac{6}{\sin 80°}\right) = 5.2763$$

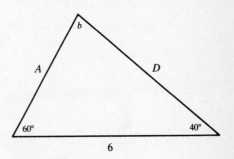

Figure 14-4

B. Given an angle, the side opposite, and one other side (Case 2)

There may be one, two, or no solution triangles for this **ambiguous case**. For example, the triangles shown in Figure 14-5 have the same values of a and B. The ambiguous case of two solutions, or of no solution, will be apparent in the calculations.

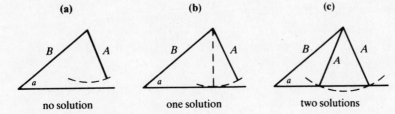

(a) no solution   **(b)** one solution   **(c)** two solutions

Figure 14-5
Ambiguous case.

Suppose you are given angle a and sides A and B. From the law of sines, $\frac{A}{\sin a} = \frac{B}{\sin b}$. Because a, A, and B are known, you can solve for $\sin b \left(\frac{B}{A} \sin a = \sin b\right)$.

- If $\sin b > 1$, then there is no triangle with the given values of a, A, and B; the sine function can never be greater than 1.
- If $\sin b = 1$, $b = 90°$.

- If $\sin b < 1$, then there are two possible values for b: One of the values will be between $0°$ and $90°$, and the other between $90°$ and $180°$. If $a > 90°$, then a value for b between $90°$ and $180°$ isn't reasonable: The sum of the angles in a triangle cannot exceed $180°$. When this happens, discard the obtuse value of b.

  If two acceptable values for b are found, proceed with separate calculations for side D and angle d corresponding to each value of b. In any case, $d = 180° - a - b$, and D is found by the law of sines in the form $\dfrac{A}{\sin a} = \dfrac{D}{\sin d}$.

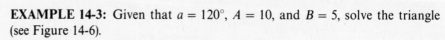

EXAMPLE 14-3: Given that $a = 120°$, $A = 10$, and $B = 5$, solve the triangle (see Figure 14-6).

Solution: From the law of sines,

$$\frac{A}{\sin a} = \frac{B}{\sin b} \quad \text{or} \quad \frac{10}{\sin 120°} = \frac{5}{\sin b}$$

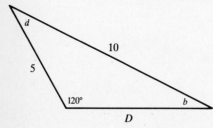

Figure 14-6

Then

$$\sin b = \frac{\sqrt{3}}{4}$$

$$b = \arcsin \frac{\sqrt{3}}{4} = 25.6589° \quad \text{or} \quad 180° - 25.6589° = 154.3411°$$

The second value of b is ignored, since the sum of the angles is greater than $180°$. Using $b = 25.6589°$, $d = 180° - a - b = 34.3411°$. Then $A/\sin a = D/\sin d$, so

$$D = A\frac{\sin d}{\sin a} = 10\frac{\sin 34.3411°}{\sin 120°} = 6.5139$$

Thus $b = 25.6589°$, $d = 34.3411°$, and $D = 6.5139$.

EXAMPLE 14-4: If $a = 45°$, $A = 8$, and $B = 12$, solve the triangle (see Figure 14-7).

Solution: From $B/\sin b = A/\sin a$, you get

$$\frac{12}{\sin b} = \frac{8}{\sin 45°}$$

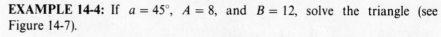

$$\sin b = \frac{3\sqrt{2}}{4} > 1$$

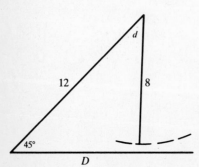

Figure 14-7

No such triangle exists.

EXAMPLE 14-5: Given that $a = 45°$, $A = 10$, and $B = 12$, solve the triangle (see Figure 14-8).

Solution: From $A/\sin a = B/\sin b$, you get

$$\frac{12}{\sin b} = \frac{10}{\sin 45°}$$

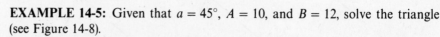

$$\sin b = \frac{3\sqrt{2}}{5} = 0.8485$$

$$b = \arcsin 0.8485 = 58.05° \quad \text{or} \quad 180° - 58.05° = 121.95°$$

Using $b = 58.05°$, $d = 180 - a - b = 76.95°$. Then $D/\sin d = A/\sin a$, or

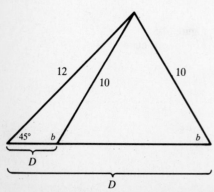

Figure 14-8

$$D = A\frac{\sin d}{\sin a} = 10\frac{\sin 76.95°}{\sin 45°} = 13.78$$

Using $b = 121.95°$, $d = 180° - a - b = 13.05°$. Then $D/\sin d = A/\sin a$, or

$$D = A\frac{\sin d}{\sin a} = 10\frac{\sin 13.05°}{\sin 45°} = 3.19$$

The solutions are $\{b = 58.05°,\ d = 76.95°,\ D = 13.78\}$, and $\{b = 121.95°,\ d = 13.05°,\ D = 3.19\}$.

EXAMPLE 14-6: Solve the triangle with $a = 30°$, $A = 8$, and $B = 4$, shown in Figure 14-9.

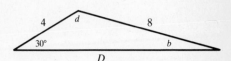

Figure 14-9

Solution: From the law of sines, $B/\sin b = A/\sin a$, or

$$\frac{4}{\sin b} = \frac{8}{\sin 30°}$$

$$\sin b = \frac{1}{4}$$

$$b = \arcsin\frac{1}{4} = 14.48° \quad \text{or} \quad (180 - 14.48)° = 165.52°$$

Discard $b = 165.52°$ since it makes the sum of the angles greater than 180°. Using $b = 14.48°$, $d = 180° - a - b = 135.52°$. Then,

$$\frac{A}{\sin a} = \frac{D}{\sin d}$$

$$D = A\frac{\sin d}{\sin a} = 8\frac{\sin 135.52°}{\sin 30°} = 11.21$$

Thus $b = 14.48°$, $d = 135.52°$, and $D = 11.21$.

14-2. Law of Cosines

Use the **law of cosines**, in any of the following forms, to solve cases 3 and 4.

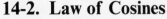

$$A^2 = B^2 + D^2 - 2BD\cos a \qquad \textbf{(14-2a)}$$

LAW OF COSINES

$$B^2 = A^2 + D^2 - 2AD\cos b \qquad \textbf{(14-2b)}$$

$$D^2 = A^2 + B^2 - 2AB\cos d \qquad \textbf{(14-2c)}$$

where A, B, and D are lengths of sides; a, b, and d are the measures of the angles opposite A, B, and D, respectively.

A. Given two sides and the angle between them (Case 3)

Assume that you know a, B, and D. From $A^2 = B^2 + D^2 - 2BD\cos a$, you find A, and from $B^2 = A^2 + D^2 - 2AD\cos b$, you find angle b. Since a is given, once b is known, d can be found from the relationship $a + b + d = 180°$.

EXAMPLE 14-7: Given that $a = 45°$, $B = 3$, and $D = 4$, solve the triangle (see Figure 14-10).

Solution: Use Equation 14-2a:

$$A^2 = B^2 + D^2 - 2BD\cos a$$
$$= 3^2 + 4^2 - 2(3)(4)\cos 45°$$
$$= 25 - 12\sqrt{2} = 8.0294$$
$$A = 2.8336$$

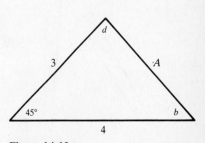

Figure 14-10

Now use Equation 14-2b:

$$B^2 = A^2 + D^2 - 2AD\cos b$$
$$3^2 = 8.0294 + 16 - (2)(2.8336)(4)\cos b$$

or
$$\cos b = \frac{9 - 8.0294 - 16}{-22.6688} = 0.6630$$

$$b = \arccos 0.6630 = 48.4710°$$

Finally, $d = 180° - a - b = 86.5290°$.

B. Given the three sides (Case 4)

If A, B, and D are known, use two forms of the law of cosines to solve for two of the angles. Then find the third angle from $a + b + d = 180°$.

EXAMPLE 14-8: Solve the triangle with $A = 5$, $B = 3$, and $D = 6$, shown in Figure 14-11.

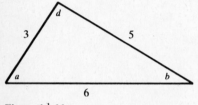

Figure 14-11

Solution: Use Equation 14-2a:

$$A^2 = B^2 + D^2 - 2BD \cos a$$

$$5^2 = 3^2 + 6^2 - 2(3)(6) \cos a$$

so
$$\cos a = \frac{25 - 9 - 36}{-36} = 0.5555$$

$$a = \arccos 0.5555 = 56.2548°$$

Now use Equation 14-2b:

$$B^2 = A^2 + D^2 - 2AD \cos b$$

$$3^2 = 5^2 + 6^2 - 2(5)(6) \cos b$$

or
$$\cos b = \frac{9 - 25 - 36}{-60} = 0.8667$$

$$b = \arccos 0.8667 = 29.9226°$$

Finally, $d = 180° - a - b = 93.8226°$.

14-3. Applications of Oblique Triangles

The examples of this Section illustrate the practical applications of oblique triangles, using the methods introduced above. Your first task in solving these examples is to determine whether you should apply the law of sines or the law of cosines.

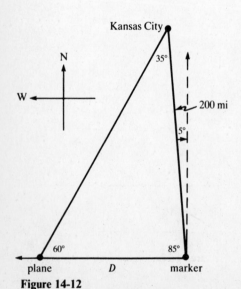

Figure 14-12

EXAMPLE 14-9: An airplane traveling due west passes a navigation marker that is directly below the plane. Kansas City is 200 miles to the northwest of the navigation marker on a line that makes a 5° angle with a north-south line through the marker. After the airplane has traveled for another 15 minutes, the angle of the line from the airplane to Kansas City measured from the east-west line is 60°. Determine the speed of the airplane.

Solution: Begin by making a diagram (see Figure 14-12). This should be your first step in solving any oblique-triangle problem. The angle at Kansas City is $(180 - 60 - 85)° = 35°$. From the law of sines,

$$\frac{D}{\sin 35°} = \frac{200}{\sin 60°}$$

$$D = \frac{200 \sin 35°}{\sin 60°} = 132.46 \text{ mi}$$

The plane covered this distance in 15 minutes (or $\frac{1}{4}$ hour), so the speed is $132.46(4) = 529.8$ miles per hour.

EXAMPLE 14-10: Two pedestrians walk from opposite ends of a city block to a point on the other side of the street (see Figure 14-13). The angle formed by their paths is 25°; one pedestrian walks 300 ft; the other, 320 ft. Determine the length of the city block.

Solution: Since you know two sides and the angle between these sides, use the law of cosines:

$$D^2 = A^2 + B^2 - 2AB\cos a$$
$$= 300^2 + 320^2 - 2(300)(320)\cos 25° = 18388.9$$
$$D = 135.6 \text{ ft}$$

EXAMPLE 14-11: Find an expression for the area of the triangle shown in Figure 14-14 given that B, D, and a are known.

Solution: Altitude H is given by $H = B\sin a$, and the area S of a triangle is given by $\frac{1}{2}$(base)(altitude). In this case,

$$S = \frac{1}{2}DB\sin a \qquad (14\text{-}3)$$

This formula requires that two sides and the angle between them be known.

EXAMPLE 14-12: Find the area of the triangle shown in Figure 14-15.

Solution: Because two sides and the included angle are given, you can use Equation 14-3:

$$S = \frac{4(5)\sin 20°}{2} = 3.4 \text{ cm}^2$$

EXAMPLE 14-13: Find an expression for the area of the triangle shown in Figure 14-14 given that angles a and b and side D are known.

Solution: First, $d = 180° - a - b$. Then, from the law of sines,

$$\frac{B}{\sin b} = \frac{D}{\sin d}$$
$$B = D\frac{\sin b}{\sin d}$$

Substituting this expression for B into Equation 14-3,

$$S = \frac{1}{2}DB\sin a$$
$$= \frac{1}{2}DD\left(\frac{\sin b}{\sin d}\right)\sin a$$
$$= \frac{1}{2}D^2\frac{\sin b\sin a}{\sin d}$$

EXAMPLE 14-14: Find the area of the triangle shown in Figure 14-16.

Solution: Find angle d: $d = (180 - 30 - 50)° = 100°$. Then, from the result of Example 14-13,

$$S = \frac{1}{2}(8^2)\frac{\sin 50°\sin 30°}{\sin 100°} = 12.4 \text{ square units}$$

EXAMPLE 14-15: Find the area of the triangle in Figure 14-14 given that the lengths of the three sides are known.

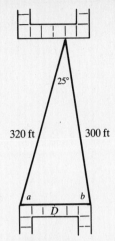

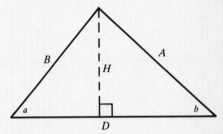

Figure 14-13

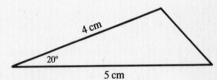

Figure 14-14

Figure 14-15

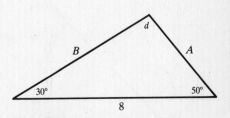

Figure 14-16

Solution: From Example 14-11,

$$S = \frac{1}{2} DB \sin a$$

$$S^2 = \frac{1}{4} D^2 B^2 \sin^2 a$$

$$= \frac{1}{4} D^2 B^2 (1 - \cos^2 a)$$

$$= \left[\frac{1}{2} DB(1 - \cos a) \right] \left[\frac{1}{2} DB(1 + \cos a) \right]$$

Use the law of cosines in the form $\cos a = (B^2 + D^2 - A^2)/2BD$:

$$\frac{1}{2} DB(1 + \cos a) = \frac{1}{2} DB \left(1 + \frac{B^2 + D^2 - A^2}{2BD} \cdot \right)$$

$$= \frac{2DB + B^2 + D^2 - A^2}{4}$$

$$= \frac{(B + D - A)(B + D + A)}{4}$$

Similarly,

$$\frac{1}{2} DB(1 - \cos a) = \frac{(A - B + D)(A + B - D)}{4}$$

Then,

$$S^2 = \frac{(B + D - A)(B + D + A)(A - B + D)(A + B - D)}{16}$$

$$S = \frac{\sqrt{(B + D - A)(B + D + A)(A - B + D)(A + B - D)}}{4}$$

You can express the area in terms of the perimeter, $P = A + B + D$, by Heron's formula:

HERON'S FORMULA $\qquad S = \frac{\sqrt{P(P - 2A)(P - 2B)(P - 2D)}}{4}$ $\qquad$ **(14-4)**

EXAMPLE 14-16: Find the area of the triangle with sides of lengths 2, 3, and 4 inches.

Solution: The perimeter of this triangle is $2 + 3 + 4 = 9$ inches. From Equation 14-4,

$$S = \frac{\sqrt{9(9 - 4)(9 - 6)(9 - 8)}}{4} = 2.9 \text{ in}^2$$

EXAMPLE 14-17: Two cars leave a point at the same time. One car travels directly east at 40 miles per hour, and the other car travels northeast at 50 miles per hour. Find the distance between the two cars after 2 hours (see Figure 14-17).

Solution: From the law of cosines,

$$D^2 = 80^2 + 100^2 - 2(80)(100) \cos 45°$$

$$= 5086.3$$

$$D = 71.3 \text{ mi}$$

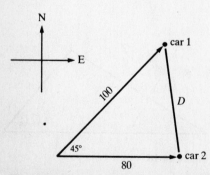

Figure 14-17

EXAMPLE 14-18: Two photographers, standing 300 feet apart, spot a lion in the distance. The lines of sight from the photographers to the lion and to each other form a triangle, as shown in Figure 14-18. Find the distance from each photographer to the lion.

Solution: The third angle is $(180 - 60 - 75)° = 45°$. Then,

$$\frac{D2}{\sin 60°} = \frac{300}{\sin 45°}$$

$$D2 = \frac{300 \sin 60°}{\sin 45°} = 367.4 \text{ ft}$$

and

$$D1 = \frac{300 \sin 75°}{\sin 45°} = 409.8 \text{ ft}$$

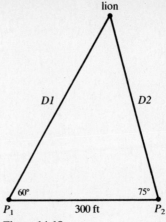

Figure 14-18

EXAMPLE 14-19: The distance from Sacramento to Davis is 16 miles and the distance from Davis to Woodland is 11 miles with the angles as shown in Figure 14-19; find the distance from Sacramento to Woodland and angle α.

Solution: From the law of cosines,

$$D^2 = 11^2 + 16^2 - 2(11)(16) \cos 80° = 315.87$$

$$D = 17.8 \text{ mi}$$

From the law of sines,

$$\frac{\sin \alpha}{11} = \frac{\sin 80°}{17.8}$$

$$\sin \alpha = \frac{11 \sin 80°}{17.8} = 0.6086$$

$$\alpha = \arcsin 0.6086 = 37.5°$$

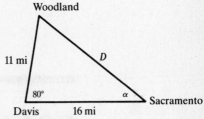

Figure 14-19

SUMMARY

1. Oblique triangles can be solved, that is, the lengths of the sides and the measure of the angles found, whenever the length of at least one side and any two other pieces of information are given.

2. The law of sines is $\dfrac{A}{\sin a} = \dfrac{B}{\sin b} = \dfrac{D}{\sin d}$ where A, B, and D are lengths of the sides and a, b, and d are the angles opposite them, respectively.

3. Oblique triangles in which two angles and any side are given can be solved by finding the third angle by subtraction and then finding the sides by using the law of sines.

4. Oblique triangles in which two sides and the angle opposite one of the sides are given can be solved by the law of sines. There may be no solutions, one solution, or two solutions in this case, as shown in the following chart.

Value of a	A and B	Value of $\sin b$	Solutions
$a \geqslant 90°$ or $\dfrac{\pi}{2}$	$A \leqslant B$	$\dfrac{B}{A} \sin a \geqslant 1$	none
	$A > B$	$\dfrac{B}{A} \sin a < 1$	one
$a < 90°$ or $\dfrac{\pi}{2}$	$A \geqslant B$	$\dfrac{B}{A} \sin a < 1$	one
	$A < B$	$\dfrac{B}{A} \sin a \begin{cases} >1 \\ =1 \\ <1 \end{cases}$	none one$(b = 90°)$ two

5. The law of cosines can be expressed in any of the following forms:

$$A^2 = B^2 + D^2 - 2BD\cos a$$

$$B^2 = A^2 + D^2 - 2AD\cos b$$

$$D^2 = A^2 + B^2 - 2AB\cos d$$

RAISE YOUR GRADES

Can you ...?

☑ use the law of sines
☑ use the law of cosines
☑ determine the number of solutions for the ambiguous case
☑ apply the law of sines and cosines to practical applications
☑ find the area of an oblique triangle
☑ use Heron's formula

SOLVED PROBLEMS

PROBLEM 14-1   Derive the law of sines.

Solution: Consider the triangle in Figure 14-20. By drawing a perpendicular (labeled *H*) from the vertex of angle *d* to the side joining vertices *a* and *b*, you form two right triangles. You see that $\sin a = H/B$, or $H = B\sin a$, and $\sin b = H/A$, or $H = A\sin b$. Equating these two expressions for *H*, you get $B\sin a = A\sin b$, or $\sin a/A = \sin b/B$. In the same way, you can show that $\sin d/D = \sin a/A$.

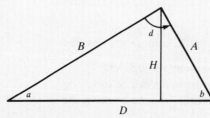

Figure 14-20

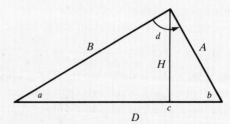

Figure 14-21

PROBLEM 14-2   Derive the law of cosines.

Solution: Consider the triangle in Figure 14-21. By drawing a perpendicular (labeled *H*) from the vertex of angle *d* to the line joining vertices *a* and *b*, you form two right triangles. Again, $\sin a = H/B$, or $H^2 = B^2\sin^2 a$. Then, using $\sin^2 a + \cos^2 a = 1$, you get $H^2 = B^2(1 - \cos^2 a)$. From right triangle *cdb*, you get $H^2 = A^2 - (D - B\cos a)^2$. Equating the right-hand sides of the two previous expressions, you get

$$A^2 - (D - B\cos a)^2 = B^2(1 - \cos^2 a)$$

$$A^2 - D^2 + 2BD\cos a - B^2\cos^2 a = B^2 - B^2\cos^2 a$$

By simplifying this expression, you get

$$A^2 = B^2 + D^2 - 2BD\cos a$$

one form of the law of cosines.

PROBLEM 14-3 A tree grows in the median of a straight divided highway. From one edge of the highway, the angle of elevation to the top of the tree is 10°; from the other edge, 8°. The highway is 300 meters wide, how tall is the tree and how far is the tree from the edges of the road?

Solution: Draw a diagram to represent the situation. We'll let x be the side of the triangle opposite the 10° angle (see Figure 14-22). The third angle of the triangle is $(180 - 10 - 8)° = 162°$. Then, from the law of sines,

$$\frac{300}{\sin 162°} = \frac{x}{\sin 10°}$$

$$x = \frac{300 \sin 10°}{\sin 162°} = 168.6 \text{ m}$$

Let d be the distance from the edge of the road at the 8° angle to the tree, and let h be the height of the tree. Then, $d/168.6 = \cos 8°$, or $d = 168.6 \cos 8° = 166.9$ m. From the same right triangle, $h/168.6 = \sin 8°$, so the height of the tree is $h = 168.6 \sin 8° = 23.5$ m. The distance of the tree from the other edge is $300 - 166.9 = 133.1$ m. [See Section 14-1.]

Figure 14-22

PROBLEM 14-4 A vertical flagpole is attached to the top edge of a building. A man stands 400 feet from the base of the building. From his viewpoint, the angle of elevation to the bottom of the flagpole is 60°; to the top, 62.5°. Determine the height of the flagpole.

Solution: Construct a diagram of the situation (see Figure 14-23). Let h represent the height of the flagpole and b the height of the building. You can see that $\tan 60° = b/400$ and $\tan 62.5° = (b + h)/400$. From the first equation, $b = 400 \tan 60° = 692.8$ ft. Then, from the second equation, $692.8 + h = 400 \tan 62.5°$, or $692.8 + h = 768.4$ and $h = 75.6$ ft. You were able to solve this problem without the law of sines or the law of cosines. [See Section 5-2.]

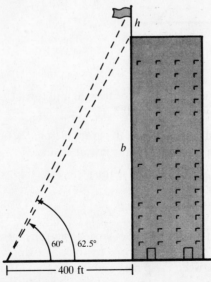

Figure 14-23

PROBLEM 14-5 A vertical diving platform is supported by two wires running from the top of the platform to the ground on opposite sides of the platform. One wire is 60 feet long and makes an angle of 36° with the ground; the second makes an angle of 40° with the ground.. Determine the length of the second wire.

Solution: Construct a diagram (see Figure 14-24). You need to find x, the length of the second wire. From the law of sines,

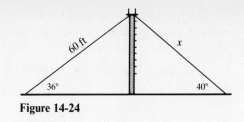

$$\frac{60}{\sin 40°} = \frac{x}{\sin 36°}$$

$$x = \frac{60 \sin 36°}{\sin 40°} = 54.9 \text{ ft} \qquad \text{[See Section 14-1.]}$$

Figure 14-24

PROBLEM 14-6 The roof of a leanto makes an angle of 80° with level ground and is supported by a pole 15 feet in length. The pole extends from the top edge of the roof and makes an angle of 53° with the ground. Determine the distance from the base of the leanto to the bottom of the pole.

Solution: In the triangle of Figure 14-25, you must find x. Since the sum of the angles of the triangle is 180°, the third angle is $(180 - 80 - 53)° = 47°$. From the law of sines,

$$\frac{x}{\sin 47°} = \frac{15}{\sin 80°}$$

$$x = \frac{15 \sin 47°}{\sin 80°} = 11.1 \text{ ft} \qquad \text{[See Section 14-1.]}$$

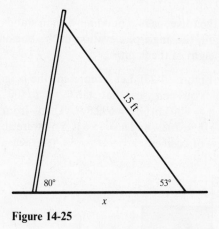

Figure 14-25

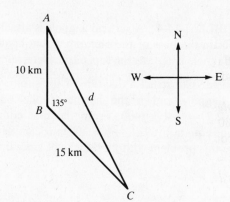

Figure 14-26

PROBLEM 14-7 Town A is 10 km north of town B and town C is 15 km southeast of town B. What is the distance from town A to town C?

Solution: Since town C is southeast of town B, the angle from the north to the line joining town B to town C is 135° (see Figure 14-26). From the law of cosines,

$$d^2 = 10^2 + 15^2 - 2(10)(15) \cos 135° = 537.13$$

so $d = \sqrt{537.13} = 23.2$ km. $\qquad$ [See Section 14-2.]

PROBLEM 14-8 A woman holds one end of a rope that runs through a pulley and has a weight attached to the other end. The section of rope between the woman and the pulley is 20 feet in length; the section of rope between the pulley and the weight is 10 feet in length. The rope bends through an angle of 32° in the pulley. How far is the woman from the weight?

Solution: You must find the length of side x of the triangle in Figure 14-27. From the law of cosines,

$$x^2 = 10^2 + 20^2 - 2(10)(20) \cos 32° = 160.78$$

so $x = \sqrt{160.78} = 12.7$ ft. $\qquad$ [See Section 14-2.]

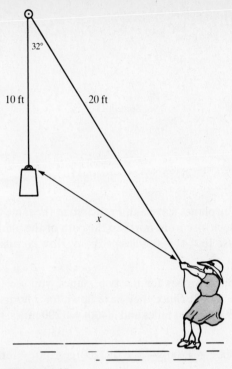

Figure 14-27

PROBLEM 14-9 A child holds two balloons on strings. The angle of elevation to the balloon on her right is 20°; the string is 60 feet in length. The angle of elevation to the balloon on her left is 26°, and the string is 75 feet in length. What is the distance between the two balloons?

Solution: The angle between the strings is $(180 - 26 - 20)° = 134°$ (see Figure 14-28). From the law of cosines,

$$d^2 = 75^2 + 60^2 - 2(75)(60) \cos 134° = 15\,476.925$$

so $d = 124.4$ ft.   [See Section 14-2.]

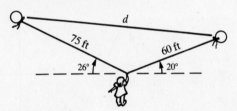

Figure 14-28

PROBLEM 14-10 Solve for side A in Figure 14-29.

Solution: The third angle is

$$(180 - 20 - 10)° = 150°.$$

From the law of sines, $8/\sin 150° = A/\sin 10°$, or $A = 8 \sin 10°/\sin 150° = 2.8$.

[See Section 14-1.]

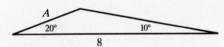

Figure 14-29

PROBLEM 14-11 Solve for angle θ in Figure 14-30.

Solution: From the law of cosines,

$$1.5^2 = 2^2 + 3^2 - 2(2)(3) \cos \theta$$

$$\cos \theta = \frac{1.5^2 - 2^2 - 3^2}{-2(2)(3)} = 0.8958$$

$$\theta = \cos^{-1}(0.8958) = 26.4°$$

[See Section 14-2.]

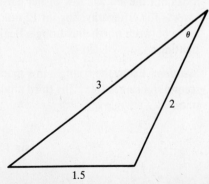

Figure 14-30

PROBLEM 14-12   Solve for side *A* in Figure 14-31.

Solution: Apply the law of cosines:

$$A^2 = 3^2 + 7^2 - 2(3)(7)\cos 40° = 25.826$$

$$A = 5.1$$

[See Section 14-2.]

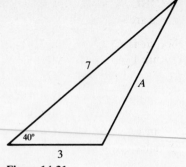

Figure 14-31

PROBLEM 14-13   Two airplanes leave San Francisco at the same time, one traveling on course 30° (the angle measured clockwise from north to the path of the plane) at 300 miles per hour and the other traveling on course 130° at 400 miles per hour. How far apart are the two airplanes after 3 hours?

Solution: By using the course values for the two planes, you see that the angle between their courses is 100° (see Figure 14-32). Since they have flown for 3 hours, the distances of the planes from San Francisco are 3(300) = 900 miles and 3(400) = 1200 miles, respectively. Apply the law of cosines:

$$d^2 = 900^2 + 1200^2 - 2(900)(1200)\cos 100° = 2\,625\,080.1$$

$$d = 1620.2 \text{ mi}$$

[See Section 14-2.]

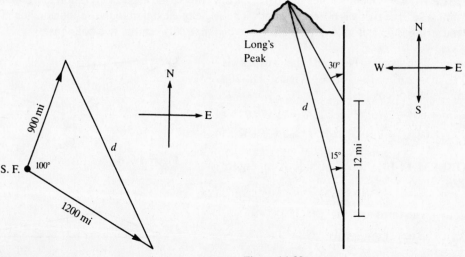

Figure 14-32   **Figure 14-33**

PROBLEM 14-14   A woman is driving north along a straight road in central Colorado. She looks out the left window of her car and sees Long's Peak. The angle between north and Long's Peak is 15°. After traveling for 12 miles, she looks out the window again and sees Long's Peak; the angle between north and Long's Peak is now 30°. How far was she from Long's Peak at the first sighting?

Solution: From the triangle in Figure 14-33, you can see that the angle inside the triangle at the second sighting is 150°. The third angle of the triangle is (180 − 150 − 15)° = 15°. From the law of sines,

$$\frac{d}{\sin 150°} = \frac{12}{\sin 15°}$$

$$d = \frac{12\sin 150°}{\sin 15°} = 23.2 \text{ mi}$$

[See Section 14-1.]

PROBLEM 14-15 A parallelogram has acute angles of 50°. The sides on either side of an acute angle are of lengths 30 and 20 centimeters. Find the lengths of the diagonals of the parallelogram.

Solution: You can apply the law of cosines to find the length, d, of the diagonal shown in Figure 14-34:

$$d^2 = 20^2 + 30^2 - 2(20)(30)\cos 50° = 528.65$$

$$d = 23 \text{ cm}$$

From geometry, the sum of the angles of a quadrilateral is 360°. The other two angles of the parallelogram are $(360 - 100)/2 = 130°$. Apply the law of cosines to find the length of side l:

$$l^2 = 20^2 + 30^2 - 2(20)(30)\cos 130° = 2071.35$$

$$l = 45.5 \text{ cm} \hspace{2cm} \text{[See Section 14-2.]}$$

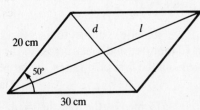

Figure 14-34

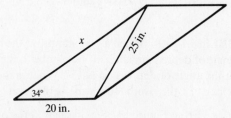

Figure 14-35

PROBLEM 14-16 A parallelogram has acute angles of 34°. The diagonal opposite these acute angles is 25 inches in length. A side of the parallelogram adjacent to the 34° angle is 20 inches long. Find the length of the other adjacent side.

Solution: Apply the law of cosines to the triangle shown in Figure 14-35:

$$25^2 = x^2 + 20^2 - 2(x)(20)\cos 34°$$

$$625 = x^2 + 400 - 33.2x$$

$$x^2 - 33.2x - 225 = 0$$

Use the quadratic formula:

$$x = \frac{-b \pm \sqrt{b^2 - 4ac}}{2a}$$

$$= \frac{-(-33.2) \pm \sqrt{(-33.2)^2 - 4(1)(-225)}}{2(1)} = 39.0 \text{ or } -5.77 \text{ in}$$

You must reject the negative value since x represents a distance. The other side is of length 39.0 inches. [See Section 14-2.]

PROBLEM 14-17 Two observers are looking at one of the World Trade Center buildings from New York Harbor. The distance between the two observers is 0.5 mi, and the angle between a line from observer 1 to observer 2 and a line from observer 1 to the World Trade Center building is 48°. Observer 1 is 3 miles from the building. How far is observer 2 from the building?

Solution: Apply the law of cosines to the triangle shown in Figure 14-36:

$$d^2 = 3^2 + 0.5^2 - 2(3)(0.5)\cos 48° = 7.24$$

$$d = 2.7 \text{ mi} \hspace{1.5cm} \text{[See Section 14-2.]}$$

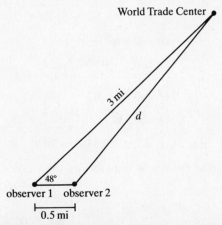

Figure 14-36

PROBLEM 14-18 The angle of elevation from an observer to the top of a building is 23°. The building is at the top of a hill that is angled upward at 21°. It is 1 mile from the observer up the hill to the bottom of the building. How tall is the building?

Solution: Let the bottom of the building be denoted by B, the top by T, the line through B parallel to the ground by L, and the position of the observer by O (see Figure 14-37). You see that angle $BOT = (23 - 21)° = 2°$, angle $LBM = 21°$, angle $MBT = (90 - 21)° = 69°$, angle $TBO = (180 - 69)° = 111°$, and angle $BTO = (180 - 2 - 111)° = 67°$. From the law of sines,

$$\frac{h}{\sin 2°} = \frac{1}{\sin 67°}$$

$$h = \frac{\sin 2°}{\sin 67°} = 0.038 \text{ mi}$$

This is approximately $0.038(5280) = 200.6$ ft.

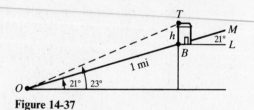

Figure 14-37

[See Section 14-1.]

PROBLEM 14-19 When a boy stands on the bank of a river and looks across to the other bank, the angle of depression is 12°. If he climbs to the top of a 10-foot tree and looks across to the other bank, the angle of depression is 15°. What is the distance from the first position of the boy to the other bank of the river? How wide is the river?

Solution: From Figure 14-38, you see that the oblique triangle formed by the tree, the line-of-sight from the top of the tree to a point on the other bank, and the line-of-sight from the bottom of the tree to the other bank has angles of 75°, 102°, and 3°. Call the distance from the boy's first position to the other side x. From the law of sines,

$$\frac{10}{\sin 3°} = \frac{x}{\sin 75°}$$

$$x = \frac{10 \sin 75°}{\sin 3°} = 184.56 \text{ ft}$$

Call the distance across the river D. Then,

$$\cos 12° = \frac{D}{184.56}$$

$$D = 184.56 \cos 12° = 180.53 \text{ ft}$$

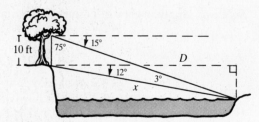

Figure 14-38

[See Section 14-2.]

PROBLEM 14-20 Two observers are watching the Goodyear Blimp as it passes overhead. The observers are 1.2 miles apart. The distance from the first observer to the blimp is 0.6 miles; the distance from the second observer to the blimp is 0.7 miles. Find the altitude of the blimp and the angles of elevation from each observer to the blimp.

Solution: The situation is shown in Figure 14-39. Use the law of cosines:

$$0.7^2 = 0.6^2 + 1.2^2 - 2(0.6)(1.2)\cos a$$

$$\cos a = 0.9097$$

$$a = \cos^{-1} 0.9097 = 24.5°$$

$$0.6^2 = 0.7^2 + 1.2^2 - 2(0.7)(1.2)\cos b$$

$$\cos b = 0.9345$$

$$b = \cos^{-1} 0.9345 = 20.9°$$

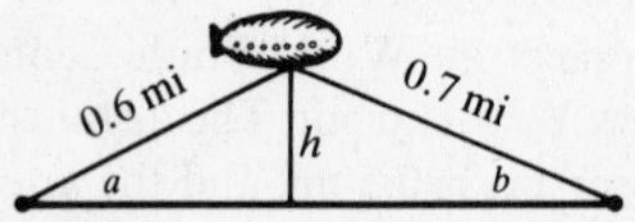

Figure 14-39

Use the law of sines to find the altitude:

$$\sin 24.5° = \frac{h}{0.6}$$

$$h = 0.6 \sin 24.5° = 0.25 \text{ mi, or } 5280(0.25) = 1320 \text{ ft} \qquad \text{[See Section 14-2.]}$$

PROBLEM 14-21   Show that the area of any quadrilateral equals one-half the product of the diagonals and the sine of either of the angles between the diagonals.

Solution: Recall that a quadrilateral is a closed figure with four sides. Rectangles and parallelograms are examples of quadrilaterals. The triangle outlined in Figure 14-40 has sides of length $L/2$ and $M/2$, where L and M are the lengths of the two diagonals, and angle θ as shown. The altitude a of the triangle is $(L/2) \sin \theta$. The area is

$$S = \frac{1}{2}\left(\frac{M}{2}\right)\left(\frac{L}{2}\right)\sin \theta = \frac{LM}{8}\sin \theta$$

There are two triangles with this area. Using the same reasoning, the other two triangles each have area

$$S = \frac{1}{2}\left(\frac{L}{2}\right)\left(\frac{M}{2}\right)\sin(180° - \theta) = \frac{LM}{8}\sin \theta.$$

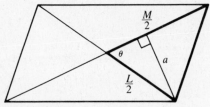

Figure 14-40

Adding the areas of the four triangles together, you get area of the quadrilateral:

$$\frac{4LM}{8}\sin \theta = \frac{LM}{2}\sin \theta \qquad \text{[See Section 14-3.]}$$

PROBLEM 14-22   Find the area of a triangular field with sides of length 150 feet, 250 feet, and 300 feet.

Solution: Recall that the area of a triangle with three known sides is given by $S = \sqrt{P(P - 2A)(P - 2B)(P - 2D)}/4$. In this problem $P = 150 + 250 + 300 = 700$ ft. Let $A = 150$, $B = 250$, and $D = 300$. Then,

$$S = \frac{\sqrt{700(700 - 2 \cdot 150)(700 - 2 \cdot 250)(700 - 2 \cdot 300)}}{4} = 18\,708.3 \text{ ft}^2$$

[See Section 14-3.]

PROBLEM 14-23   A motorcyclist drives 52.34 miles due south. At that point he makes a turn and drives for another 30.72 miles. Find the angle between the initial and final directions of the motorcyclist if the angle measured clockwise from due south to the line running from the starting position to the ending position of the motorcyclist is 32°.

Solution: The sides of the triangle in Figure 14-41 labeled A and B are the lines of travel of the motorcyclist. The side labeled C is the line from the starting position to the ending position of the motorcyclist. Apply the law of sines to find θ:

$$\frac{\sin \theta}{52.34} = \frac{\sin 32°}{30.72}$$

$$\sin \theta = \frac{52.34 \sin 32°}{30.72} = 0.9029$$

$$\theta = \sin^{-1} 0.9029 = 64.5°$$

Angle α is the third angle of the triangle which has measure $(180 - 32 - 64.5)° = 83.5°$.

[See Section 14-1.]

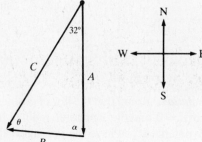

Figure 14-41

PROBLEM 14-24   In Dodger Stadium the straight-line distance from home plate through second base to seat A in the first row of the stands is 400 feet. The bases lie on a square, 90 feet on each side; find the distance from seat A to third base.

Solution: The situation is shown in Figure 14-42. From the law of cosines,

$$x^2 = 90^2 + 400^2 - 2(90)(400)\cos 45° = 117\,188.31$$

$$x = 342.33 \text{ ft} \qquad \text{[See Section 14-2.]}$$

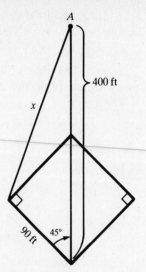

Figure 14-42

PROBLEM 14-25 Consider Figure 14-43. Find *D*.

Solution: Apply the law of cosines:

$$D^2 = 20^2 + 21^2 - 2(20)(21)\cos 100° = 986.9$$

$$D = 31.4$$

[See Section 14-2.]

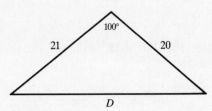

Figure 14-43

PROBLEM 14-26 Solve the triangle in Figure 14-44.

Solution: You see that $d = (180 - 40 - 85)° = 55°$. Apply the law of sines:

$$\frac{7}{\sin 40°} = \frac{D}{\sin 55°}$$

$$D = \frac{7\sin 55°}{\sin 40°} = 8.9$$

Apply the law of sines again:

$$\frac{7}{\sin 40°} = \frac{B}{\sin 85°}$$

$$B = 10.8$$

[See Section 14-1.]

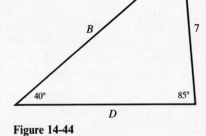

Figure 14-44

PROBLEM 14-27 Show that the law of cosines reduces to the Pythagorean Theorem in the case of a right triangle.

Solution: Recall that the law of cosines can be stated as $D^2 = A^2 + B^2 - 2AB\cos d$. For $d = 90°$, $\cos d = 0$, and the expression reduces to $D^2 = A^2 + B^2$. Since side *D*, opposite the 90° angle, is the hypotenuse, you have the Pythagorean theorem. [See Section 14-2.]

Supplementary Exercises

PROBLEM 14-28 Cleveland is 390 miles from Evansville. At Evansville, the clockwise angle between due north and the line running from Evansville to Cleveland is 51°. Lexington is 280 miles due east of Evansville. How far is Lexington from Cleveland?

PROBLEM 14-29   A boat traveling 20 miles per hour leaves the shore of a lake and travels for 1 hour. The boat then turns through a counterclockwise angle of 10° and proceeds at the same speed on this path for 15 minutes before stopping. What is the distance from the starting position of the boat to its ending position?

PROBLEM 14-30   A vacationer observes that the angle of elevation of a mountain peak from his cottage is 18°. He leaves the cottage and walks 2000 feet up a slope of 10° toward the mountain and then finds the angle of elevation of the peak to be 31°. What is the distance from the cottage to the mountain peak?

PROBLEM 14-31   Two angles of a triangle are 58° 46′ and 57° 18′. The largest side is 93.63 feet long; find the length of the smallest side.

PROBLEM 14-32   One diagonal of a parallelogram makes angles of 28° 19′ and 32° 41′ with the sides. The length of the diagonal is 84.56 inches; how long are the sides?

For Problems 14-33 through 14-37, assume that the triangle has angles a, b, and d, and sides of length A, B, and D, respectively.

PROBLEM 14-33 $A = 60$ feet, $B = 70$ feet, and $D = 90$ feet; find angle d.

PROBLEM 14-34 $A = 634.9$ inches, $a = 37° 47′$, and $d = 38° 47′$; find D.

PROBLEM 14-35 $a = 60°$, $A = 6$ inches, and $B = 4$ inches; solve the triangle.

PROBLEM 14-36 $a = 75°$, $B = 5$ meters, and $D = 8$ meters; solve the triangle.

PROBLEM 14-37 $A = 7$ cm, $B = 4$ cm, and $D = 10$ centimeters; solve the triangle.

PROBLEM 14-38   Two planes take off from the same airport. The first plane flies on a course of 234°. The second plane flies on a course of 178°. After the first plane flies 478 kilometers, the bearing from the first plane to the airport is 30° and to the second plane is 40°. How far is the second plane from the airport?

PROBLEM 14-39   A diagonal of a parallelogram is 50 centimeters long and forms angles of 35° and 18° with the sides at one end. Find the lengths of the sides.

PROBLEM 14-40   Two sides of a parallelogram are 9 inches and 13 inches in length and have an included angle of 72°. Find the lengths of the diagonals.

PROBLEM 14-41   A man standing at the bottom of a hill of slope 17° sights a tower on top of the hill. The angle of elevation from the man to the top of the tower is 37°. At a point 1500 feet up the side of the hill the angle of elevation to the top of the tower is 50°. Find the distance from the bottom of the hill to the top of the tower (see Figure 14-45).

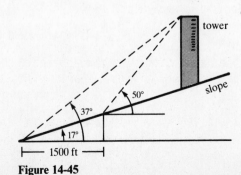

Figure 14-45

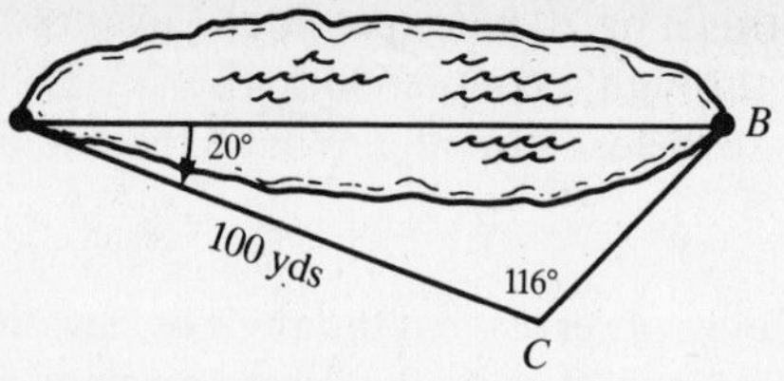

Figure 14-46

PROBLEM 14-42 The opposite ends of a lake are marked as points *A* and *B* in Figure 14-46. A man on the side of the lake makes the following measurements: the distance from *A* to *C* is 100 yards, angle *CAB* is 20°, and angle *ACB* is 116°. Find the length of the lake.

PROBLEM 14-43 A woman wants to measure the width of a river. She sees a tree directly across the river. She then walks 1200 feet along the straight bank of the river to a point from which the angle from the bank to the tree is 40°. Find the width of the river.

PROBLEM 14-44 A pole 35 feet tall makes an angle of 15° with the vertical. What is the minimum length of rope that will reach from the top of the pole to the ground?

PROBLEM 14-45 Find the area of a triangular field having angles 112° 20′ and 47° 30′ with included side of 140 feet.

PROBLEM 14-46 Find the lengths of the sides of a parallelogram with diagonals of length 40 centimeters and 70 centimeters and an included angle of 108°.

PROBLEM 14-47 Two airplanes leave an airport one hour apart. One flies northeast at 500 miles per hour, and the other flies directly south at 400 miles per hour. How far apart are they after the first plane has traveled for 3 hours?

PROBLEM 14-48 How many triangles can be found given that $A = 4$, $B = 10$, and $a = 30°$?

PROBLEM 14-49 Solve the model oblique triangle with $b = 134°$, $B = 526$, and $D = 481$.

PROBLEM 14-50 Solve the model oblique triangle with $A = 9$, $B = 10$, and $d = 60°$.

Answers to Supplementary Exercises

14-28: 246.52 mi

14-29: 24.94 mi

14-30: 3186.2 ft

14-31: 87.72 ft

14-32: 45.86 and 52.21 in

14-33: 87.3°

14-34: 649.1

14-35: $d = 84.7°$, $b = 35.3°$, $D = 6.90$

14-36: $A = 8.3$ m, $b = 35.6°$, $d = 69.4°$

14-37: $b = 18.2°$, $a = 33.1°$, $d = 128.7°$

14-38: 90.9 km

14-39: 19.3 and 35.9

14-40: 13.3 and 18.0

14-41: 3631.7 ft

14-42: 129.4 yd

14-43: 1006.9 ft

14-44: 33.8 ft

14-45: 19 385.9 ft²

14-46: 34.5 cm and 45.4 cm

14-47: 2141.7 mi

14-48: None—sin $b > 1$

14-49: $d = 41.1°$, $a = 4.9°$, $A = 62.5$

14-50: $D = 9.5$, $a = 55.13°$, $b = 64.88°$

15 *POLAR COORDINATES*

THIS CHAPTER IS ABOUT

☑ **The Polar Coordinate System**
☑ **Relating Polar and Cartesian Coordinates**
☑ **Polar Equations and Graphs**

15-1. The Polar Coordinate System

So far we've used only Cartesian coordinates to locate a point in a plane. Another important method is the **polar coordinate system**, defined by a fixed point O, the **origin** or **pole**, and a fixed initial ray, the **polar axis** or **polar line**, with its vertex at point O. You represent a point P by an ordered pair of real numbers (r, θ), where r is the **radius vector**—the directed distance from the pole to point P—and θ is the **vectorial angle**—the directed angle in radians from the polar axis to line segment OP (see Figure 15-1a). You write $P = (r, \theta)$ or $P(r, \theta)$ to specify the **polar coordinates** of P and you locate them as follows:

1. If $r > 0$, place the point at distance r from the pole on the angle θ ray.
2. If $r < 0$, place the point at distance r from the pole on the angle $\theta + \pi$ ray (see Figure 15-1b).
3. If $r = 0$, then P is the pole and θ is arbitrary.

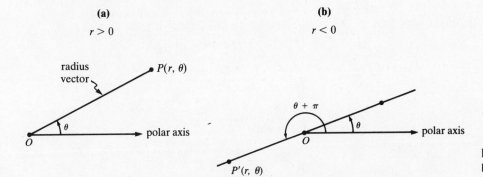

(a)
$r > 0$

(b)
$r < 0$

Figure 15-1
Polar coordinates.

Note that the radius vector is *positive* when measured on the terminal side of the vectorial angle (line segment OP) and *negative* when measured on the terminal side of angle $\theta + \pi$ (line segment OP'). The vectorial angle is positive when generated by counterclockwise rotation from the polar axis and negative when generated by clockwise rotation. Figure 15-2 shows equivalent ways of describing the position of point P, depending on the direction of measure of angle θ.

You can describe all of the possible polar coordinates of $P(r, \theta)$, including those that result from the periodic nature of θ, as follows:

$$P(r, \theta) = \begin{cases} (r, \theta + 2n\pi) \\ \text{and} \\ (-r, \theta + (2n+1)\pi) \end{cases} \Bigg| \; n \text{ any integer} \Bigg\} \qquad \textbf{(15-1)}$$

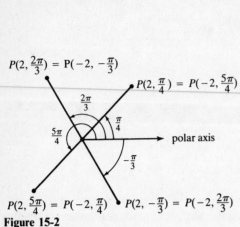

Figure 15-2
Polar coordinates of $P(2, \theta)$.

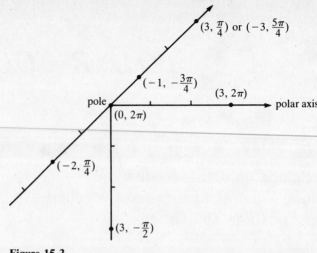

Figure 15-3

EXAMPLE 15-1: Plot the points in the $r\theta$-plane whose polar coordinates are
(a) $(3, 2\pi)$; **(b)** $\left(3, \frac{1}{4}\pi\right)$; **(c)** $\left(3, -\frac{1}{2}\pi\right)$; **(d)** $\left(-1, -\frac{3}{4}\pi\right)$; **(e)** $\left(-2, \frac{1}{4}\pi\right)$; **(f)** $(0, 2\pi)$;
(g) $\left(-3, \frac{5}{4}\pi\right)$.

Solution: The points are shown in Figure 15-3. Note that $\left(3, \frac{1}{4}\pi\right)$ and $\left(-3, \frac{5}{4}\pi\right)$ coincide.

EXAMPLE 15-2: Write all possible polar coordinates for $P\left(2, \frac{1}{4}\pi\right)$. Plot the points for which $-2\pi \leqslant \theta \leqslant 2\pi$ on the $r\theta$-plane.

Solution: From Equation 15-1,

$$P\left(2, \frac{\pi}{4}\right) = \left\{ \begin{array}{l} \left(2, \dfrac{\pi}{4} + 2n\pi\right) \\[2ex] \text{and} \\[2ex] \left(-2, \dfrac{\pi}{4} + (2n+1)\pi\right) \end{array} \right| \left. n \text{ any integer} \right\}$$

$$= \left\{ \begin{array}{l} \left(2, \dfrac{\pi}{4} + \dfrac{8n\pi}{4}\right) \\[2ex] \text{and} \\[2ex] \left(-2, \dfrac{\pi}{4} + \dfrac{4\pi(2n+1)}{4} \cdots \right. \end{array} \right.$$

$$= \left\{ \begin{array}{l} \left(2, \dfrac{8n\pi + \pi}{4}\right) \\[2ex] \text{and} \\[2ex] \left(-2, \dfrac{8n\pi + 4\pi + \pi}{4}\right) \cdots \end{array} \right.$$

$$= \left\{ \begin{array}{l} \left(2, \dfrac{\pi(8n+1)}{4}\right) \\[2ex] \text{and} \\[2ex] \left(-2, \dfrac{\pi(8n+5)}{4}\right) \end{array} \right| \left. n \text{ any integer} \right\}$$

For $n = 0$, $P(2,\tfrac{1}{4}\pi) = (2,\tfrac{1}{4}\pi)$ and $(-2,\tfrac{5}{4}\pi)$.

For $n = -1$, $P(2,\tfrac{1}{4}\pi) = (2, -\tfrac{7}{4}\pi)$ and $(-2, -\tfrac{3}{4}\pi)$.

Figure 15-4 shows the points plotted on a polar coordinate system. You should note that they all coincide.

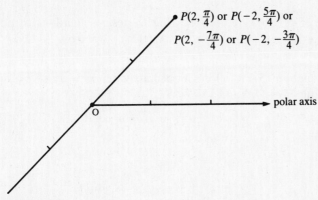

$P(2, \tfrac{\pi}{4})$ or $P(-2, \tfrac{5\pi}{4})$ or

$P(2, -\tfrac{7\pi}{4})$ or $P(-2, -\tfrac{3\pi}{4})$

polar axis

Figure 15-4

15-2. Relating Polar and Cartesian Coordinates

Every point in a plane can be described by both Cartesian and polar coordinates. If point P has Cartesian coordinates (x, y) and polar coordinates (r, θ) and the positive x axis of the Cartesian system and the polar axis of the polar coordinate system coincide, then the coordinates in the two systems can be interconverted as follows:

POLAR TO CARTESIAN $\quad x = r\cos\theta \qquad y = r\sin\theta$ **(15-2)**

CARTESIAN TO POLAR $\quad r^2 = x^2 + y^2 \qquad \theta = \tan^{-1}\dfrac{y}{x}$ **(15-3)**

EXAMPLE 15-3: Find the Cartesian coordinates (x, y) of the point P with polar coordinates $(2,\tfrac{1}{3}\pi)$.

Solution: From Equations 15-2,

$$x = r\cos\theta = 2\cos\frac{\pi}{3} = 2\left(\frac{1}{2}\right) = 1$$

$$y = r\sin\theta = 2\sin\frac{\pi}{3} = \frac{2\sqrt{3}}{2} = \sqrt{3}$$

The relationship is shown in Figure 15-5.

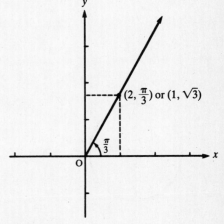

$(2, \tfrac{\pi}{3})$ or $(1, \sqrt{3})$

Figure 15-5

EXAMPLE 15-4 Find two sets of polar coordinates for the point with Cartesian coordinates $(-1, 1)$.

Solution: You want values of r and θ such that

$$r^2 = x^2 + y^2 = (-1)^2 + 1^2 = 1 + 1 = 2$$

$$r = \pm\sqrt{2}$$

and

$$\theta = \tan^{-1}\frac{y}{x} = \tan^{-1}\frac{1}{-1} = -\frac{\pi}{4}$$

Two sets of polar coordinates for $(-1, 1)$ are $(-\sqrt{2}, -\tfrac{1}{4}\pi)$ and $(\sqrt{2}, \pi - \tfrac{1}{4}\pi) = (\sqrt{2}, \tfrac{3}{4}\pi)$ (see Figure 15-6).

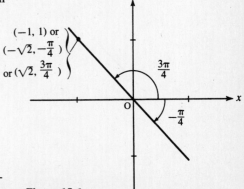

$(-1, 1)$ or
$(-\sqrt{2}, -\tfrac{\pi}{4})$
or $(\sqrt{2}, \tfrac{3\pi}{4})$

$\tfrac{3\pi}{4}$

$-\tfrac{\pi}{4}$

Figure 15-6

15-3. Polar Equations and Graphs

A. Polar equations

An equation in r and θ of the form $r = f(\theta)$ is a **polar equation**. You can interconvert simple polar and Cartesian equations with Equations 15-2 and 15-3.

EXAMPLE 15-5: Convert the following Cartesian equations to polar equations: **(a)** $x^2 + y^2 = a^2$; **(b)** $x^2 + y^2 = ax$; **(c)** $x^2 + y^2 = by$.

Solution:

(a) From Equation 15-3 it follows immediately that $r^2 = a^2$, so $r = a$.

(b) From Equations 15-2 and 15-3,

$$r^2 = x^2 + y^2 = ax = ar\cos\theta$$
$$r = a\cos\theta$$

(c) As in **b**,

$$r^2 = x^2 + y^2 = by = br\sin\theta$$
$$r = b\sin\theta$$

EXAMPLE 15-6: Convert the following polar equations to Cartesian equations: **(a)** $\theta = 1$; **(b)** $r = 2$; **(c)** $r = \cos\theta$.

Solution: Use Equations 15-3.

(a) The equation $\theta = 1$ can be written $\tan^{-1}(y/x) = 1$, or $y/x = \tan 1 = 1.557$. This can be written $y = 1.557x$, which is the equation of a straight line through the origin with slope 1.557.

(b) The equation $r = 2$ can be written $\sqrt{x^2 + y^2} = 2$ or $x^2 + y^2 = 4$, the equation of a circle with radius 2 and centered at the origin.

(c) Multiplying $r = \cos\theta$ by r, you get $r^2 = r\cos\theta$. This can be written $x^2 + y^2 = x$, or $x^2 - x + \frac{1}{4} + y^2 = \frac{1}{4}$. The first three terms factor as $\left(x - \frac{1}{2}\right)^2$, so the equation can be written $\left(x - \frac{1}{2}\right)^2 + y^2 = \left(\frac{1}{2}\right)^2$. This is the equation of a circle with center at $\left(\frac{1}{2}, 0\right)$ and radius $\frac{1}{2}$.

B. Graphs of polar equations

The graph of the polar equation $r = f(\theta)$ is the set of all points $P(r, \theta)$ that satisfy the given equation.

EXAMPLE 15-7: Graph $r = 4$ where θ is arbitrary.

Solution: A few points on the graph of $r = 4$ are $(4, 0)$, $\left(4, \frac{1}{2}\pi\right)$, $(4, \pi)$, and $\left(4, \frac{3}{2}\pi\right)$. The set of all points $(4, \theta)$, where θ is arbitrary, forms a circle of radius 4 centered at the origin (see Figure 15-7).

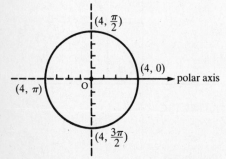

Figure 15-7

EXAMPLE 15-8: Graph $\theta = \frac{1}{4}\pi$ where r is arbitrary.

Solution: Some points on the graph of $\theta = \frac{1}{4}\pi$ are $\left(-2, \frac{1}{4}\pi\right)$, $\left(-1, \frac{1}{4}\pi\right)$, $\left(0, \frac{1}{4}\pi\right)$, and $\left(1, \frac{1}{4}\pi\right)$. The set of all points $\left(r, \frac{1}{4}\pi\right)$, r arbitrary, is a line through the origin making an angle of $\frac{1}{4}\pi$ with the polar axis (see Figure 15-8).

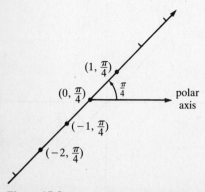

Figure 15-8

C. Symmetry

There are many symmetry rules for polar equations. The three that follow are among the easiest to apply to polar equations of the form $r = f(\theta)$:

1. If $f(-\theta) = f(\theta)$ for any θ, then the graph is symmetric about the polar axis (see Figure 15-9a).

2. If $f(\pi - \theta) = f(\theta)$ for any θ, then the graph is symmetric about the line $\theta = \frac{1}{2}\pi$ (see Figure 15-9b).

3. If $f(\theta + \pi) = f(\theta)$ for any θ, then the graph is symmetric about the pole (see Figure 15-9c).

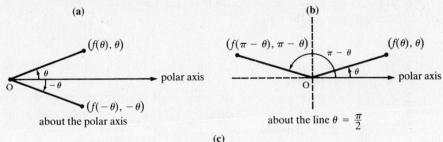

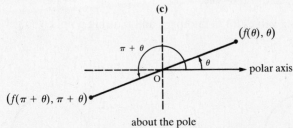

about the pole

Figure 15-9

Note: These three symmetry tests are not double-valued. If the conditions are met, then symmetry exists; if a test fails, you cannot reach any conclusion about the shape of the curve. Symmetry may exist even if a test fails.

EXAMPLE 15-9: Sketch the graph of $r = \sin \theta$.

Solution: Begin by using the three tests for symmetry:

1. $f(-\theta) = \sin(-\theta) = -\sin \theta \neq f(\theta)$, so you have no information about polar-axis symmetry.
2. $f(\pi - \theta) = \sin(\pi - \theta) = \sin \theta = f(\theta)$, so the graph is symmetric about $\theta = \frac{1}{2}\pi$.
3. $f(\theta + \pi) = \sin(\theta + \pi) = -\sin \theta \neq f(\theta)$, so you have no information on symmetry about the pole.

Because of the symmetry about $\theta = \frac{1}{2}\pi$, you need only find coordinates of points for θ on the interval $\left[0, \frac{1}{2}\pi\right]$. Points on the interval $\left[\frac{1}{2}\pi, \pi\right]$ will be reflections of the points on $\left[0, \frac{1}{2}\pi\right]$ through the line $\theta = \frac{1}{2}\pi$. Selected co-ordinates follow:

θ	0	$\frac{\pi}{6}$	$\frac{\pi}{4}$	$\frac{\pi}{3}$	$\frac{\pi}{2}$
r	0	$\frac{1}{2}$	$\frac{\sqrt{2}}{2}$	$\frac{\sqrt{3}}{2}$	1

If you plot these points on the polar axis, you get a semicircle of radius $\frac{1}{2}$ with center at $r = \frac{1}{2}$, and $\theta = \frac{1}{2}\pi$ as shown in Figure 15-10. The curve from $\frac{1}{2}\pi$ to π is a reflection of this semicircle. As θ varies from π to 2π, you can check that points on the circle are specified a second time.

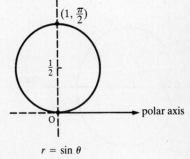

$r = \sin \theta$

Figure 15-10

EXAMPLE 15-10: Graph the three polar equations from the results of Example 15-5.

Solution:

(a) $r = a$: θ is arbitrary and r is a constant. This equation yields a circle centered at the pole of radius a.
(b) $r = a \cos \theta$:

    1. $f(-\theta) = a \cos(-\theta) = a \cos \theta = f(\theta)$, so the graph is symmetric about the polar axis.

2. $f(\pi - \theta) = a\cos(\pi - \theta) = -a\cos\theta \neq f(\theta)$; you have no information on symmetry about $\theta = \frac{1}{2}\pi$.
3. $f(\theta + \pi) = a\cos(\theta + \pi) = -a\cos\theta \neq f(\theta)$; you have no information on symmetry about the pole.

(c) $r = b\sin\theta$:

1. $f(-\theta) = b\sin(-\theta) = -b\sin\theta \neq f(\theta)$; you have no information about polar-axis symmetry.
2. $f(\pi - \theta) = b\sin(\pi + \theta) = b\sin\theta$, so the graph is symmetric about $\theta = \frac{1}{2}\pi$.
3. $f(\theta + \pi) = b\sin(\theta + \pi) = -b\sin\theta \neq f(\theta)$; you have no information on symmetry about the pole.

After substituting values for θ, finding r, and plotting points, you get the graphs shown in Figure 15-11.

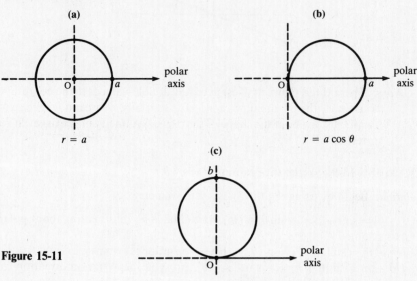

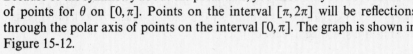

Figure 15-11

$r = b\sin\theta$

EXAMPLE 15-11: Sketch the graph of $r = 1 + \cos\theta$.

Solution:

1. $f(-\theta) = 1 + \cos(-\theta) = 1 + \cos\theta = f(\theta)$, so the graph is symmetric about the polar axis.
2. $f(\pi - \theta) = 1 + \cos(\pi - \theta) = 1 - \cos\theta \neq f(\theta)$; you have no information on symmetry about $\theta = \frac{1}{2}\pi$.
3. $f(\theta + \pi) = 1 + \cos(\theta + \pi) = 1 - \cos\theta \neq f(\theta)$; you have no information on symmetry about the pole.

Because of the symmetry about the polar axis, you need only find coordinates of points for θ on $[0, \pi]$. Points on the interval $[\pi, 2\pi]$ will be reflections through the polar axis of points on the interval $[0, \pi]$. The graph is shown in Figure 15-12.

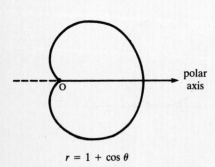

$r = 1 + \cos\theta$

Figure 15-12

EXAMPLE 15-12: Sketch the graph of $r = \sin 2\theta$.

Solution:

1. $f(-\theta) = \sin(-2\theta) = -\sin 2\theta \neq f(\theta)$; you have no information about polar-axis symmetry.
2. $f(\pi - \theta) = \sin[2(\pi - \theta)] = -\sin(2\theta) \neq f(\theta)$; you have no information on symmetry about $\theta = \frac{1}{2}\pi$.
3. $f(\theta + \pi) = \sin[2(\theta + \pi)] = \sin(2\theta) = f(\theta)$, so the graph is symmetric about the pole.

Since there is symmetry about the pole, you need plot only the coordinates of points for θ in the range 0 to $\frac{1}{2}\pi$. Points in the other quadrants of the plane are reflections through the pole, the polar axis, and the line $\theta = \frac{1}{2}\pi$, respectively, of first-quadrant points. The results are shown in Figure 15-13. Although two of the symmetry tests failed, the graph is indeed symmetric about the pole, the polar axis, and the line $\theta = \frac{1}{2}\pi$.

EXAMPLE 15-13: Sketch the graph of $r = \theta$ for $\theta \geqslant 0$.

Solution:

1. Since θ is nonnegative, this test for symmetry is not applicable.
2. $f(\pi - \theta) = \pi - \theta \neq f(\theta)$; you have no information on symmetry about $\theta = \frac{1}{2}\pi$.
3. $f(\theta + \pi) = \theta + \pi \neq f(\theta)$; you have no information on symmetry about the pole.

Upon plotting the pairs (r, θ) for several points with positive values of θ, you get the graph shown in Figure 15-14.

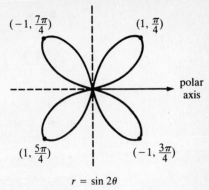

$r = \sin 2\theta$

Figure 15-13

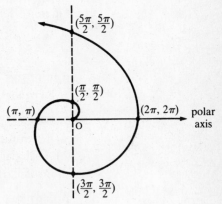

Figure 15-14

SUMMARY

1. A polar coordinate system is defined by a pole and a polar axis.
2. The polar coordinates (r, θ) of a point are given by the angle θ measured from the polar axis to a line from the pole through the point and by the distance r along this line from the pole to the point.
3. If x and y are the Cartesian coordinates of a point, the polar coordinates (r, θ) are given by $r = x^2 + y^2$ and $\theta = \tan^{-1}(y/x)$.
4. If r and θ are the polar coordinates of a point, the Cartesian coordinates (x, y) are given by $x = r\cos\theta$ and $y = r\sin\theta$.
5. Graphs of polar equations can be sketched by using the rules of symmetry and plotting points or by converting to Cartesian coordinates and plotting the points.
6. Symmetry

   (a) If $f(-\theta) = f(\theta)$ for any θ, then the graph of a polar equation is symmetric about the polar axis.
   (b) If $f(\pi - \theta) = f(\theta)$ for any θ, then the graph of a polar equation is symmetric about the line $\theta = \frac{1}{2}\pi$.
   (c) If $f(\theta + \pi) = f(\theta)$ for any θ, then the graph of a polar equation is symmetric about the pole.

RAISE YOUR GRADES
Can you ...?

☑ plot any point given its polar coordinates
☑ find the polar coordinates of any point given the Cartesian coordinates of that point
☑ find the Cartesian coordinates of any point given the polar coordinates of that point
☑ find the polar equation of any Cartesian equation
☑ find the Cartesian equation of any polar equation
☑ sketch the graph of any polar equation

SOLVED PROBLEMS

PROBLEM 15-1 Plot the following points given by polar coordinates (r, θ): (a) $(1, \frac{1}{3}\pi)$; (b) $(-1, \frac{1}{3}\pi)$; (c) $(1, -\frac{1}{3}\pi)$; (d) $(1, \frac{4}{3}\pi)$; (e) $(-1, -\frac{1}{3}\pi)$.

Solution: You plot each of these points by first using the given angle to find the line on which each point will lie. If r is negative, use angle $\theta + \pi$ rather than θ. In either case, you then move out distance $|r|$ along the line and plot the point. Figure 15-15 shows the points.

[See Section 15-1.]

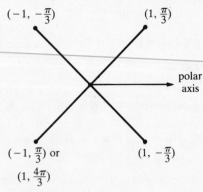

$(-1, -\frac{\pi}{3})$ $(1, \frac{\pi}{3})$

polar axis

$(-1, \frac{\pi}{3})$ or $(1, -\frac{\pi}{3})$

$(1, \frac{4\pi}{3})$

Figure 15-15

PROBLEM 15-2 Find the polar coordinates with $r > 0$ and $0 \leqslant \theta < 2\pi$ for the following points given by their Cartesian coordinates: (a) $(3, 4)$; (b) $(1, 0)$; (c) $(1, -\sqrt{3})$.

Solution:

(a) From Equation 15-3, $r = \sqrt{x^2 + y^2} = \sqrt{3^2 + 4^2} = 5$ and $\theta = \tan^{-1}(\frac{4}{3}) = 0.93$ rad.
(b) As in part **a**, $r = \sqrt{1^2 + 0^2} = 1$ and $\theta = \tan^{-1}(\frac{0}{1}) = 0$.
(c) As in part **a**, $r = \sqrt{1^2 + (-\sqrt{3})^2} = 2$ and $\theta = \tan^{-1}(\frac{-\sqrt{3}}{1}) = -\frac{1}{3}\pi$.   [See Section 15-2.]

PROBLEM 15-3 Find the Cartesian coordinates for the following points given by their polar coordinates (r, θ): (a) $(2, 45°)$; (b) $(1, \frac{5}{6}\pi)$; (c) $(-3, \frac{1}{6}\pi)$.

Solution:

(a) From Equation 15-2,
$$x = r \cos \theta = 2 \cos 45° = \frac{2\sqrt{2}}{2} = \sqrt{2}$$
$$y = r \sin \theta = 2 \sin 45° = \frac{2\sqrt{2}}{2} = \sqrt{2}$$

(b) As in part **a**, $x = 1 \cos \frac{5}{6}\pi = \frac{-\sqrt{3}}{2}$, and $y = 1 \sin \frac{5}{6}\pi = \frac{1}{2}$.
(c) Since r is negative, convert the angle to $\pi + \frac{1}{6}\pi = \frac{7}{6}\pi$ and use $r = 3$. Then the coordinates are given by $x = 3 \cos \frac{7}{6}\pi = 3(\frac{-\sqrt{3}}{2}) = \frac{-3\sqrt{3}}{2}$ and $y = 3 \sin \frac{7}{6}\pi = 3(-\frac{1}{2}) = -\frac{3}{2}$.   [See Section 15-2.]

PROBLEM 15-4 Find the polar equation for $x^2 + x + y^2 = 1$.

Solution: Write the equation as $x^2 + y^2 + x = 1$ and use the relationships $x^2 + y^2 = r^2$ and $x = r \cos \theta$ to get the polar equation $r^2 + r \cos \theta = 1$.   [See Section 15-3.]

PROBLEM 15-5 Find the rectangular coordinate equation for $r = \sin 3\theta$.

Solution: Use the identities $\sin(\alpha + \theta) = \sin \alpha \cos \theta + \cos \alpha \sin \theta$, $\sin 2\theta = 2 \sin \theta \cos \theta$, and $\cos 2\theta = \cos^2\theta - \sin^2\theta$:

$$r = \sin 3\theta = \sin(2\theta + \theta)$$
$$= \sin 2\theta \cos \theta + \cos 2\theta \sin \theta$$
$$= 2 \sin \theta \cos^2\theta + \cos^2\theta \sin \theta - \sin^3\theta$$

The solution is simplified if you multiply both sides of this equation by r^3:

$$r^4 = 2r \sin \theta \, r^2\cos^2\theta + r^2\cos^2\theta \, r \sin \theta - r^3\sin^3\theta$$

Now use $r^2 = x^2 + y^2$, $r \sin \theta = y$, and $r \cos \theta = x$:

$$(x^2 + y^2)^2 = 2x^2y + x^2y - y^3$$
$$(x^2 + y^2)^2 = 3x^2y - y^3$$ [See Section 15-3.]

PROBLEM 15-6   Graph $r = 3 \sec \theta$.

Solution: You could substitute values for θ and plot the resulting points, but it is easier to multiply both sides of the equation by $\cos \theta$ to get $r \cos \theta = 3$. Then, since $r \cos \theta = x$, the equation is $x = 3$, a vertical line passing through the point $(3, 0)$ (see Figure 15-16).   [See Section 15-3.]

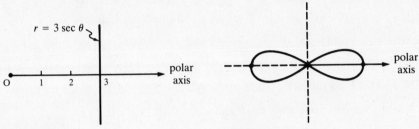

Figure 15-16   **Figure 15-17**

PROBLEM 15-7   Graph $r^2 = \cos 2\theta$.

Solution: This problem demonstrates another symmetry principle. Since r is squared, replacing r by $-r$ does not change the equation. This means that the graph is symmetric about the pole. Also, replacing θ by $-\theta$ does not change the equation, so the graph is also symmetric about the polar axis. Any curve that is symmetric about the pole and the polar axis must also be symmetric about the line $\theta = \frac{1}{2}\pi$. Symmetry about the pole and both lines means that you have to plot only points for θ running from 0 to $\frac{1}{2}\pi$. The graph in the other four quadrants is a reflection of the graph in the first quadrant (see Figure 15-17).   [See Section 15-3.]

PROBLEM 15-8   Graph $r = \theta + 1$, for $\theta > 0$.

Solution: Using the symmetry tests, you have no information on symmetry about the polar axis, the pole, or the line $\theta = \frac{1}{2}\pi$. Find the coordinates of a number of points. In this case the r coordinate is simply the angle plus 1, so as the angle increases, r increases. When you plot points, you will get the graph shown in Figure 15-18.   [See Section 15-3.]

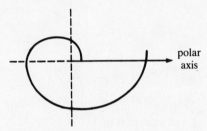

Figure 15-18

PROBLEM 15-9   Graph $r = 2 \cos \theta$.

Solution: Use the symmetry tests:

1. $f(-\theta) = 2 \cos(-\theta) = 2 \cos \theta = f(\theta)$, so the graph is symmetric about the polar axis.
2. $f(\pi - \theta) = 2 \cos(\pi - \theta) = 2(\cos \pi \cos \theta + \sin \pi \sin \theta) = -2 \cos \theta \neq f(\theta)$. You have no information on symmetry about the line $\theta = \frac{1}{2}\pi$; as shown in the text, the negative sign on the right side also implies that the graph is symmetric about the polar axis.
3. $f(\theta + \pi) = 2 \cos(\theta + \pi) = 2(\cos \theta \cos \pi - \sin \theta \sin \pi) = -2 \cos \theta \neq f(\theta)$. You have no information on symmetry about the pole.

Using the symmetry about the polar axis, you need plot only points for θ varying from $-\frac{1}{2}\pi$ to $\frac{1}{2}\pi$:

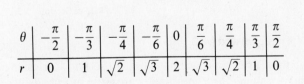

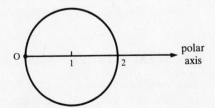

Figure 15-19

Plotting these points, you get a circle of radius 1 with center at $(1, 0)$ as shown in Figure 15-19.

[See Section 15-3.]

PROBLEM 15-10   Find the Cartesian coordinates for the following points given by their polar coordinates (r, θ): **(a)** $\left(4, \frac{1}{6}\pi\right)$; **(b)** $\left(-4, \frac{9}{4}\pi\right)$; **(c)** $\left(4, -\frac{1}{2}\pi\right)$; **(d)** $\left(-4, -\frac{1}{3}\pi\right)$.

Solution:

(a) From Equation 15-2, $x = r\cos\theta = 4\cos\frac{1}{6}\pi = \frac{4\sqrt{3}}{2} = 2\sqrt{3}$ and $y = r\sin\theta = 4\sin\frac{1}{6}\pi = 4\left(\frac{1}{2}\right) = 2$.

(b) As above, $x = -4\cos\frac{9}{4}\pi = \frac{-4\sqrt{2}}{2} = -2\sqrt{2}$ and $y = -4\sin\frac{9}{4}\pi = \frac{-4\sqrt{2}}{2} = -2\sqrt{2}$.

(c) As above, $x = 4\cos\left(-\frac{1}{2}\pi\right) = -4(0) = 0$ and $y = 4\sin\left(-\frac{1}{2}\pi\right) = 4(-1) = -4$.

(d) As above, $x = -4\cos\left(-\frac{1}{3}\pi\right) = -4\left(\frac{1}{2}\right) = -2$ and $y = -4\sin\left(-\frac{1}{3}\pi\right) = -4\left(\frac{-\sqrt{3}}{2}\right) = 2\sqrt{3}$.

[See Section 15-2.]

PROBLEM 15-11   Find the polar equation for $x^2/9 + y^2 = 1$.

Solution: Use the substitutions $x = r\cos\theta$ and $y = r\sin\theta$:

$$\frac{(r\cos\theta)^2}{9} + (r\sin\theta)^2 = 1$$

$$r^2\left(\frac{\cos^2\theta}{9} + \sin^2\theta\right) = 1$$

$$r^2 = \left(\frac{\cos^2\theta}{9} + \sin^2\theta\right)^{-1} \qquad \text{[See Section 15-3.]}$$

PROBLEM 15-12   Find the polar equation for $2x - y = 0$.

Solution: Replace x by $r\cos\theta$ and y by $r\sin\theta$:

$$2r\cos\theta - r\sin\theta = 0$$

$$r = \frac{\sin\theta}{2\cos\theta} = \frac{\tan\theta}{2} \qquad \text{[See Section 15-3.]}$$

PROBLEM 15-13   Find the rectangular equation for $r = \sin\theta + 2\cos\theta$.

Solution: Since $x = r\cos\theta$, $y = r\sin\theta$, and $r^2 = x^2 + y^2$, a first step is to multiply both sides of the given equation by r:

$$r^2 = r\sin\theta + 2r\cos\theta$$

$$x^2 + y^2 = y + 2x$$

$$x^2 - 2x + y^2 - y = 0$$

By completing squares on the left side, you get

$$(x - 1)^2 + \left(y - \frac{1}{2}\right)^2 = \left(\frac{\sqrt{5}}{2}\right)^2$$

which is the equation of a circle with center at $\left(1, \frac{1}{2}\right)$ and radius $\frac{\sqrt{5}}{2}$.   [See Section 15-3.]

PROBLEM 15-14   Find the rectangular equation for $4\theta = 3$.

Solution: Replace θ by $\tan^{-1}(y/x)$ to get $4\tan^{-1}(y/x) = 3$, or $\tan^{-1}(y/x) = \frac{4}{3}$. Then $y/x = \tan\frac{4}{3} = 4.13$, or $y = 4.13x$, the equation of a straight line through the origin with slope 4.13.

[See Section 15-3.]

PROBLEM 15-15   Find the distance between $\left(2, \frac{1}{4}\pi\right)$ and $\left(-2, \frac{1}{4}\pi\right)$.

Solution: You know a distance formula for the Cartesian coordinate system, so find Cartesian coordinates of the points and use that distance formula. The x coordinates are given by $r\cos\theta$ and the y coordinates by $r\sin\theta$: $\left(2, \frac{1}{4}\pi\right)$ has $x = 2\cos\frac{1}{4}\pi = \sqrt{2}$ and $y = 2\sin\frac{1}{4}\pi = \sqrt{2}$; $\left(-2, \frac{1}{4}\pi\right)$ has $x = -2\cos\frac{1}{4}\pi = -\sqrt{2}$ and $y = -2\sin\frac{1}{4}\pi = -\sqrt{2}$. Then,

$$d = \sqrt{(x_1 - x_2)^2 + (y_1 - y_2)^2}$$

$$= \sqrt{(\sqrt{2} - (-\sqrt{2}))^2 + (\sqrt{2} - (-\sqrt{2}))^2} = \sqrt{16} = 4 \qquad \text{[See Section 15-2.]}$$

PROBLEM 15-16 Find the distance between (r, θ) and (s, α).

Solution: The x coordinates are given by $r\cos\theta$ and the y coordinates by $r\sin\theta$: (r, θ) has $x = r\cos\theta$ and $y = r\sin\theta$; (s, α) has $x = s\cos\alpha$ and $y = s\sin\alpha$. Then,

$$d = \sqrt{(r\cos\theta - s\cos\alpha)^2 + (r\sin\theta - s\sin\alpha)^2} = \sqrt{r^2 + s^2 - 2rs\cos(\theta - \alpha)}$$

This should look familiar. It is one form of the law of cosines. [See Section 15-2.]

PROBLEM 15-17 Write two sets of polar coordinates for $(3, 4)$ and for $(-3, -4)$.

Solution: The distance of $(3, 4)$ from the pole is given by $\sqrt{3^2 + 4^2} = 5$ and the angle with the positive x axis is $\theta = \tan^{-1}(\frac{4}{3}) = 53.1°$. Then $(3, 4) = (5, 53.1°)$. You can find another expression for $(3, 4)$ by adding $360°$ to the angle. Then $(3, 4) = (5, 413.1°)$.

The distance of $(-3, -4)$ from the pole is given by $\sqrt{(-3)^2 + (-4)^2} = 5$ and the angle with the positive x axis is $\theta = \tan^{-1}(\frac{-4}{-3}) + 180° = 233.1°$ ($180°$ was added because the point lies in quadrant III). Then $(-3, -4) = (5, 233.1°)$. You can find another expression for $(-3, -4)$ by subtracting $180°$ from the angle to get $53.1°$ and by changing the sign of r. Then $(-3, -4) = (-5, 53.1°)$. [See Section 15-2.]

PROBLEM 15-18 Write two sets of polar coordinates for $(4, -3)$ and $(-4, 3)$.

Solution: The distance of $(4, -3)$ from the pole is given by $\sqrt{4^2 + (-3)^2} = 5$ and the angle with the positive x axis is $\theta = \tan^{-1}(-\frac{3}{4}) = -36.9°$. Then $(4, -3) = (5, -36.9°)$. You can find another expression for $(4, -3)$ by adding $360°$ to the angle to get $323.1°$. Then $(4, -3) = (5, 323.1°)$.

The distance of $(-4, 3)$ from the pole is given by $\sqrt{(-4)^2 + 3^2} = 5$ and the angle with the positive x axis is $\theta = \tan^{-1}(\frac{3}{-4}) + 180° = 143.1°$ ($180°$ was added because the point lies in quadrant II). Then $(-4, 3) = (5, 143.1°)$. You can find another expression for $(-4, 3)$ by subtracting $180°$ from the angle to get $-36.9°$ and by changing the sign of r. Then $(-4, 3) = (-5, -36.9°)$. [See Section 15-2.]

PROBLEM 15-19 Plot each of the following points given by their polar coordinates: (a) $(2, \frac{1}{4}\pi)$; (b) $(3, \frac{11}{3}\pi)$; (c) $(1, \frac{7}{12}\pi)$; (d) $(4, \frac{17}{4}\pi)$.

Solution: Figure 15-20 shows the plotted points. [See Section 15-1.]

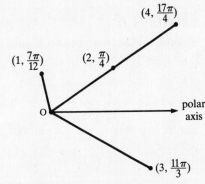

Figure 15-20

PROBLEM 15-20 Write all possible polar coordinates for the point whose Cartesian coordinates are $(12, 5)$.

Solution: The given point has $r = \sqrt{12^2 + 5^2} = 13$ and $\theta = \tan^{-1}(\frac{5}{12})$. Then any multiple of 2π added to $\tan^{-1}(\frac{5}{12})$ will give a representation for the point. Also, by making r negative and at the same time adding π to $\tan^{-1}(\frac{5}{12})$, you get another representation for $(12, 5)$. Now, multiples of 2π can be added to this representation: $(12, 5) = (13, \tan^{-1}(\frac{5}{12}) + 2n\pi) = (-13, \tan^{-1}(\frac{5}{12}) + \pi + 2n\pi)$, for n any integer. [See Section 15-1.]

PROBLEM 15-21 Sketch the curves whose equations are: (a) $r = 2$; (b) $\theta = \frac{1}{3}\pi$; (c) $\theta^2 = \frac{1}{9}\pi^2$; (d) $r^2 - 3r + 2 = 0$.

Solution:

(a) This curve has constant radius—thus it is a circle with radius 2 and center at the pole.
(b) The angle is constant: the graph is a line through the pole making an angle of $\frac{1}{3}\pi$ with the positive x axis.

(c) By taking square roots of both sides of this equation, you see that the angle can be either $\frac{1}{3}\pi$ or $-\frac{1}{3}\pi$. These are two straight lines through the origin with angles of $\frac{1}{3}\pi$ and $-\frac{1}{3}\pi$, respectively, with the positive x axis.

(d) You can factor this equation as $(r-2)(r-1)=0$, so either $r=2$ or $r=1$. These are concentric circles centered at the origin with radii 2 and 1, respectively (see Figure 15-21).

[See Section 15-3.]

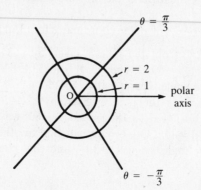

Figure 15-21

PROBLEM 15-22 Convert $x^2 + y^2 - 3x = 0$ to a polar equation.

Solution: Use the substitutions $x^2 + y^2 = r^2$ and $x = r\cos\theta$ to get $r^2 - 3r\cos\theta$ or $r = 3\cos\theta$.

[See Section 15-3.]

PROBLEM 15-23 Convert $y^2 = 4x$ to a polar equation.

Solution: Use the substitutions $y = r\sin\theta$ and $x = r\cos\theta$ in the given equation: $r^2\sin^2\theta = 4r\cos\theta$. Divide both sides by $r\sin^2\theta$: $r = 4\cot\theta\csc\theta$.

[See Section 15-3.]

PROBLEM 15-24 Convert $r = 1 + 2\cos\theta$ to a Cartesian equation.

Solution: First, multiply both sides of the given equation by r. Then replace r by $\sqrt{x^2 + y^2}$ and $r\cos\theta$ by x:

$$r^2 = r + 2r\cos\theta$$
$$x^2 + y^2 = \sqrt{x^2 + y^2} + 2x$$

Subtract $2x$ from both sides, square both sides to get rid of the radical, and move all terms to the left side:

$$x^4 + 2x^2y^2 + y^4 - 4x^3 - 4xy^2 + 3x^2 - y^2 = 0$$

[See Section 15-3.]

PROBLEM 15-25 Convert $r = 4/(1 + \cos\theta)$ to a Cartesian equation.

Solution: Multiply both sides by $1 + \cos\theta$ to get $r + r\cos\theta = 4$. Then replace r by $\sqrt{x^2 + y^2}$ and $r\cos\theta$ by x:

$$\sqrt{x^2 + y^2} + x = 4$$

Subtract x from both sides and square:

$$x^2 + y^2 = 16 - 8x + x^2$$
$$y^2 = 16 - 8x$$

[See Section 15-3.]

PROBLEM 15-26 Give the polar coordinates of two points on the curve $r^2 = 8\cos\theta$.

Solution: Replace θ by 0. Then $r^2 = 8$, or $r = \pm\sqrt{8}$. Two points on the curve are $(\sqrt{8}, 0)$ and $(-\sqrt{8}, 0)$.

[See Section 15-3.]

PROBLEM 15-27 Give the polar coordinates of two points on the curve $r\theta = 6$.

Solution: If you replace r by 1, then $\theta = 6$. Conversely, if you replace θ by 1, then $r = 6$. Two points on the curve are $(1, 6)$ and $(6, 1)$.   [See Section 15-3.]

PROBLEM 15-28 Find the Cartesian coordinates of the points whose polar coordinates are $\left(-2, -\frac{3}{2}\pi\right)$ and $\left(2, -\frac{3}{2}\pi\right)$.

Solution: You use the expression $x = r\cos\theta$ and $y = r\sin\theta$ to get $x = -2\cos\left(-\frac{3}{2}\pi\right) = 0$ and $y = -2\sin\left(-\frac{3}{2}\pi\right) = -2$ in the first instance and $x = 2\cos\left(-\frac{3}{2}\pi\right) = 0$ and $y = 2\sin\left(-\frac{3}{2}\pi\right) = 2$ in the second instance. The points are $(0, -2)$ and $(0, 2)$, respectively.   [See Section 15-2.]

PROBLEM 15-29 Find the distance between $\left(3, \frac{1}{4}\pi\right)$ and $\left(3, \frac{5}{4}\pi\right)$.

Solution: The distance formula for the distance between (r, θ) and (s, α) is $d = \sqrt{r^2 + s^2 - 2rs\cos(\theta - \alpha)}$ (see Problem 15-16). Replace r by 3, θ by $\frac{5}{4}\pi$, s by 3, and α by $\frac{1}{4}\pi$:

$$d = \sqrt{3^2 + 3^2 - 2(3)(3)\cos\left(\frac{5\pi}{4} - \frac{\pi}{4}\right)} = \sqrt{36} = 6 \qquad \text{[See Section 15-2.]}$$

Supplementary Exercises

PROBLEM 15-30 Express in polar coordinates with $r > 0$ and $0 \leqslant \theta < 2\pi$: **(a)** $(3, 4)$; **(b)** $(-5, 12)$; **(c)** $(-1, -\sqrt{3})$; **(d)** $(4, -3)$.

PROBLEM 15-31 Graph $r = 2 + \cos\theta$.

PROBLEM 15-32 Plot the following points given by their polar coordinates: **(a)** $\left(5, \frac{11}{6}\pi\right)$; **(b)** $\left(-5, \frac{13}{6}\pi\right)$; **(c)** $\left(4, -\frac{5}{6}\pi\right)$; **(d)** $(2, 17\pi)$; **(e)** $(2, 68°)$; **(f)** $(-3, 45°)$; **(g)** $(-2, 180°)$.

PROBLEM 15-33 Plot each of the following points given by their polar coordinates: **(a)** $\left(2, \frac{13}{4}\pi\right)$; **(b)** $\left(2, -\frac{13}{4}\pi\right)$; **(c)** $\left(-2, \frac{13}{4}\pi\right)$; **(d)** $\left(-2, -\frac{13}{4}\pi\right)$.

PROBLEM 15-34 Sketch the curve whose equation is $r = \sin\theta$.

PROBLEM 15-35 Sketch the curve whose equation is $r = \cos 2\theta$.

PROBLEM 15-36 Sketch the curve whose equation is $r = 1 - \sin\theta$.

PROBLEM 15-37 Express in polar coordinates with $r > 0$ and $0 \leqslant \theta < 2\pi$: **(a)** $(5, 12)$; **(b)** $(-2, -1)$; **(c)** $(-4, 4)$; **(d)** $(7, -3)$.

PROBLEM 15-38 Find the Cartesian coordinates for the following points given by their polar coordinates (r, θ): **(a)** $(3, 60°)$; **(b)** $\left(-3, \frac{5}{6}\pi\right)$; **(c)** $(4, 120°)$; **(d)** $\left(5, \frac{1}{2}\pi\right)$.

PROBLEM 15-39 Find the polar equation for $x^2 + y^2 = 16$.

PROBLEM 15-40 Find the rectangular equation for $r = -\csc\theta$.

PROBLEM 15-41 Sketch the curves whose equations are **(a)** $r = 3\sin\theta$; **(b)** $r = 3\cos\theta$; **(c)** $r = \theta + 2$; **(d)** $\sin\theta - r\cos^2\theta = 0$.

PROBLEM 15-42 Write all possible polar coordinates for the point with Cartesian coordinates $(3, 4)$.

PROBLEM 15-43 Find the Cartesian coordinates of each of the following points given in polar coordinates: **(a)** $\left(2, \frac{7}{12}\pi\right)$; **(b)** $\left(-2, \frac{19}{12}\pi\right)$; **(c)** $\left(-2, -\frac{5}{12}\pi\right)$; **(d)** $\left(2, \frac{55}{12}\pi\right)$

Convert each of the following to polar equations.

PROBLEM 15-44 $x + 2y = 3$

PROBLEM 15-45 $x^2 + 2y^2 = 2$

PROBLEM 15-46 $x^2 - y^2 = 1$

Convert each of the following equations to Cartesian coordinates.

PROBLEM 15-47 $r = 6$

PROBLEM 15-48 $r = 5\sin\theta$

PROBLEM 15-49 $r = 3(\sin\theta - \cos\theta)$

PROBLEM 15-50 $r\sin\theta = 3$

Give the polar coordinates of two points on each of the following.

PROBLEM 15-51 $r = \tan\theta$

PROBLEM 15-52 $r = 6\cos 2\theta$

Find the Cartesian coordinates for the following points given by their polar coordinates (r, θ).

PROBLEM 15-53 $\left(1, \frac{1}{3}\pi\right), \left(-1, -\frac{1}{3}\pi\right)$, and $\left(1, -\frac{1}{3}\pi\right)$

PROBLEM 15-54 $\left(3, \frac{7}{6}\pi\right)$ and $\left(3, \frac{1}{6}\pi\right)$

PROBLEM 15-55 Find the distance between $\left(1, \frac{1}{3}\pi\right)$ and $\left(2, \frac{1}{3}\pi\right)$.

PROBLEM 15-56 Write two sets of polar coordinates for $(-2, -3)$ and $(-2, 3)$.

Answers to Supplementary Exercises

15-30: (a) $(5, 0.93)$; (b) $(13, 1.97)$; (c) $\left(2, \frac{4}{3}\pi\right)$;
(d) $(5, 5.64)$

15-31:

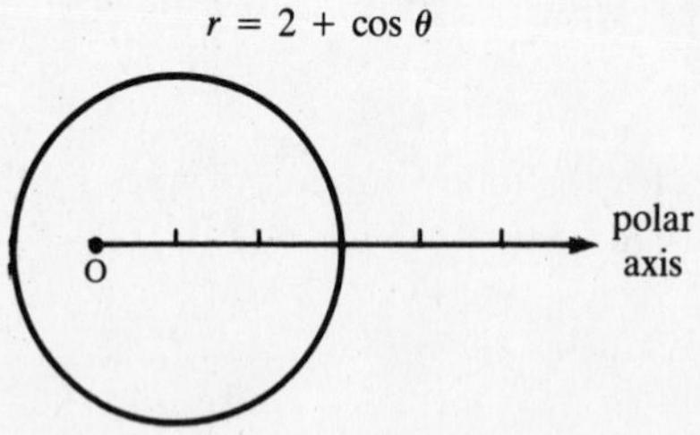

$r = 2 + \cos\theta$

Figure 15-22

15-32:

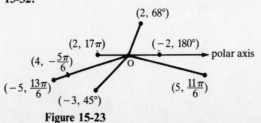

Figure 15-23

15-33:

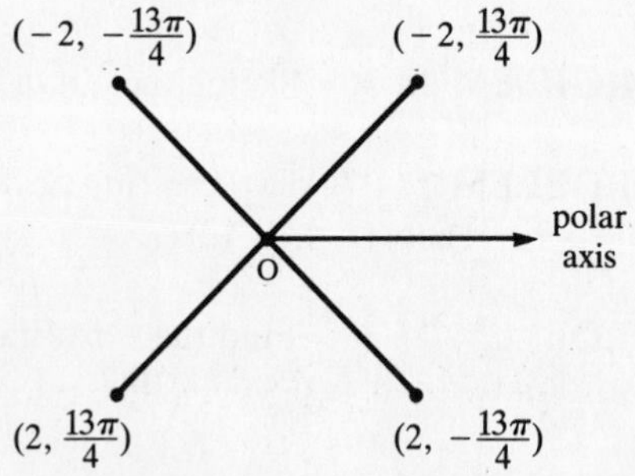

Figure 15-24

15-34:

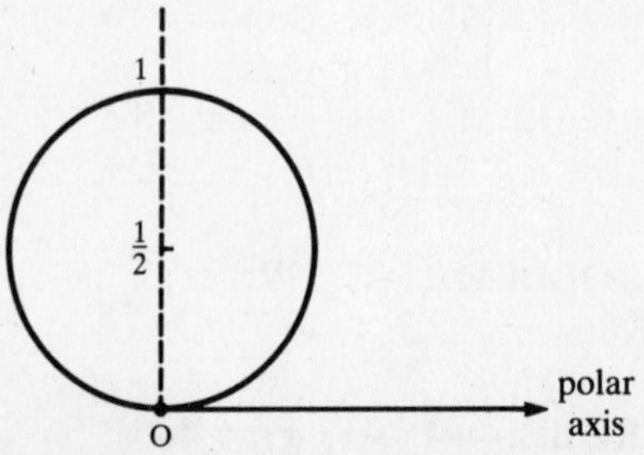

Figure 15-25

15-35:

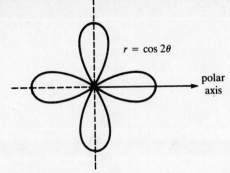

$r = \cos 2\theta$

polar axis

Figure 15-26

15-36:

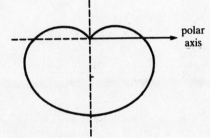

polar axis

Figure 15-27

15-37: **(a)** $(13, 1.18)$; **(b)** $(\sqrt{5}, 3.61)$; **(c)** $(\sqrt{32}, 2.36)$; **(d)** $(\sqrt{58}, -0.40)$

15-38: **(a)** $\left(\dfrac{3}{2}, \dfrac{3\sqrt{3}}{2}\right)$; **(b)** $\left(\dfrac{3\sqrt{3}}{2}, \dfrac{-3}{2}\right)$;

(c) $(-2, 2\sqrt{3})$; **(d)** $(0, 5)$

15-39: $r = 4$

15-40: $y = -1$

15-41: See Figure 15-28

15-42: $\left(5, \tan^{-1}\!\left(\tfrac{4}{3}\right) + 2n\pi\right)$, $\left(-5, \tan^{-1}\!\left(\tfrac{4}{3}\right) + \pi + 2n\pi\right)$

15-43: **(a)** $(-0.52, 1.93)$; **(b)** $(-0.52, 1.93)$; **(c)** $(-0.52, 1.93)$; **(d)** $(-0.52, 1.93)$

15-44: $r = 3/(\cos\theta + 2\sin\theta)$

15-45: $r^2 = 2/(1 + \sin^2\theta)$

15-46: $r^2 = 1/\cos 2\theta$

15-47: $x^2 + y^2 = 6^2$

15-48: $x^2 + y^2 = 5y$

15-49: $x^2 + y^2 = 3y - 3x$

15-50: $y = 3$

15-51: $(0, 0)$ and $\left(1, \tfrac{1}{4}\pi\right)$

15-52: $(6, 0)$ and $\left(0, \tfrac{1}{4}\pi\right)$

15-53: $\left(\dfrac{1}{2}, \dfrac{\sqrt{3}}{2}\right), \left(-\dfrac{1}{2}, \dfrac{\sqrt{3}}{2}\right)$, and $\left(\dfrac{1}{2}, \dfrac{-\sqrt{3}}{2}\right)$

15-54: $\left(\dfrac{-3\sqrt{3}}{2}, -\dfrac{3}{2}\right)$ and $\left(\dfrac{3\sqrt{3}}{2}, \dfrac{3}{2}\right)$

15-55: 1

15-56: $(\sqrt{13}, 236.3°)$ or $(-\sqrt{13}, 56.3°)$ and $(\sqrt{13}, 123.7°)$ or $(-\sqrt{13}, -56.3°)$

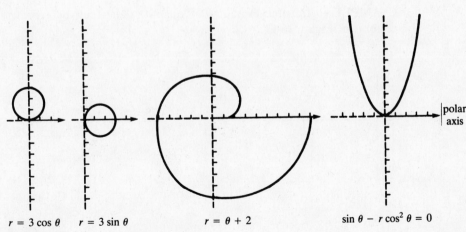

$r = 3\cos\theta$ $r = 3\sin\theta$ $r = \theta + 2$ $\sin\theta - r\cos^2\theta = 0$

Figure 15-28

16 *VECTORS*

THIS CHAPTER IS ABOUT

☑ **Vectors and Scalars**
☑ **Cartesian Representations of Vectors**
☑ **Arithmetic Operations on Vectors**
☑ **Resolving Vectors**
☑ **Course, Bearing, and Heading**

16-1. Vectors and Scalars

Vectors represent physical quantities that require a *magnitude* (how much) and a *direction* (which way) for their description. Vector quantities are often represented by arrows (directed line segments): The head of the arrow indicates direction and the shaft length indicates magnitude. Velocity, acceleration, and force all act in particular directions with specific magnitudes, so they are represented by vectors. Mass, volume, and temperature are physical quantities defined by magnitude alone; they are represented by **scalars** (real numbers).

• Two vectors are equal if and only if they have the same magnitude and direction.

Vectors are denoted in printed material by boldface letters such as **v**, **a**, or **F**, or by letters with arrows over them ($\vec{v}$, $\vec{a}$, and $\vec{F}$), especially in handwritten work. Boldface notation is used in this book. The magnitude of vector **v** is denoted by |**v**|. The magnitude of the *zero vector* **0** is zero and its direction is arbitrary.

EXAMPLE 16-1: Some examples of quantities that can be represented by vectors (see Figure 16-1):

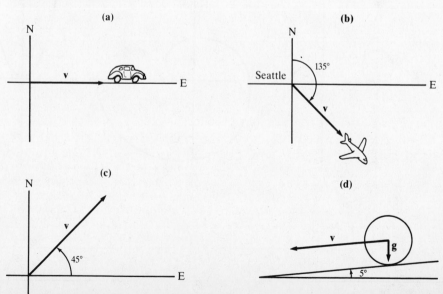

Figure 16-1
Vector quantities.

234

(a) A car travels east from a point at 30 kilometers per hour. Its *velocity* is **v**; its *speed* is |**v**|.

(b) An airplane flies from Seattle to Denver at 450 miles per hour on a course of 135°. (Course is the clockwise angle from north to the actual line of travel of the airplane.)

(c) The wind is blowing from the southwest at 30 miles per hour.

(d) A ball is rolling down a 5° incline at 1.0 ft/s. Its velocity is **v**; its acceleration due to gravity is **g**.

EXAMPLE 16-2: A boat is crossing from the south bank of a river to the parallel north bank at 3 miles per hour. The current is flowing at 2 miles per hour to the east (see Figure 16-2). Find **r**, the **resultant** velocity of the boat; that is, the *vector sum* of its velocity and that of the current. Determine the course of the boat.

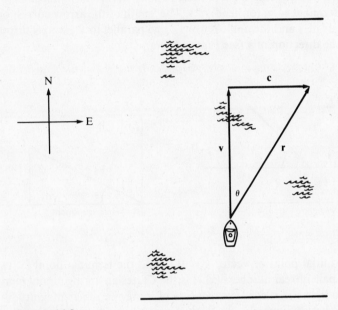

Figure 16-2

Solution: From the triangle in Figure 16-2, the magnitude of the resultant velocity vector is

$$|\mathbf{r}| = \sqrt{2^2 + 3^2} = \sqrt{13}$$

and the course is

$$\tan\theta = \frac{2}{3}$$

$$\theta = \tan^{-1}\frac{2}{3} = 33.7° \text{ (east of north)}$$

EXAMPLE 16-3: A farmer pulls a loaded wagon up a hill. The hill has an angle of elevation of 8° with the horizontal. It requires 300 pounds of force to pull the wagon; what is the force in the horizontal direction (see Figure 16-3)?

Solution: From the triangle in Figure 16-3, you see that the magnitude of the vector in the horizontal direction |**F**$_h$| is

$$\cos 8° = \frac{|\mathbf{F}_h|}{300}$$

$$|\mathbf{F}_h| = 300\cos 8° = 297.1 \text{ pounds}$$

Figure 16-3

16-2. Cartesian Representations of Vectors

In the Cartesian plane, vectors are again represented by directed line segments (arrows). If vector **v** runs from point *A* to point *B*, *A* is the **initial point** (or tail) and *B* is the **terminal point** (or head) of **v**. The length of the arrow corresponds to the magnitude of **v** and the angle θ from a line parallel to the *x* axis through *A* to **v** defines the direction of **v** (see Figure 16-4a).

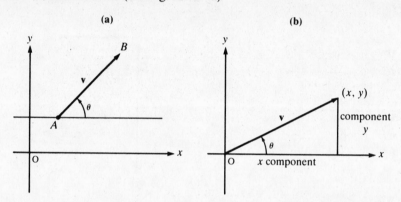

Figure 16-4
Cartesian vector notation.

If the initial point of vector **v** is $(0, 0)$ and the terminal point (x, y), then the ordered pair of real numbers (x, y) is the **Cartesian** (or algebraic) **representation** of **v**. The **first** and **second components** of **v** are the scalar magnitudes *x* and *y*, respectively (see Figure 16-4b). Both components of the zero vector, **0**, equal zero.

If $\mathbf{v} = (x, y)$, then the magnitude, $|\mathbf{v}| = \sqrt{x^2 + y^2}$. The direction of **v** is defined as the smallest nonnegative angle measured from the positive *x* axis, or a line parallel to the positive *x* axis if the base of the vector is not at the origin, to the vector. You find the direction of the vector by using $\theta = \tan^{-1}|y/x|$ as the reference angle for the vector (if $x = 0$, reference angle θ is $90°$). Then find the direction by using the principles of reference angles discussed in Chapters 6 and 7.

EXAMPLE 16-4: Find the magnitudes and directions of the following vectors:
(a) $(1, 3)$; **(b)** $(-2, -5)$; **(c)** $(0, -10)$.

Solution:

(a) The magnitude is $\sqrt{1^2 + 3^2} = \sqrt{10}$. The reference angle for direction is $\tan^{-1}\left|\frac{3}{1}\right| = 71.6°$. Since the head of the vector lies in quadrant I, the direction is $71.6°$.

(b) The magnitude is $\sqrt{(-2)^2 + (-5)^2} = \sqrt{29}$. The reference angle for direction is $\tan^{-1}\left|\frac{-5}{-2}\right| = 68.2°$. Since the head of the vector lies in quadrant III, the direction is $(180 + 68.2)° = 248.2°$.

(c) The magnitude is $\sqrt{0^2 + (-10)^2} = 10$. Since $x = 0$, the reference angle for direction is $90°$, and since the head of the vector lies on the negative *y* axis, the direction is $270°$.

• Vectors $\mathbf{v} = (x, y)$ and $\mathbf{w} = (u, v)$ are equal if and only if $x = u$ and $y = v$.

EXAMPLE 16-5: Each of the following pairs of vectors are equal; find *x* and *y*:
(a) $(x, x - y)$ and $(3, 4)$, **(b)** $(x - y, 4)$ and $(3, y + x)$.

Solution:

(a) From the definition of vector equality, $x = 3$ and $x - y = 4$. Therefore, $3 - y = 4$, or $y = 3 - 4 = -1$.

(b) From the definition of vector equality, $x - y = 3$ and $4 = y + x$. Adding these two equations, you get $2x = 7$. Thus $x = \frac{7}{2}$. Replacing x by $\frac{7}{2}$ in either equation, you get $y = \frac{1}{2}$.

16-3. Arithmetic Operations on Vectors

If k is a real number and $\mathbf{v} = (x, y)$ and $\mathbf{w} = (u, v)$ are vectors, then we can define the following arithmetic operations:

1. SCALAR MULTIPLICATION: Each component of the vector is multiplied by the scalar. In mathematical notation,

$$kv = (kx, ky) \tag{16-1}$$

2. VECTOR ADDITION: Corresponding components of the two vectors are added. In mathematical notation,

$$\mathbf{v} + \mathbf{w} = (x + u, y + v) \tag{16-2}$$

3. VECTOR SUBTRACTION: Corresponding components of the two vectors are subtracted. In mathematical notation,

$$\mathbf{v} - \mathbf{w} = (x - u, y - v) \tag{16-3}$$

4. VECTOR MULTIPLICATION: Corresponding components of the two vectors are multiplied and added. The result is a scalar quantity called the **dot product**. In mathematical notation,

$$\mathbf{v} \cdot \mathbf{w} = xu + yv \qquad \text{or} \qquad |\mathbf{v}||\mathbf{w}| \cos \theta \tag{16-4}$$

where θ is the angle between the two vectors.

EXAMPLE 16-6: Let $k = 3$, $\mathbf{v} = (2, 4)$ and $\mathbf{w} = (-1, 2)$. Find $\mathbf{v} + \mathbf{w}$, $\mathbf{v} - \mathbf{w}$, $k\mathbf{v}$, $k\mathbf{w}$, and $\mathbf{v} \cdot \mathbf{w}$.

Solution: From Equations 16-1 through 16-3,

$$\mathbf{v} + \mathbf{w} = (2, 4) + (-1, 2) = (2 - 1, 4 + 2) = (1, 6)$$

$$\mathbf{v} - \mathbf{w} = (2, 4) - (-1, 2) = (2 - (-1), 4 - 2) = (3, 2)$$

$$k\mathbf{v} = 3(2, 4) = (3 \times 2, 3 \times 4) = (6, 12)$$

$$k\mathbf{w} = 3(-1, 2) = (3 \times (-1), 3 \times 2) = (-3, 6)$$

$$\mathbf{v} \cdot \mathbf{w} = 2(-1) + 4(2) = 6$$

A. Parallelogram rule

You can represent $\mathbf{v} + \mathbf{w}$ graphically as the diagonal of the parallelogram with $\mathbf{v}$ and $\mathbf{w}$ as adjacent sides; $\mathbf{v} - \mathbf{w}$ can be represented by the diagonal of the parallelogram from the head of $\mathbf{w}$ to the head of $\mathbf{v}$ (see Figure 16-5). This representation of sums and differences of vectors is called the **parallelogram rule**.

The scalar product $k\mathbf{v}$ is a vector with the same direction as $\mathbf{v}$ if $k > 0$ and the opposite direction if $k < 0$. The magnitude of $k\mathbf{v}$ is $|k||\mathbf{v}|$. If $|k| > 1$, $k\mathbf{v}$ has greater magnitude than $\mathbf{v}$, while if $|k| < 1$, the magnitude of $k\mathbf{v}$ is smaller than the magnitude of $\mathbf{v}$.

EXAMPLE 16-7: Let $k = 3$, $\mathbf{v} = (2, 4)$ and $\mathbf{w} = (-1, 2)$. Sketch $\mathbf{v}$, $\mathbf{w}$, $\mathbf{v} + \mathbf{w}$, $\mathbf{v} - \mathbf{w}$, $k\mathbf{v}$, and $k\mathbf{w}$.

Solution: See Figure 16-6.

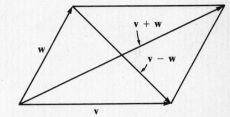

Figure 16-5
Parallelogram Rule.

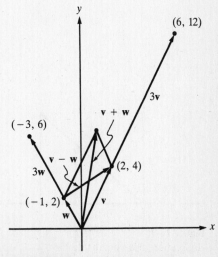

Figure 16-6

B. Unit vectors

Any vector with magnitude 1 is called a **unit vector**. If **v** is not the zero vector, then $(1/|\mathbf{v}|)\mathbf{v}$ is a unit vector.

EXAMPLE 16-8: Let $k = 3$, $\mathbf{v} = (2, 4)$ and $\mathbf{w} = (-1, 2)$. Find the following: **(a)** $\mathbf{v} + \mathbf{w}$, $\mathbf{v} - \mathbf{w}$, $k\mathbf{v}$, and $k\mathbf{w}$; **(b)** the magnitudes of **v**, **w**, and the vectors of part **a**; **(c)** unit vectors in the same direction as **v**, **w**, $\mathbf{v} + \mathbf{w}$, $\mathbf{v} - \mathbf{w}$, $k\mathbf{v}$, and $k\mathbf{w}$.

Solution:

(a) From Example 16-6,

$$\mathbf{v} + \mathbf{w} = (1, 6)$$
$$\mathbf{v} - \mathbf{w} = (3, 2)$$
$$k\mathbf{v} = (6, 12)$$
$$k\mathbf{w} = (-3, 6)$$

(b) You find magnitudes by taking the square root of the sum of the squares of the components as follows:

$$|\mathbf{v}| = \sqrt{2^2 + 4^2} = \sqrt{20}$$
$$|\mathbf{w}| = \sqrt{(-1)^2 + 2^2} = \sqrt{5}$$
$$|\mathbf{v} + \mathbf{w}| = \sqrt{1^2 + 6^2} = \sqrt{37}$$
$$|\mathbf{v} - \mathbf{w}| = \sqrt{3^2 + 2^2} = \sqrt{13}$$
$$|k\mathbf{v}| = \sqrt{6^2 + 12^2} = \sqrt{180}$$
$$|k\mathbf{w}| = \sqrt{(-3)^2 + 6^2} = \sqrt{45}$$

(c) To find a unit vector in the direction of a given vector, you divide the components of the vector by the magnitude of the vector:

$$\text{Unit vector in direction of } \mathbf{v} = \left(\frac{2}{\sqrt{20}}, \frac{4}{\sqrt{20}}\right)$$

$$\text{Unit vector in direction of } \mathbf{w} = \left(\frac{-1}{\sqrt{5}}, \frac{2}{\sqrt{5}}\right)$$

$$\text{Unit vector in direction of } \mathbf{v} + \mathbf{w} = \left(\frac{1}{\sqrt{37}}, \frac{6}{\sqrt{37}}\right)$$

$$\text{Unit vector in direction of } \mathbf{v} - \mathbf{w} = \left(\frac{3}{\sqrt{13}}, \frac{2}{\sqrt{13}}\right)$$

$$\text{Unit vector in direction of } k\mathbf{v} = \left(\frac{6}{\sqrt{180}}, \frac{12}{\sqrt{180}}\right)$$

$$\text{Unit vector in direction of } k\mathbf{w} = \left(\frac{-3}{\sqrt{45}}, \frac{6}{\sqrt{45}}\right)$$

16-4. Resolving Vectors

You can write vector $\mathbf{v} = (x, y)$ in the Cartesian plane in terms of the unit vectors $\mathbf{i} = (1, 0)$ and $\mathbf{j} = (0, 1)$ as

$$\mathbf{v} = x\mathbf{i} + y\mathbf{j} \qquad \text{where} \qquad \begin{aligned} x &= |\mathbf{v}| \cos \theta \\ y &= |\mathbf{v}| \sin \theta \end{aligned} \qquad \textbf{(16-5)}$$

The first or **horizontal component** of **v** is x and the second or **vertical component** of **v** is y (see Figure 16-7). Expressing a vector in terms of its components is called **resolution** of the vector into its components.

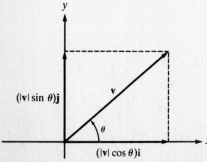

Figure 16-7
Resolution of a vector into its components.

EXAMPLE 16-9: If vector **F** represents a force of 10 pounds acting in the direction of 60° to the horizontal, resolve **F** into horizontal and vertical components (see Figure 16-8).

Solution: The force in the horizontal direction is

$$10\cos 60° = 10\left(\frac{1}{2}\right) = 5 \text{ lb}$$

The force in the vertical direction is

$$10\sin 60° = 10\frac{\sqrt{3}}{2} = 8.7 \text{ lb}$$

You can then write **F** = 5**i** + 8.7**j**.

EXAMPLE 16-10: The components of vector **v** are 3**i** and 4**j**. Find the magnitude and direction of **v** (see Figure 16-9).

Solution: From the right triangle in the figure,

$$|\mathbf{v}| = \sqrt{3^2 + 4^2} = 5$$

The direction of **v** is

$$\theta = \tan^{-1}\left(\tfrac{4}{3}\right) = 53.13°$$

16-5. Course, Bearing, and Heading

In navigation problems you will often see the terms course, bearing, and heading. They are defined as follows (see Figure 16-10):

1. **Course:** the angle measured *clockwise* from due north to the actual direction of movement of an object.
2a. **Bearing** of *A* from *B*: The angle measured *clockwise* from a north-south line through *B* to the line segment joining *B* and *A*.
2b. **Bearing** of *B* from *A*: The angle measured *clockwise* from a north-south line through *A* to the line segment joining *A* and *B*.
3. **Heading:** The heading is the direction in which a moving object is pointed. Winds or current will translate the heading into the course followed by the object; that is, course is the vector sum of heading and external physical forces.

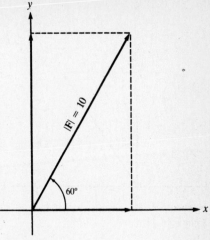

Figure 16-8

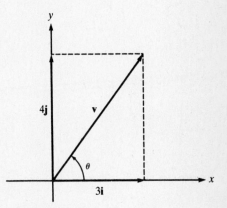

Figure 16-9

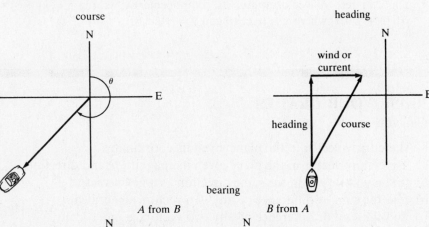

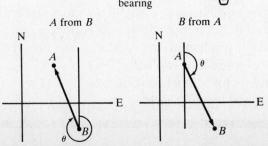

Figure 16-10
Course, bearing, and heading.

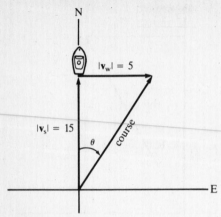

Figure 16-11

EXAMPLE 16-11: A ship is traveling at 15 miles per hour on a heading of 0° and the wind is blowing from the west at 5 miles per hour (see Figure 16-11). **(a)** Find the course of the ship and its speed. **(b)** If the ship maintains the constant speed of part **a**, what should its heading be if its destination is due north?

Solution:

(a) The velocity vector of the ship $\mathbf{v}_s$ has magnitude 15 and the velocity vector of the wind $\mathbf{v}_w$ has magnitude 5. From the right triangle formed by these vectors, you see that the magnitude of the course vector, the ship's speed, is $\sqrt{15^2 + 5^2} = 15.8$ miles per hour. The course of the ship is

$$\tan \theta = \frac{5}{15} = \frac{1}{3}$$

$$\theta = \arctan \frac{1}{3} = 18.4°$$

(b) The heading required for a course of 0° is $(360 - 18.4)° = 341.6°$.

SUMMARY

1. A vector quantity has magnitude and direction; a scalar has magnitude only.
2. If (x, y) is the algebraic representation of vector $\mathbf{v}$, its magnitude is $|\mathbf{v}| = \sqrt{x^2 + y^2}$ and the reference angle for its direction is given by $\theta = \tan^{-1}(y/x)$.
3. A unit vector in the direction of vector $\mathbf{v}$ is given by $\mathbf{v}/|\mathbf{v}|$.
4. Two vectors $\mathbf{v} = (x, y)$ and $\mathbf{w} = (u, v)$ are equal if $x = u$ and $v = y$.
5. If $\mathbf{v} = (x, y)$ and $\mathbf{w} = (u, v)$,

$$k\mathbf{v} = (kx, ky)$$

$$\mathbf{v} + \mathbf{w} = (x + u, y + v)$$

$$\mathbf{v} - \mathbf{w} = (x - u, y - v)$$

$$\mathbf{v} \cdot \mathbf{w} = xu + yv = |\mathbf{v}||\mathbf{w}| \cos \theta$$

6. The sum and difference of two vectors can be represented geometrically by the diagonals of the parallelogram with the two vectors as sides.
7. You can resolve vector $\mathbf{v}$ into its components by writing $\mathbf{v} = x\mathbf{i} + y\mathbf{j} = x(1, 0) + y(0, 1)$ where $x = |\mathbf{v}| \cos \theta$ and $y = |\mathbf{v}| \sin \theta$.

RAISE YOUR GRADES

Can you...?

☑ sketch any vector in the plane given its coordinates
☑ sketch any vector in the plane given its magnitude and direction
☑ find a unit vector in the same direction as a given vector
☑ find the sum or difference of any two vectors algebraically
☑ find the sum or difference of any two vectors geometrically
☑ find the scalar product of any number k and any vector
☑ find the dot product of two vectors
☑ resolve any vector into its components
☑ find course and heading vectors

SOLVED PROBLEMS

PROBLEM 16-1 A force of 160 pounds acts perpendicularly to a force of 120 pounds. Find the magnitude of the resultant and find the angle between the resultant and the direction of the 120-pound force.

Solution: Figure 16-12 is a diagram of the situation. The magnitude of the resultant force **R** is $\sqrt{160^2 + 120^2} = 200$ pounds, and the angle between the resultant and the 120-pound force, angle θ, is $\tan^{-1}\left(\frac{160}{120}\right) = 53.13°$ [See Section 16-3.]

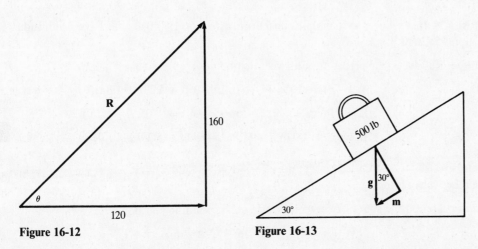

Figure 16-12   **Figure 16-13**

PROBLEM 16-2 A package weighing 500 pounds rests on a frictionless ramp. The ramp makes a 30° angle with the horizontal. How much force parallel to the ramp will be needed to keep the package from slipping?

Solution: The package exerts a force of 500 pounds in the downward direction (vector **g** in Figure 16-13). This force can be resolved into a force **m** parallel to the ramp and a force perpendicular to the ramp. By congruent triangles; the angle opposite **m** is 30°. Then,

$$\sin 30° = \frac{|\mathbf{m}|}{500}$$

$$|\mathbf{m}| = 500\sin 30° = 250 \text{ lb}$$ [See Section 16-4.]

PROBLEM 16-3 The airspeed of an airplane is 300 miles per hour. A wind blows from east to west at 40 miles per hour. In what direction should the plane head in order to travel due north? What will the groundspeed be?

Solution: In Figure 16-14, **h** is the heading of the plane and **c** is the desired course of the plane. You can see that $\sin \theta = 40/|\mathbf{h}| = 40/300$, or $\theta = \sin^{-1}(40/300) = 7.66°$. The plane should then head 7.66° east of north. The ground speed is the magnitude of **c**. This is given by $\sqrt{300^2 - 40^2} = 297.32$ miles per hour.

[See Section 16-5.]

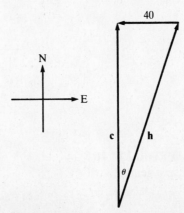

Figure 16-14

PROBLEM 16-4 Find the vector with twice the magnitude of $(3, -4)$ in the direction opposite that of vector $(1, -4)$.

Solution: The magnitude of $(3, -4)$ is $\sqrt{3^2 + (-4)^2} = 5$. By sketching vector $(-1, 4)$, you can see that it has direction opposite that of vector $(1, -4)$. You need a vector with magnitude 10 in the same direction as $(-1, 4)$. The magnitude of $(-1, 4)$ is $\sqrt{(-1)^2 + 4^2} = \sqrt{17}$. A unit vector in the direction of $(-1, 4)$ is $\left(\frac{-\sqrt{17}}{17}, \frac{4\sqrt{17}}{17}\right)$. If you multiply this unit vector by 10 you get the desired vector, $\left(\frac{-10\sqrt{17}}{17}, \frac{40\sqrt{17}}{17}\right)$. [See Section 16-3.]

PROBLEM 16-5 A vector has magnitude 25 and direction 15.4°. Find its x and y components.

Solution: The x component is $25 \cos 15.4° = 24.1$, and the y component is $25 \sin 15.4° = 6.6$.

[See Section 16-4.]

PROBLEM 16-6 Let **A** have magnitude 5.6 and direction 19° 10′, and **B** have magnitude 0.9 and direction 95° 30′. Find **A** + **B**.

Solution: Resolve each vector into unit vectors and add:

$$\mathbf{A} = 5.6(\cos 19° 10′ \mathbf{i} + \sin 19° 10′ \mathbf{j}) = 5.29\mathbf{i} + 1.84\mathbf{j}$$

$$\mathbf{B} = 0.9(\cos 95° 30′ \mathbf{i} + \sin 95° 30′ \mathbf{j}) = -0.09\mathbf{i} + 0.90\mathbf{j}$$

$$\mathbf{A} + \mathbf{B} = (5.29 - 0.09)\mathbf{i} + (1.84 + 0.90)\mathbf{j} = 5.20\mathbf{i} + 2.74\mathbf{j} \quad \text{[See Section 16-4.]}$$

PROBLEM 16-7 Let $\mathbf{v}_1 = 2\mathbf{i} - 4\mathbf{j}$ and $\mathbf{v}_2 = -\mathbf{i} + 5\mathbf{j}$. Find $\mathbf{v}_1 + \mathbf{v}_2$, $\mathbf{v}_1 - \mathbf{v}_2$, $|\mathbf{v}_1|$, $|\mathbf{v}_2|$, and the angle θ between $\mathbf{v}_1$ and $\mathbf{v}_2$.

Solution: To find the sum of two vectors, add corresponding components:

$$\mathbf{v}_1 + \mathbf{v}_2 = (2 + (-1))\mathbf{i} + (-4 + 5)\mathbf{j} = \mathbf{i} + \mathbf{j}$$

To find the difference, subtract corresponding components:

$$\mathbf{v}_1 - \mathbf{v}_2 = (2 - (-1))\mathbf{i} + (-4 - 5)\mathbf{j} = 3\mathbf{i} - 9\mathbf{j}$$

The magnitude of $\mathbf{v}_1$ is

$$|\mathbf{v}_1| = \sqrt{2^2 + (-4)^2} = 2\sqrt{5}$$
$$|\mathbf{v}_2| = \sqrt{(-1)^2 + 5^2} = \sqrt{26}$$

The angle that $\mathbf{v}_1$ makes with the positive x axis is $\tan^{-1}\left(-\frac{4}{2}\right) = -63.4°$, and the angle that $\mathbf{v}_2$ makes with the positive x axis is $\tan^{-1}\left(\frac{5}{-1}\right) + 180° = 101.3°$. The difference between these angles is $(101.3 - (-63.4))° = 164.7°$ (see Figure 16-15). [See Section 16-3.]

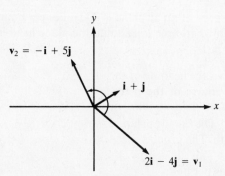

Figure 16-15

PROBLEM 16-8 For $\mathbf{v}_1 = \mathbf{i} - \mathbf{j}$ and $\mathbf{v}_2 = 5\mathbf{i} + 3\mathbf{j}$, find $\mathbf{v}_1 + \mathbf{v}_2$ and $\mathbf{v}_1 - \mathbf{v}_2$, both algebraically and geometrically.

Solution: To find the sum of two vectors, add corresponding components:

$$\mathbf{v}_1 + \mathbf{v}_2 = (1 + 5)\mathbf{i} + (-1 + 3)\mathbf{j} = 6\mathbf{i} + 2\mathbf{j}$$

To find the difference of two vectors, subtract corresponding components:

$$\mathbf{v}_1 - \mathbf{v}_2 = (1 - 5)\mathbf{i} + (-1 - 3)\mathbf{j} = -4\mathbf{i} - 4\mathbf{j}$$

See Figure 16-16 for the geometric solution. [See Section 16-3.]

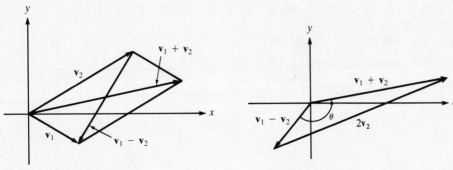

Figure 16-16	**Figure 16-17**

PROBLEM 16-9 Let θ be the angle measured from $\mathbf{v}_1 + \mathbf{v}_2 = 6\mathbf{i} + 2\mathbf{j}$ to $\mathbf{v}_1 - \mathbf{v}_2 = -4\mathbf{i} - 4\mathbf{j}$. Find $\cos \theta$.

Solution: From the Figure 16-17, you see that the vector

$$\mathbf{v}_1 + \mathbf{v}_2 - (\mathbf{v}_1 - \mathbf{v}_2) = 2\mathbf{v}_2 = 10\mathbf{i} + 6\mathbf{j}$$

is opposite the desired angle. The magnitudes of the three vectors as shown in the figure are

$$|\mathbf{v}_1 + \mathbf{v}_2| = \sqrt{40}, |\mathbf{v}_1 - \mathbf{v}_2| = \sqrt{32} \text{ and } |2\mathbf{v}_2| = \sqrt{136}$$

Apply the law of cosines to the triangle:

$$(\sqrt{136})^2 = (\sqrt{40})^2 + (\sqrt{32})^2 - 2\sqrt{40}(\sqrt{32})\cos \theta$$

$$\cos \theta = -\frac{2\sqrt{5}}{5}$$

[See Section 16-3.]

PROBLEM 16-10 A force of 60 pounds act horizontally on an object. Another force of 75 pounds acts on the object at an angle of 55° above the horizontal. What is the resultant of these forces?

Solution: You can represent the force of 60 pounds by the vector $60\mathbf{i}$, and the 75 pound force by the vector $75\mathbf{i} \cos 55° + 75\mathbf{j} \sin 55° = 43\mathbf{i} + 61.4\mathbf{j}$. The sum of these vectors is $103\mathbf{i} + 61.4\mathbf{j}$, a vector with magnitude $\sqrt{103^2 + 61.4^2} = 119.9$ pounds with angle equal to $\tan^{-1}(61.4/103) = 30.8°$ above the horizontal. [See Section 16-4.]

In Problems 16-11 and 16-12, find a vector (**a**) in the same direction as the given vector, (**b**) in the opposite direction to the given vector, and (**c**) perpendicular to the given vector.

PROBLEM 16-11 $\mathbf{v} = \mathbf{i} + \mathbf{j}$

Solution: You can find a vector in the same direction as the given vector by multiplying by the scalar 2, in the opposite direction by multiplying by the scalar -1, and you can find a perpendicular vector by reversing the coefficients and changing one sign (see Figure 16-18). The vectors $2\mathbf{i} + 2\mathbf{j}$, $-\mathbf{i} - \mathbf{j}$, and $-\mathbf{i} + \mathbf{j}$, respectively, satisfy the given conditions. [See Section 16-3.]

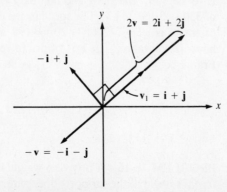

Figure 16-18

PROBLEM 16-12 $v = -3i + 4j$

Solution: Apply the steps outlined in Problem 16-11. You get $2(-3i + 4j) = -6i + 8j$ as a vector in the same direction, $-1(-3i + 4j) = 3i - 4j$ as a vector in the opposite direction, and $4i + 3j$ as a perpendicular vector.   [See Section 16-3.]

PROBLEM 16-13   An airplane travels on a course of 220° with a speed of 600 miles per hour. How far south does the plane fly in 2 hours?

Solution: From the triangle in Figure 16-19, you see that the distance south d_s flown in two hours is $1200 \cos 40° = 919.3$ mi.   [See Section 16-5.]

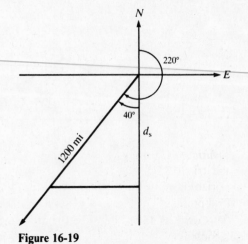

Figure 16-19

PROBLEM 16-14   An artillery cannon is inclined at 15° above the horizontal. The initial speed of a projectile fired from the cannon is 2000 feet per second; find the magnitudes of the horizontal and vertical components of the initial velocity.

Solution: From the triangle in Figure 16-20, you see that the horizontal component is $2000 \cos 15° = 1931.9$ feet per second; the vertical component is $2000 \sin 15° = 517.6$ feet per second.   [See Section 16-4.]

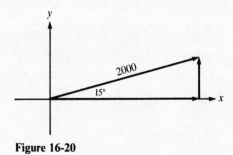

Figure 16-20

PROBLEM 16-15   A pilot wants to fly on course 350° from Los Angeles to San Francisco, a distance of 300 miles. If the plane cruises at 200 miles per hour and there is a steady wind of 20 miles per hour from the west, what heading should the pilot take? How long will the trip take?

Solution: From the triangle shown in Figure 16-21, you see that angle θ between the heading vector **h** with $|h| = 200$ and course vector **c** is $\sin^{-1}(20/200) = 5.7°$. If the desired course is to be 350°, the pilot should take a heading of $350 - 5.7 = 344.3°$. The speed in the direction of travel is $|c| = \sqrt{200^2 - 20^2} = 199$ miles per hour. It will take $\frac{300}{199} = 1.51$ hours to reach San Francisco.   [See Section 16-5.]

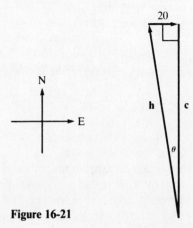

Figure 16-21

PROBLEM 16-16   A boy who weighs 80 pounds sits on a swing. His father pushes the swing so that the angle of the support ropes with the vertical is 10°; what is the force on each of the support ropes?

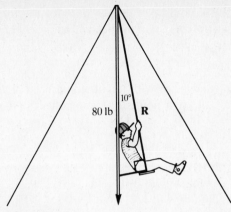

Figure 16-22

Solution: The resultant force vector on the support ropes is denoted by **R** in Figure 16-22. From the triangle of vectors formed by the boy's downward force of magnitude 80 pounds, the horizontal force, and the resultant force, you see that $\cos 10° = |\mathbf{R}|/80$ and $|\mathbf{R}| = 80 \cos 10° = 78.8$ pounds. The force on each rope is $78.8/2 = 39.4$ pounds.   [See Section 16-4.]

PROBLEM 16-17   Find the magnitude and angle with the positive x axis of $-6\mathbf{i} + 4\mathbf{j}$.

Solution: The magnitude is $\sqrt{(-6)^2 + 4^2} = \sqrt{52}$. The angle with the positive x axis is $\tan^{-1}\left(\frac{4}{-6}\right) + 180° = 146.3°$.

PROBLEM 16-18   Find the magnitude and angle with the positive x axis of $-3\mathbf{i} - 4\mathbf{j}$.

Solution: The magnitude is $\sqrt{(-3)^2 + (-4)^2} = \sqrt{25} = 5$. The angle with the positive x axis is $\tan^{-1}\left(\frac{-4}{-3}\right) + 180° = 233.1°$.   [See Section 16-4.]

PROBLEM 16-19   Let **u** and **v** be vectors oriented as shown in Figure 16-23. Find $|\mathbf{v}|$ when $|\mathbf{u}| = 5$ and $|\mathbf{u} + \mathbf{v}| = 13$.

Solution: Apply the law of cosines:

$$13^2 = 5^2 + |\mathbf{v}|^2 - 2(5)(|\mathbf{v}|)\cos 160°$$

$$|\mathbf{v}|^2 + 9.4|\mathbf{v}| - 144 = 0$$

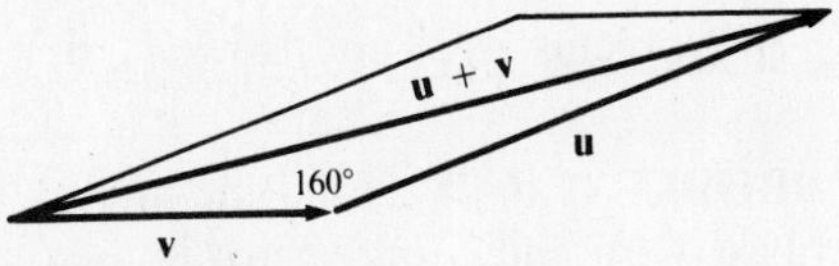

Figure 16-23

Use the quadratic formula:

$$|\mathbf{v}| = \frac{-9.4 \pm \sqrt{9.4^2 - 4(1)(-144)}}{2} = 8.2$$

(The negative root is discarded.)   [See Section 16-4.]

For Problems 16-20 through 16-22, find a representation of the form $a\mathbf{i} + b\mathbf{j}$ for each vector, where θ is the angle that the vector makes with the positive x axis.

PROBLEM 16-20 $|\mathbf{v}| = 5$ and $\theta = 90°$.

Solution: The horizontal component is $5 \cos 90° = 0$ and the vertical component is $5 \sin 90° = 5$, so $\mathbf{v} = 0\mathbf{i} + 5\mathbf{j}$.

PROBLEM 16-21 $|\mathbf{v}| = 5$ and $\theta = 270°$.

Solution The horizontal component is $5 \cos 270° = 0$ and the vertical component is $5 \sin 270° = -5$, so $\mathbf{v} = 0\mathbf{i} - 5\mathbf{j}$.

PROBLEM 16-22 $|\mathbf{v}| = 10$ and $\theta = 320°$.

Solution: The horizontal component is $10 \cos 320° = 7.7$ and the vertical component is $10 \sin 320° = -6.4$, so $\mathbf{v} = 7.7\mathbf{i} - 6.4\mathbf{j}$.

For Problems 16-23 through 16-25, find a representation of the form $a\mathbf{i} + b\mathbf{j}$ for each vector, where the tail of the vector is at point A and the head at point B. [See Section 16-4.]

PROBLEM 16-23 $A(8, 10)$ and $B(-3, -4)$

Solution: The components of the vector with tail at A and head at B are found by subtracting corresponding components of the two points: the component of $\mathbf{i} = -3 - 8 = -11$ and the component of $\mathbf{j} = -4 - 10 = -14$, so $\mathbf{v} = -11\mathbf{i} - 14\mathbf{j}$.

PROBLEM 16-24 $A(8, -10)$ and $B(-3, -4)$

Solution: Subtract corresponding components of the two points: the component of $\mathbf{i} = -3 - 8 = -11$ and the component of $\mathbf{j} = -4 - (-10) = 6$, so $\mathbf{v} = -11\mathbf{i} + 6\mathbf{j}$.

PROBLEM 16-25 $A(1, -2)$ and $B(-3, -4)$

Solution: Subtract corresponding components of the two points: the component of $\mathbf{i} = -3 - 1 = -4$ and the component of $\mathbf{j} = -4 - (-2) = -2$, so $\mathbf{v} = -4\mathbf{i} - 2\mathbf{j}$.

PROBLEM 16-26 Find a representation of the form $a\mathbf{i} + b\mathbf{j}$ for the unit vector that makes a $120°$ angle with the positive x axis.

Solution: Since the magnitude of a unit vector is 1, the components in the horizontal and vertical directions are $\cos 120° = -0.5$ and $\sin 120° = \frac{\sqrt{3}}{2}$, respectively; $\mathbf{v} = -0.5\mathbf{i} + \frac{\sqrt{3}}{2}\mathbf{j}$.
[See Section 16-4.]

PROBLEM 16-27 Find a representation of the form $a\mathbf{i} + b\mathbf{j}$ for the unit vector that makes a $225°$ angle with the positive x axis.

Solution: The components in the horizontal and vertical directions are $\cos 225° = \frac{-\sqrt{2}}{2}$ and $\sin 225° = \frac{-\sqrt{2}}{2}$, respectively; $\mathbf{v} = -\sqrt{2}\mathbf{i}/2 - \sqrt{2}\mathbf{j}/2$.
[See Section 16-4.]

For Problems 16-28 and 16-29, let $\mathbf{u} = 2\mathbf{i} + 3\mathbf{j}$, $\mathbf{v} = -3\mathbf{i} + 4\mathbf{j}$, and $\mathbf{w} = 12\mathbf{i} + 5\mathbf{j}$.

PROBLEM 16-28 Find the angle that $2\mathbf{u} + 3\mathbf{v}$ makes with the positive x axis, the magnitude of $2\mathbf{u} + 3\mathbf{v}$, and a unit vector in the direction of $2\mathbf{u} + 3\mathbf{v}$.

Solution: Adding components, $2\mathbf{u} + 3\mathbf{v} = -5\mathbf{i} + 18\mathbf{j}$. The angle with the positive x axis is $\tan^{-1}\left(\frac{18}{-5}\right) + 180° = 105.5°$, and the magnitude is $|2\mathbf{u} + 3\mathbf{v}| = \sqrt{(-5)^2 + 18^2} = 18.7$. A unit vector in the same direction as $2\mathbf{u} + 3\mathbf{v}$ is found by dividing $2\mathbf{u} + 3\mathbf{v}$ by 18.7. The result is $\left(\frac{-5}{18.7}\right)\mathbf{i} + \left(\frac{18}{18.7}\right)\mathbf{j}$.
[See Section 16-4.]

PROBLEM 16-29 Find the angle that $3\mathbf{u} - 4\mathbf{w}$ makes with the positive x axis, the magnitude of $3\mathbf{u} - 4\mathbf{w}$, and a unit vector in the direction of $3\mathbf{u} - 4\mathbf{w}$.

Solution: Adding components, $3\mathbf{u} - 4\mathbf{w} = -42\mathbf{i} - 11\mathbf{j}$. The angle with the positive x axis is $\tan^{-1}\left(\frac{-11}{-42}\right) + 180° = 194.7°$ and the magnitude is $|3\mathbf{u} - 4\mathbf{w}| = \sqrt{(-42)^2 + (-11)^2} = 43.4$. A unit vector in the same direction as $3\mathbf{u} - 4\mathbf{w}$ is found by dividing $3\mathbf{u} - 4\mathbf{w}$ by 43.4. The result is $\left(\frac{-42}{43.4}\right)\mathbf{i} + \left(\frac{-11}{43.4}\right)\mathbf{j}$.
[See Section 16-4.]

Supplementary Exercises

PROBLEM 16-30 A person slides a 500-pound package up a frictionless ramp. The person can push with a force of 200 pounds; what is the greatest angle formed by the ramp and the horizontal such that the person doesn't lose ground?

PROBLEM 16-31   A package is strung up by two ropes. The package exerts a force of 200 pounds against one rope, which is at an angle of 30° to the vertical. The other rope is at an angle of 45° to the vertical. What is the weight of the package?

PROBLEM 16-32   Find the vector of unit length in the direction of the vector $(5, -12)$.

PROBLEM 16-33   Consider the points $A = (1, 3)$, $B = (-2, 7)$, $C = (-3, -2)$, and $D = (2, 5)$. Let v_1 be the vector with tail at A and head at B, v_2 be the vector with tail at B and head at C, v_3 be the vector with tail at C and head at D. Find $v_1 + v_2 + v_3$.

PROBLEM 16-34   What angle does vector $(7, 4)$ make with the x axis?

PROBLEM 16-35   For $v_1 = (1, 5)$ and $v_2 = (-1, 3)$, find $2v_1 - 3v_2$.

PROBLEM 16-36   An airplane is headed due north at 300 miles per hour. The wind is from the east at 50 miles per hour; find the course and velocity of the airplane.

PROBLEM 16-37   An object is acted upon by a force of 200 pounds at an angle of 35° to the horizontal. What is the horizontal component of the force?

PROBLEM 16-38   A bullet is fired from an airplane at an angle of 20° below the horizontal and in the direction the airplane is moving. The bullet leaves the muzzle of the gun with a speed of 1200 feet per second and the airplane is flying 500 feet per second; what is the resultant speed of the bullet?

PROBLEM 16-39   An airplane flies due west with an airspeed of 150 miles per hour. The wind is blowing from the southwest (45° south of west) at 35 miles per hour; find the resultant speed of the airplane and the direction it is traveling.

In the following problems find a vector (**a**) in the same direction as the given vector, (**b**) in the opposite direction to the given vector, and (**c**) perpendicular to the given vector.

PROBLEM 16-40 $v = 5i + (-3j)$

PROBLEM 16-41 $v = i - j$.

PROBLEM 16-42   A man rows straight north at 4.5 miles per hour. The current is carrying the boat to the west at 5 miles per hour. Find the course of the boat.

PROBLEM 16-43   An airplane heads east at 200 miles per hour. The wind is blowing from south to north at 40 miles per hour. Find the course and speed of the airplane.

PROBLEM 16-44   A 4000 pound car is parked on a hill that is inclined at a 5° angle to the horizontal. Find the minimum force required to keep the car from rolling down the hill.

PROBLEM 16-45   An airplane travels on a course of 120° with a speed of 200 miles per hour. How far east does the plane fly in 1 hour?

PROBLEM 16-46   An artillery cannon is inclined at 10° above the horizontal. The initial speed of a projectile fired from the cannon is 1500 feet per second; find the magnitudes of the horizontal and vertical components of the initial velocity.

PROBLEM 16-47   A rocket rises at 500 miles per second at an angle of 72° above the horizontal. How fast is the altitude of the rocket increasing?

PROBLEM 16-48   A boat can travel at 4 kilometers per hour in still water. It heads from one bank of a stream directly to the other bank. The current is 2 kilometers per hour. Find the resultant speed and direction of the boat.

PROBLEM 16-49 Given that the boat in Problem 16-48 is supposed to reach a point directly opposite on the other bank. Find the direction in which it should head. What is the resultant speed of the boat if it heads in this direction?

PROBLEM 16-50 An airplane is flying at 600 miles per hour with an 80 mile per hour wind at right angles to the airplane. Find the angle between the course and heading of the airplane.

PROBLEM 16-51 A ship is sailing due west at a constant speed. At 3 PM a lighthouse is observed on a bearing of 310° at a distance of 20 nautical miles. At 4 PM the bearing is 320°. Find the speed of the ship.

PROBLEM 16-52 Find the magnitude and angle with the positive x axis of $3\mathbf{i} + 2\mathbf{j}$.

PROBLEM 16-53 Find the magnitude and angle with the positive x axis is $2\mathbf{i} - 4\mathbf{j}$.

PROBLEM 16-54 Find the horizontal and vertical components of a vector of magnitude 5 whose angle with the positive x axis is 58°.

PROBLEM 16-55 Find the horizontal and vertical components of a vector of magnitude 5 whose angle with the positive x axis is 135°.

PROBLEM 16-56 Let $\mathbf{u}$ and $\mathbf{v}$ be vectors with $\mathbf{u}$ oriented straight north and $\mathbf{v}$ oriented straight east. Find $|\mathbf{u} + \mathbf{v}|$ and the angle between $\mathbf{u}$ and $\mathbf{u} + \mathbf{v}$ when (a) $|\mathbf{u}| = 2$ and $|\mathbf{v}| = 3$ and (b) $|\mathbf{u}| = 12$ and $|\mathbf{v}| = 5$.

Find a representation of the form $a\mathbf{i} + b\mathbf{j}$ for each of the following vectors, where θ is the angle that the vector makes with the positive x axis.

PROBLEM 16-57 $|\mathbf{v}| = 8$ and $\theta = 30°$

PROBLEM 16-58 $|\mathbf{v}| = 12$ and $\theta = 120°$

PROBLEM 16-59 $|\mathbf{v}| = 18$ and $\theta = 200°$

Find a representation of the form $a\mathbf{i} + b\mathbf{j}$ for each of the following vectors, where the tail of the vector is point A and the head of the vector is point B.

PROBLEM 16-60 $A(2, 3)$ and $B(3, 4)$

PROBLEM 16-61 $A(-2, 3)$ and $B(3, 4)$

PROBLEM 16-62 $A(8, 10)$ and $B(3, 4)$

PROBLEM 16-63 Find a representation of the form $a\mathbf{i} + b\mathbf{j}$ for the unit vector that makes a 55° angle with the positive x axis.

Let $\mathbf{u} = 2\mathbf{i} + 3\mathbf{j}$, $\mathbf{v} = -3\mathbf{i} + 4\mathbf{j}$, and $\mathbf{w} = 12\mathbf{i} + 5\mathbf{j}$.

PROBLEM 16-64 Find the magnitudes of $\mathbf{u}$, $\mathbf{v}$, and $\mathbf{w}$.

PROBLEM 16-65 Find unit vectors in the same directions as $\mathbf{u}$, $\mathbf{v}$, and $\mathbf{w}$.

PROBLEM 16-66 Find the angle that $\mathbf{u} + \mathbf{v} + \mathbf{w}$ makes with the positive x axis, the magnitude of $\mathbf{u} + \mathbf{v} + \mathbf{w}$, and a unit vector in the direction of $\mathbf{u} + \mathbf{v} + \mathbf{w}$.

PROBLEM 16-67 Find the angle that $\mathbf{v} - 4\mathbf{w}$ makes with the positive x axis, the magnitude of $\mathbf{v} - 4\mathbf{w}$, and a unit vector in the direction of $\mathbf{v} - 4\mathbf{w}$.

PROBLEM 16-68 Are the vectors $\mathbf{i} + \mathbf{j}$ and $\mathbf{i} - \mathbf{j}$ perpendicular to each other?

PROBLEM 16-69 Are the vectors $\sqrt{2}\mathbf{i} + \sqrt{2}\mathbf{j}$ and $-\sqrt{2}\mathbf{i} + \sqrt{2}\mathbf{j}$ perpendicular to each other?

PROBLEM 16-70 Are the vectors $\mathbf{i}$ and $-\mathbf{j}$ perpendicular to each other?

Answers to Supplementary Exercises

16-30: 23.58°

16-31: $100(1 + \sqrt{3})$ lb

16-32: $\left(\frac{5}{13}, \frac{-12}{13}\right)$

16-33: $\mathbf{i} + 2\mathbf{j}$

16-34: 29.74°

16-35: $5\mathbf{i} + \mathbf{j}$

16-36: 350.54° and 304.14 mi/h

16-37: 163.8 lb

16-38: 1627.6 ft/s

16-39: speed = 127.7 mi/h and course = 281.2°

16-40: $10\mathbf{i} - 6\mathbf{j}, -5\mathbf{i} + 3\mathbf{j}, 3\mathbf{i} + 5\mathbf{j}$

16-41: $2\mathbf{i} - 2\mathbf{j}, -\mathbf{i} + \mathbf{j}, \mathbf{i} + \mathbf{j}$

16-42: course = 312°

16-43: course = 78.7°, speed = 204 mi/h

16-44: 348.6 lb

16-45: 173.2 mi

16-46: horizontal = 1477.2, vertical = 260.5

16-47: 475.5 mi/s

16-48: speed = $\sqrt{20}$ km/h, angle between a line directly across the stream and the direction vector is 26.6°

16-49: angle between the heading vector and a line directly across the stream is 26.6° in the upstream direction; speed = $\sqrt{12}$ km/h

16-50: 7.6°

16-51: 4.5 nautical miles per hour

16-52: magnitude = $\sqrt{13}$, angle = $\tan^{-1}\left(\frac{2}{3}\right) =$ 33.7°

16-53: magnitude = $\sqrt{20}$, angle = $\tan^{-1}\left(\frac{-4}{2}\right) =$ $-63.4°$

16-54: horizontal = 2.6, vertical = 4.2

16-55: horizontal = -3.5, vertical = 3.5

16-56: **(a)** $|\mathbf{u} + \mathbf{v}| = \sqrt{13}; \theta = 56.3°;$ **(b)** $|\mathbf{u} + \mathbf{v}| =$ 13, $\theta = 22.6°$

16-57: $4\sqrt{3}\mathbf{i} + 4\mathbf{j}$

16-58: $-6\mathbf{i} + 6\sqrt{3}\mathbf{j}$

16-59: $-16.9\mathbf{i} - 6.2\mathbf{j}$

16-60: $\mathbf{i} + \mathbf{j}$

16-61: $5\mathbf{i} + \mathbf{j}$

16-62: $-5\mathbf{i} - 6\mathbf{j}$

16-63: $0.57\mathbf{i} + 0.82\mathbf{j}$

16-64: $\sqrt{13}$, 5, and 13

16-65: $\frac{2}{\sqrt{13}}\mathbf{i} + \frac{3}{\sqrt{13}}\mathbf{j}, -\left(\frac{3}{5}\right)\mathbf{i} + \left(\frac{4}{5}\right)\mathbf{j}$, and $\left(\frac{12}{13}\right)\mathbf{i} + \left(\frac{5}{13}\right)\mathbf{j}$

16-66: angle = 47.5°, magnitude = 16.3, $\left(\frac{11}{16.3}\right)\mathbf{i} + \left(\frac{12}{16.3}\right)\mathbf{j}$

16-67: angle = 197.4°, magnitude = 53.45, $\left(\frac{-51}{53.45}\right)\mathbf{i} + \left(\frac{-16}{53.45}\right)\mathbf{j}$

16-68: yes

16-69: yes

16-70: yes

17 COMPLEX NUMBERS

THIS CHAPTER IS ABOUT

☑ **Algebraic Representations of Complex Numbers**
☑ **Arithmetic Operations on Complex Numbers**
☑ **Geometric Representations of Complex Numbers**
☑ **Geometric Operations on Complex Numbers**
☑ **Polar Form of a Complex Number**

17-1. Algebraic Representations of Complex Numbers

A. Imaginary numbers

The equation $x^2 + 1 = 0$ has no solution in the set of real numbers. The **imaginary number**, i, is defined as the number that satisfies $x^2 + 1 = 0$. Note that $i^2 = -1$ or $i = \sqrt{-1}$.

EXAMPLE 17-1: **(a)** Find the first five powers of i; **(b)** Evaluate i^{53}; **(c)** Evaluate i^{78}.

Solution:

(a) $i^1 = i$
$i^2 = -1$
$i^3 = i^2(i) = -i$
$i^4 = i^3(i) = -i(i) = -(i^2) = -(-1) = 1$
$i^5 = i^4(i) = 1(i) = i$
Because the fifth power of i equals i, the first four powers of i display the only possible values for powers of i. This means that

$$i^{4n} = 1$$
$$i^{4n+1} = i$$
POWERS OF i $\qquad\qquad\qquad\qquad$ **(17-1)**
$$i^{4n+2} = -1$$
$$i^{4n+3} = -i$$

where n is any integer.

(b) To evaluate i^n, divide n by 4:

If the remainder is	$i^n =$
0	1
1	i
2	-1
3	$-i$

For i^{53}, $\frac{53}{4}$ equals 13 with a remainder of 1, so $i^{53} = i$.
(c) Dividing 78 by 4, you get 19 with a remainder of 2, so $i^{78} = -1$.

B. Complex numbers

If x and y are two real numbers, then $z = x + yi$ is called a **complex number** in standard form. The number x is the **real part** of z, denoted by Re(z), and y is the **imaginary part** of z, denoted by Im(z). In standard form, $z = x + yi = \text{Re}(z) + \text{Im}(z)i$.

EXAMPLE 17-2: Find the real and imaginary parts of $2 + 3i$.

Solution: If $z = 2 + 3i$, then Re(z) = Re($2 + 3i$) = 2 and Im(z) = Im($2 + 3i$) = 3.

- Complex numbers of the form $x + 0i = x$ are called **real** and those of the form $z = 0 + yi = yi$ are called **pure imaginary**. The zero complex number is $0 + 0i = 0$. The **complex conjugate** of $z = x + yi$, is written as follows:

COMPLEX **CONJUGATE**	$\bar{z} = x - yi = x + (-y)i$	**(17-2)**

Note: The sum and product of conjugate complex numbers—$z + \bar{z}$ and $z\bar{z}$—are real numbers.

EXAMPLE 17-3: Find the conjugates of $-6 + 2i$, 3, and $-8i$.

Solution: The conjugate of $-6 + 2i$ is $\bar{z} = -6 - 2i$; the conjugate of 3 ($3 + 0i$) is $\bar{z} = 3 - 0i = 3$; the conjugate of $-8i$ ($0 - 8i$) is $\bar{z} = 0 + 8i = 8i$.

- Complex number z_1 equals complex number z_2, that is, $z_1 = z_2$, if and only if the real part of z_1 equals the real part of z_2 and the imaginary part of z_1 equals the imaginary part of z_2. In symbols, $x_1 + y_1 i = x_2 + y_2 i$, if and only if $x_1 = x_2$ and $y_1 = y_2$.

EXAMPLE 17-4: Are $\frac{4}{2} + \sqrt{9}i$ and $2 + 3i$ equal?

Solution: Re($\frac{4}{2} + \sqrt{9}i$) = $\frac{4}{2}$ = 2; Re($2 + 3i$) = 2. Im($\frac{4}{2} + \sqrt{9}i$) = $\sqrt{9}$ = 3; Im($2 + 3i$) = 3. Because the real and imaginary parts of both numbers are equal, the numbers are equal.

EXAMPLE 17-5: Are $5 + 7i$ and $2 + 7i$ equal?

Solution: Because Re($5 + 7i$) = $5 \neq 2$ = Re($2 + 7i$), the numbers aren't equal.

Note: The complex number $z = x + yi$ can also be written as the ordered pair (x, y). All of the standard-form properties of complex numbers can be translated to this **ordered pair representation**.

EXAMPLE 17-6: Write $-6 + 2i$, 3, and $-8i$ and their conjugates using ordered pair notation.

Solution: If $z = -6 + 2i = (-6, 2)$, then $\bar{z} = -6 - 2i = (-6, -2)$. If $z = 3 + 0i = (3, 0)$, then $\bar{z} = 3 - 0i = (3, 0)$. If $z = 0 - 8i = (0, -8)$, then $\bar{z} = 0 + 8i = (0, 8)$.

17-2. Arithmetic Operations on Complex Numbers

The operations of addition, subtraction, multiplication, and division for real numbers can be extended to the complex numbers. In this section, $z_1 = x_1 + y_1 i$ and $z_2 = x_2 + y_2 i$.

A. Addition

The sum of two complex numbers is the complex number whose real part is the sum of the two real parts and whose imaginary part is the sum of the two imaginary parts. This takes the algebraic form

ADDITION $\qquad z_1 + z_2 = (x_1 + x_2) + (y_1 + y_2)i$ $\qquad$ **(17-3)**

EXAMPLE 17-7:

$$(6 + 4i) + (4 + 3i) = (6 + 4) + (4 + 3)i = 10 + 7i$$

$$\left(\frac{1}{2} + \sqrt{2}i\right) + (2 + i) = \left(\frac{1}{2} + 2\right) + (\sqrt{2} + 1)i = \frac{5}{2} + (1 + \sqrt{2})i$$

B. Subtraction

The difference of two complex numbers is the complex number whose real part is the difference of the two real parts and whose imaginary part is the difference of the two imaginary parts. This takes the algebraic form

SUBTRACTION $\qquad z_1 - z_2 = (x_1 - x_2) + (y_1 - y_2)i$ $\qquad$ **(17-4)**

EXAMPLE 17-8:

$$(6 + 4i) - (4 + 3i) = (6 - 4) + (4 - 3)i = 2 + i$$

$$\left(\frac{1}{2} + \sqrt{2}i\right) - (2 + i) = \left(\frac{1}{2} - 2\right) + (\sqrt{2} - 1)i = -\frac{3}{2} + (\sqrt{2} - 1)i$$

C. Multiplication

The product of two complex numbers is the complex number obtained by multiplying each term in the first factor by each term in the second factor, replacing i^2 by -1, and combining like terms:

MULTIPLICATION $\qquad \begin{aligned} z_1 z_2 &= x_1 x_2 + y_1 y_2 i^2 + x_1 y_2 i + y_1 x_2 i \\ &= (x_1 x_2 - y_1 y_2) + (x_1 y_2 + x_2 y_1)i \end{aligned}$ $\qquad$ **(17-5)**

EXAMPLE 17-9:

$$(2 + 4i)(-6 + 3i) = [2(-6) - (4)(3)] + [2(3) + (4)(-6)]i = -24 - 18i$$

Note: The product of real number r and complex number z is a special case of the definition of multiplication. Because $r = r + 0i$,

$$rz = (r + 0i)(x + yi) = (rx - 0y) + (0x + ry)i = rx + ryi. \quad \textbf{(17-6)}$$

In other words, you multiply a complex number by a real number by multiplying the real and imaginary parts of the complex number by the real number.

EXAMPLE 17-10: Perform the following multiplications: **(a)** $3(2 - i)$; **(b)** $3(2i)$; **(c)** $-2(4 + i)$

Solution:

(a) $3(2 - i) = 6 - 3i$
(b) $3(2i) = 6i$
(c) $-2(4 + i) = -8 - 2i$

D. Division

The division formula for two complex numbers is derived by multiplying the quotient of the two complex numbers by the complex conjugate of the denominator:

$$\frac{z_1}{z_2} = \frac{x_1 + y_1 i}{x_2 + y_2 i} = \frac{x_1 + y_1 i}{x_2 + y_2 i}\left(\frac{x_2 - y_2 i}{x_2 - y_2 i}\right) = \frac{(x_1 x_2 + y_1 y_2) + (y_1 x_2 - x_1 y_2)i}{x_2^2 + y_2^2}$$

In standard form,

DIVISION $$\frac{z_1}{z_2} = \frac{x_1 x_2 + y_1 y_2}{x_2^2 + y_2^2} + \frac{(y_1 x_2 - x_1 y_2)}{x_2^2 + y_2^2} i \qquad \textbf{(17-7)}$$

EXAMPLE 17-11: Divide $z_1 = 2 + 2i$ by $z_2 = 4 - 2i$.

Solution: Substitute $x_1 = 2$, $y_1 = 2$, $x_2 = 4$, and $y_2 = -2$ into Equation 17-7:

$$\frac{z_1}{z_2} = \frac{2(4) + 2(-2)}{4^2 + (-2)^2} + \frac{(2(4) - 2(-2))}{4^2 + (-2)^2} i$$

$$= \frac{4}{20} + \frac{12}{20} i = \frac{1}{5} + \frac{3}{5} i$$

In summary: For complex numbers in standard form, you add, subtract, multiply, and divide expressions by using the ordinary rules of algebra, replacing i^2 by -1.

17-3. Geometric Representations of Complex Numbers

You can represent any complex number $z = x + yi$ geometrically in two ways: as the point $P(x, y)$, or as a vector **r**. Both representations are made on the **complex**, or **Gaussian**, plane where the x axis is the real number axis and the y axis is the imaginary axis (see Figure 17-1).

EXAMPLE 17-12: Plot a point in the complex plane that corresponds to: (**a**) $3 + 2i$; (**b**) $-4i$; (**c**) $2 - i$; (**d**) $2 + i$; (**e**) $i - 2$.

Solution: Figure 17-2 shows these points.

For $z = x + yi$, any vector **r** with initial point at an arbitrary point (a, b) and terminal point at $(a + x, b + y)$ is an equivalent vector representation of z.

EXAMPLE 17-13: Sketch the vector **r** representing $z = 3 + 4i$ with initial point at $(-2, 1)$.

Solution: The terminal point is at $(-2 + 3, 1 + 4) = (1, 5)$. The vector is shown in Figure 17-3.

For $z = x + yi$, the magnitude of the vector **r** representing z is the **modulus** or **absolute value** of z, denoted $|z|$:

MODULUS $$|z| = |\mathbf{r}| = \sqrt{x^2 + y^2} \qquad \textbf{(17-8)}$$

The smallest nonnegative angle measured from the positive x axis, or a line parallel to the positive x axis if the initial point of the vector isn't the origin, to **r** is the **argument** of z, denoted $\arg z$. The reference angle corresponding to this argument is $\theta = \arctan |y/x|$ if $x \neq 0$, and $\theta = 90°$ if $x = 0$. You find the argument by using the principles of reference angles developed in Chapters 6 and 7. The arguments of several complex numbers are shown in Figure 17-4.

EXAMPLE 17-14: Find the modulus and argument of $z = -\sqrt{3} + i$.

Solution: The modulus of z is

$$|z| = \sqrt{(-\sqrt{3})^2 + 1^2} = 2$$

The argument of z is

$$\arg z = \pi - \frac{\pi}{6} = \frac{5\pi}{6}$$

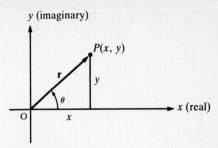

Figure 17-1
Complex plane.

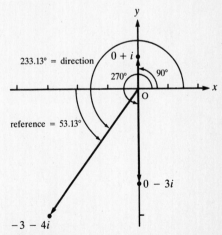

Figure 17-2

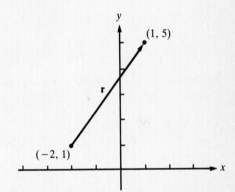

Figure 17-3

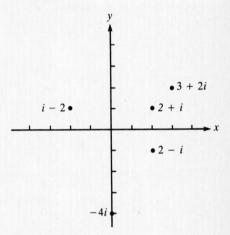

Figure 17-4
Arguments of selected complex numbers.

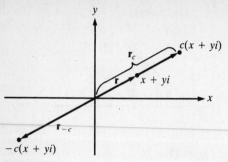

Figure 17-5
Scalar multiplication of a complex numbers.

17-4. Geometric Operations on Complex Numbers

A. Scalar multiplication

For any real number c and complex number z, cz is a complex number with magnitude $|c|$ times the modulus of z. Thus $|cz| = |c||z|$. An argument of cz is $\arg z$ if $c > 0$, and $\pi + \arg z$ if $c < 0$.

If $cz \neq 0$, then $\mathbf{r}_c$, a vector representing cz, is parallel to $\mathbf{r}$, the vector representing z. The vector $\mathbf{r}_c$ points in the same direction as $\mathbf{r}$ if $c > 0$, and in the opposite direction if $c < 0$. This is shown in Figure 17-5.

B. Vector addition

Let vectors $\mathbf{r}_1, \mathbf{r}_2, \mathbf{r}_3 \cdots \mathbf{r}_n$ represent complex numbers $z_1, z_2, z_3 \cdots z_n$. The resultant vector $\mathbf{R}$ that represents complex number $z_1 + z_2 + \cdots + z_n$ is found by placing the vectors head to tail and drawing a new vector from the tail of the first to the head of the last (see Figure 17-6).

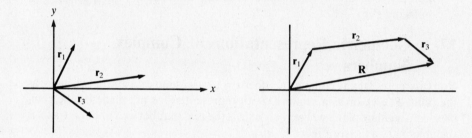

Figure 17-6
Vector addition.

EXAMPLE 17-15: Find the sum of $2 + i$ and $1 + 3i$ by the vector method.

Solution: Figure 17-7 shows the solution.

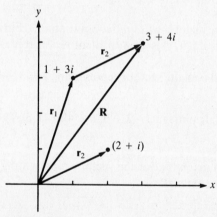

Figure 17-7

C. Vector subtraction

The difference $z_2 - z_1$ can be written as $z_2 + (-1)z_1$, and this sum can be found by the methods of section **A**. But recall from Chapter 16 that there is another representation of differences of vectors, the parallelogram rule: If z_1 and z_2 are represented by vectors that form adjacent sides of a parallelogram, then the sum $z_1 + z_2$ and difference $z_1 - z_2$ are represented by vectors along the diagonals of the parallelogram (see Figure 16-5).

17-5. Polar Form of a Complex Number

A. Definition

Look again at Figure 17-1. We can readily express the complex number $z = x + yi$ in terms of polar coordinates. The point $P(x, y)$ has polar coordinates (r, θ), where $r = |\mathbf{r}|$, and so $r > 0$, and where

$$x = r \cos \theta \quad \text{and} \quad y = r \sin \theta \qquad \textbf{(17-9)}$$

Substituting these values into z,

$$z = x + yi = r \cos \theta + (r \sin \theta)i$$
$$= r(\cos \theta + i \sin \theta) \qquad \textbf{(17-10)}$$

We introduce the notation

$$\operatorname{cis} \theta = \cos \theta + i \sin \theta \qquad \textbf{(17-11)}$$

derived from Equation 17-10 by letting $r = 1$. You can think of $\operatorname{cis} \theta$ as a unit vector that makes an angle θ with the positive x axis.

Note: You can also write

$$\cos \theta + i \sin \theta = e^{i\theta} \qquad \textbf{(17-12)}$$

where e is a constant ($e \cong 2.71728$).

EXAMPLE 17-16:

$$\operatorname{cis} \frac{\pi}{3} = \cos \frac{\pi}{3} + i \sin \frac{\pi}{3} = \frac{1}{2} + \frac{i\sqrt{3}}{2}$$

$$\operatorname{cis} \pi = \cos \pi + i \sin \pi = -1 + 0i = -1$$

$$\operatorname{cis} \frac{\pi}{4} = \cos \frac{\pi}{4} + i \sin \frac{\pi}{4} = \frac{\sqrt{2}}{2} + \frac{i\sqrt{2}}{2}$$

B. Properties of cis θ

The complex function $\operatorname{cis} \theta$ has several interesting properties. If θ_1 and θ_2 are real, then

$$\operatorname{cis} \theta_1 \operatorname{cis} \theta_2 = \operatorname{cis}(\theta_1 + \theta_2) \qquad \textbf{(17-13)}$$

PROPERTIES OF CIS θ

$$\frac{\operatorname{cis} \theta_1}{\operatorname{cis} \theta_2} = \operatorname{cis}(\theta_1 - \theta_2) \qquad \textbf{(17-14)}$$

$$\frac{1}{\operatorname{cis} \theta} = \operatorname{cis}(-\theta) \qquad \textbf{(17-15)}$$

EXAMPLE 17-17: Verify Equations 17-13 and 17-14 with $\theta_1 = \frac{1}{3}\pi$ and $\theta_2 = \frac{1}{2}\pi$.

Solution:

$$\operatorname{cis} \frac{\pi}{3} = \cos \frac{\pi}{3} + i \sin \frac{\pi}{3} = \frac{1}{2} + \frac{i\sqrt{3}}{2}$$

$$\operatorname{cis} \frac{\pi}{2} = \cos \frac{\pi}{2} + i \sin \frac{\pi}{2} = 0 + 1i = i$$

From Equation 17-13,

$$\operatorname{cis} \frac{\pi}{3} \operatorname{cis} \frac{\pi}{2} = \operatorname{cis}\left(\frac{\pi}{3} + \frac{\pi}{2}\right)$$

$$\left(\frac{1}{2} + \frac{i\sqrt{3}}{2}\right)i = \operatorname{cis} \frac{5\pi}{6}$$

$$\frac{1}{2}i + \frac{i^2 \sqrt{3}}{2} = \cos \frac{5\pi}{6} + i \sin \frac{5\pi}{6}$$

$$\frac{-\sqrt{3}}{2} + \frac{1}{2}i = -\frac{\sqrt{3}}{2} + \frac{1}{2}i$$

From Equation 17-14,

$$\frac{\operatorname{cis}\dfrac{\pi}{3}}{\operatorname{cis}\dfrac{\pi}{2}} = \operatorname{cis}\left(\frac{\pi}{3} - \frac{\pi}{2}\right)$$

$$\frac{\dfrac{1}{2} + \dfrac{i\sqrt{3}}{2}}{i} = \operatorname{cis}\left(-\frac{\pi}{6}\right)$$

$$-\left[\frac{1}{2}i + \frac{\sqrt{3}}{2}i^2\right] = \cos\left(-\frac{\pi}{6}\right) + i\sin\left(-\frac{\pi}{6}\right)$$

$$-\frac{1}{2}i + \frac{\sqrt{3}}{2} = \cos\frac{\pi}{6} - i\sin\frac{\pi}{6}$$

$$\frac{\sqrt{3}}{2} - \frac{1}{2}i = \frac{\sqrt{3}}{2} - \frac{1}{2}i$$

C. DeMoivre's Theorem

DeMoivre's Theorem allows you to find powers of complex numbers expressed in polar form:

DEMOIVRE'S THEOREM
$$(r\operatorname{cis}\theta)^n = r^n\operatorname{cis} n\theta$$

$$[r(\cos\theta + i\sin\theta)]^n = r^n(\cos n\theta + i\sin n\theta) \tag{17-16}$$

EXAMPLE 17-18: Find $[2\,(\cos\frac{1}{24}\pi + i\sin\frac{1}{24}\pi)]^8$

Solution: From DeMoivre's Theorem,

$$\left[2\left(\cos\frac{\pi}{24} + i\sin\frac{\pi}{24}\right)\right]^8 = 2^8\left(\cos\frac{8\pi}{24} + i\sin\frac{8\pi}{24}\right)$$

$$= 256\left(\cos\frac{\pi}{3} + i\sin\frac{\pi}{3}\right)$$

$$= 256\left(\frac{1}{2} + \frac{i\sqrt{3}}{2}\right)$$

D. Polar form of z

For the complex number $z = x + yi$, $r = |z|$ and $\theta = \arg z$. You can write the polar (trigonometric) form of $z = x + yi$ as

$$z = |z|\operatorname{cis}(\arg z) = r\operatorname{cis}\theta \tag{17-17}$$

EXAMPLE 17-19: Find the polar form of

$$z = -\frac{1}{2} - \frac{i\sqrt{3}}{2}$$

Solution:

$$r = |z| = \sqrt{\left(-\frac{1}{2}\right)^2 + \left(-\frac{\sqrt{3}}{2}\right)^2} = \sqrt{\frac{1}{4} + \frac{3}{4}} = 1$$

$$\arg z = \pi + \arctan\frac{(-\sqrt{3}/2)}{(-1/2)} = \pi + \arctan\sqrt{3} = \pi + \frac{\pi}{3} = \frac{4\pi}{3}$$

$$z = |z|\operatorname{cis}(\arg z) = 1\operatorname{cis}\left(\frac{4\pi}{3}\right)$$

E. Arithmetic operations on polar forms

Let $z_1 = x_1 + iy_1 = r_1 \operatorname{cis}\theta_1$ and $z_2 = x_2 + iy_2 = r_2 \operatorname{cis}\theta_2$. It follows that

$$z_1 z_2 = (r_1 \operatorname{cis}\theta_1)(r_2 \operatorname{cis}\theta_2) = r_1 r_2 \operatorname{cis}(\theta_1 + \theta_2) \qquad \textbf{(17-18)}$$

From this equality and the fact that $|z_1| = r_1$ and $|z_2| = r_2$, you see that

- the modulus of a product is the product of the moduli. In symbols,

$$|z_1 z_2| = |z_1||z_2|$$

- the argument of a product is the sum of the arguments. In symbols,

$$\arg z_1 z_2 = \arg z_1 + \arg z_2$$

With a similar procedure you can show that

- the modulus of a quotient is the quotient of the moduli. In symbols,

$$\left|\frac{z_1}{z_2}\right| = \frac{|z_1|}{|z_2|}$$

- The argument of a quotient is the difference of the arguments. In symbols,

$$\arg \frac{z_1}{z_2} = \arg z_1 - \arg z_2$$

EXAMPLE 17-20: Find $z_1 z_2$ and z_1/z_2 if $z_1 = 3\operatorname{cis}\frac{5}{24}\pi$ and $z_2 = 4\operatorname{cis}\frac{1}{24}\pi$.

Solution: From Equation 17-17, $|z_1| = 3$, $|z_2| = 4$, $\theta_1 = \frac{5}{24}\pi$, and $\theta_2 = \frac{1}{24}\pi$. So,

$$|z_1||z_2| = 3(4) = 12$$

$$\theta_1 + \theta_2 = \frac{5\pi}{24} + \frac{\pi}{24} = \frac{6\pi}{24} = \frac{\pi}{4}$$

and

$$z_1 z_2 = 12\left(\cos\frac{\pi}{4} + i\sin\frac{\pi}{4}\right) = 12\left(\frac{\sqrt{2}}{2} + \frac{i\sqrt{2}}{2}\right)$$

$$= 6\sqrt{2} + 6i\sqrt{2}$$

Also,

$$\frac{|z_1|}{|z_2|} = \frac{3}{4}$$

$$\theta_1 - \theta_2 = \frac{5\pi}{24} - \frac{\pi}{24} = \frac{4\pi}{24} = \frac{\pi}{6}$$

so,

$$\frac{z_1}{z_2} = \frac{3}{4}\operatorname{cis}\frac{\pi}{6} = \frac{3}{4}\left(\frac{\sqrt{3}}{2} + \frac{1}{2}i\right) = \frac{3\sqrt{3}}{8} + \frac{3}{8}i$$

F. Roots

If n is any positive integer and z a complex number, define the nth root of z to be the complex number w such that $w^n = z$. By raising both sides of $w^n = z$ to the $1/n$ power and using DeMoivre's Theorem, you get a formula for an $\boldsymbol{n^{\text{th}}}$ **root of** z:

n^{th} ROOT OF z
$$w = \sqrt[n]{r\operatorname{cis}\theta} = |z|^{1/n}\operatorname{cis}\frac{\arg z}{n} \qquad \textbf{(17-19)}$$

where $\arg z$ is any argument of z.

EXAMPLE 17-21: Find a 4th root of $z = \sqrt{3} + i$.

Solution:

$$|z| = \sqrt{(\sqrt{3})^2 + 1^2} = \sqrt{4} = 2$$

$$\arg z = \arg(\sqrt{3} + i) = \arctan\frac{1}{\sqrt{3}}$$

$$= \arctan\frac{\sqrt{3}}{3} = \frac{\pi}{6}$$

From Equation 17-19, a 4th root is

$$w = |z|^{1/4}\text{cis}\frac{\arg z}{4}$$

$$= 2^{1/4}\text{cis}\frac{\pi/6}{4} = 2^{1/4}\text{cis}\frac{\pi}{24}$$

DeMoivre's Theorem shows that each nonzero complex number has exactly n distinct n^{th} roots, which are given by

ROOTS OF z $$w_k = |z|^{1/n}\text{cis}\left(\frac{\arg z}{n} + k\frac{2\pi}{n}\right), \qquad k = 0, 1, 2, \ldots, n-1 \qquad \textbf{(17-20)}$$

where $|z|^{1/n}$ is any n^{th} root of $|z|$ and $\arg z$ is any fixed argument of z.

EXAMPLE 17-22: Find the four 4th roots of $z = \sqrt{3} + i$.

Solution: From Example 17-21, $|z| = 2$ and $\arg z = \frac{1}{6}\pi$. Substituting into Equation 17-20 you get

$$w_k = |2|^{1/4}\text{cis}\left(\frac{\pi/6}{4} + k\frac{2\pi}{4}\right), \qquad \text{for } k = 0, 1, 2, 3$$

and

$$w_0 = 2^{1/4}\text{cis}\frac{\pi}{24}$$

$$w_1 = 2^{1/4}\text{cis}\frac{13\pi}{24}$$

$$w_2 = 2^{1/4}\text{cis}\frac{25\pi}{24}$$

$$w_3 = 2^{1/4}\text{cis}\frac{37\pi}{24}$$

• Each of the nth roots of $z = x + yi$ is a vector with tail at the origin and head on the circle of radius $|z|^{1/n}$ with center at the origin. The angle between any two n^{th} roots of z is a multiple of $2\pi/n$.

EXAMPLE 17-23: Draw a diagram of the 8th roots of $z = 256$.

Solution: Since $|z| = 256$ and $\arg z = 0$, you can find the 8th roots of 256 from

$$w_k = 256^{1/8}\text{cis}\left(0 + k\frac{2\pi}{8}\right) = 2\,\text{cis}\left(k\frac{\pi}{4}\right) \quad \text{for } k = 0, 1, \ldots, 7$$

You can represent each root by a vector with tail at the origin and head on the circle with center at the origin and radius 2. The angle between successive vectors is $\frac{1}{4}\pi$. The vectors are shown in Figure 17-8.

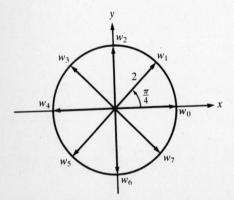

Figure 17-8

Note: The sum of the 8th roots of 256 is zero. It can be shown that the sum of the nth roots of any complex number is zero.

SUMMARY

1. The imaginary number i is the solution of $x^2 + 1 = 0$.
2. The powers of i repeat in the pattern i, -1, $-i$, 1.
3. Complex numbers can be represented in standard form as $z = x + yi$ or as ordered pairs of real numbers.
4. The sum of two complex numbers is a complex number whose real part is the sum of the two real parts and whose imaginary part is the sum of the two imaginary parts.
5. The difference of two complex numbers is a complex number whose real part is the difference of the two real parts and whose imaginary part is the difference of the two imaginary parts.
6. The product of two complex numbers is found by multiplying each term by each of the other terms and simplifying by using $i^2 = -1$.
7. The quotient of two complex numbers is found by multiplying the quotient of the two complex numbers by the complex conjugate of the denominator.
8. Complex numbers can be represented as vectors with the x component equal to the real part and the y component equal to the imaginary part.
9. The polar form of a complex number is $r \operatorname{cis} \theta$ or $r(\cos \theta + i \sin \theta)$.
10. Multiply complex numbers in polar form by multiplying the moduli and adding the angles, that is,

$$(r \operatorname{cis} \theta)(s \operatorname{cis} \alpha) = rs \operatorname{cis}(\theta + \alpha)$$

11. Divide complex numbers in polar form by dividing the moduli and subtracting the angles, that is,

$$\frac{r \operatorname{cis} \theta}{s \operatorname{cis} \alpha} = \frac{r}{s} \operatorname{cis}(\theta - \alpha)$$

12. If $r \operatorname{cis} \theta$ is the polar form of a complex number, DeMoivre's Theorem says that the nth power of the number is $r^n \operatorname{cis} n\theta$.
13. Find the nth roots of the complex number $r \operatorname{cis} \theta$ in polar form by

$$w_k = r^{1/n} \operatorname{cis}\left(\frac{\theta}{n} + k\,\frac{2\pi}{n}\right) \text{ for } k = 0, 1, 2, \dots, n - 1.$$

RAISE YOUR GRADES
Can you...?

☑ find any power of i
☑ sketch any complex number in the complex plane
☑ find the magnitude and angle of any complex number
☑ find the sum, difference, product, scalar product, or quotient of any two complex numbers algebraically or geometrically
☑ find the complex conjugate of any complex number
☑ find the polar form of any complex number from the standard form
☑ find the standard form of any complex number from the polar form
☑ use DeMoivre's Theorem to multiply and divide complex numbers in polar form
☑ find the nth roots of any complex number

SOLVED PROBLEMS

PROBLEM 17-1 Express $\sqrt{-4}$ in terms of i.

Solution: You know that $i = \sqrt{-1}$, so $\sqrt{-4} = \sqrt{4(-1)} = \sqrt{4}(\sqrt{-1}) = 2i$.   [See Section 17-1.]

PROBLEM 17-2 Express $-\sqrt{-12}$ in terms of i.

Solution: Since $i = \sqrt{-1}$, $-\sqrt{-12} = -\sqrt{12(-1)} = -\sqrt{4(3)}(\sqrt{-1}) = -\sqrt{4}(\sqrt{3})i = -2i\sqrt{3}$.

[See Section 17-1.]

For Problems 17-3 and 17-4, simplify and leave your answers in terms of i.

[See Section 17-1.]

PROBLEM 17-3 $\sqrt{-24}/\sqrt{-12}$

Solution: $\sqrt{-24}/\sqrt{-12} = \sqrt{-24/-12} = \sqrt{2}$.

PROBLEM 17-4 $-\sqrt{6}/\sqrt{-9}$

Solution: You can write $\sqrt{-9}$ as $3i$. Then $-\sqrt{6}/\sqrt{-9} = -\sqrt{6}/3i = (-\sqrt{6}/3i)(i/i) = -i\sqrt{6}/3i^2 = -i\sqrt{6}/-3 = i\sqrt{6}/3$.

For Problems 17-5 through 17-12, perform the indicated operations and simplify.

[See Section 17-2.]

PROBLEM 17-5 $(2 + i) + (2 - i)$

Solution: Add the real components to each other and the complex components to each other: $(2 + i) + (2 - i) = (2 + 2) + (i + (-i)) = 4 + 0i = 4$.

PROBLEM 17-6 $(2 + i)(2 - i)$

Solution: Multiply the two binomials and then combine terms; simplify with $i^2 = -1$: $(2 + i)(2 - i) = 2(2) - 2i + 2i - i^2 = 4 - i^2 = 4 - (-1) = 5$.

PROBLEM 17-7 $(2 + i)/(2 - i)$

Solution: Multiply numerator and denominator by the complex conjugate of the denominator:

$$\frac{2 + i}{2 - i} = \frac{2 + i}{2 - i}\left(\frac{2 + i}{2 + i}\right) = \frac{(2 + i)(2 + i)}{(2 + i)(2 - i)}$$

$$= \frac{4 + 4i + i^2}{4 - i^2} = \frac{3 + 4i}{5} = \frac{3}{5} + \frac{4}{5}i$$

PROBLEM 17-8 $(4 + i)^2$

Solution: You raise a complex number to a power as you would a binomial, and then use the simplification $i^2 = -1$:

$$(4 + i)^2 = 4^2 + 2(4)i + i^2 = 16 + 8i + i^2 = 15 + 8i$$

PROBLEM 17-9 $(2i)^8$

Solution: Since i to any multiple of the 4th power is 1, $(2i)^8 = 2^8 i^8 = 2^8 = 256$.

PROBLEM 17-10 i^{77}

Solution: Divide the exponent by 4. The remainder is 1, so $i^{77} = i$.   [See Section 17-1.]

PROBLEM 17-11 $(4 + i)/i$

Solution: Multiply numerator and denominator by the conjugate of i, $-i$:

$$\frac{4 + i}{i}\left(\frac{-i}{-i}\right) = \frac{-4i - i^2}{-i^2} = 1 - 4i$$

PROBLEM 17-12 $(3 + i)/(1 - i) + (6 - i)/(1 + i)$

Solution: Begin by performing the two divisions:

$$\frac{3 + i}{1 - i} = \frac{3 + i}{1 - i}\left(\frac{1 + i}{1 + i}\right) = \frac{3 + 4i + i^2}{1 - i^2} = 1 + 2i$$

$$\frac{6 - i}{1 + i} = \frac{6 - i}{1 + i}\left(\frac{1 - i}{1 - i}\right) = \frac{6 - 7i + i^2}{1 - i^2} = \frac{5}{2} - \frac{7}{2}i$$

Now add the results:

$$1 + 2i + \frac{5}{2} - \frac{7}{2}i = \frac{7}{2} - \frac{3}{2}i$$

For Problems 17-13 through 17-17, perform the indicated operations both graphically and algebraically.   [See Sections 17-2 and 17-4.]

PROBLEM 17-13 $(3 + i) - (2 - i)$

Solution: Subtracting, you get $3 - 2 + (i + i) = 1 + 2i$. Figure 17-9 shows the result.

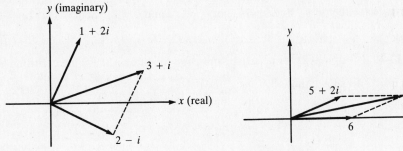

Figure 17-9   **Figure 17-10**

PROBLEM 17-14 $(5 + 2i) + 6$

Solution: Adding, you get $5 + 6 + 2i = 11 + 2i$. Figure 17-10 shows the result.

PROBLEM 17-15 $i - (2 - i)$

Solution: Subtracting, you get $-2 + i + i = -2 + 2i$. Figure 17-11 shows the result.

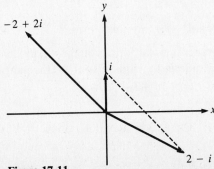

Figure 17-11

PROBLEM 17-16 $(\sqrt{5} + i) + (\sqrt{5} + 2i)$

Solution: Adding, you get $\sqrt{5} + \sqrt{5} + i + 2i = 2\sqrt{5} + 3i$. Figure 17-12 shows the result.

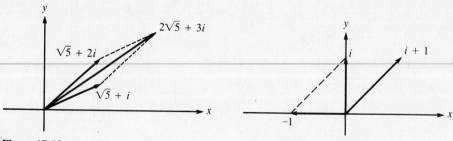

Figure 17-12   **Figure 17-13**

PROBLEM 17-17 $i - (i + 1)$

Solution: Subtracting, you get $-1 + i - i = -1$. Figure 17-13 shows the result.

PROBLEM 17-18 Simplify $\sqrt{5 - 12i}$.

Solution: If you let $\sqrt{5 - 12i} = a + bi$, then $5 - 12i = (a + bi)^2 = (a^2 - b^2) + 2abi$. If two complex numbers are equal, then their magnitudes are equal and their real parts are equal. In the present situation this gives the two equations

$$a^2 + b^2 = \sqrt{169} = 13$$
$$a^2 - b^2 = 5$$

Solving, you get $a^2 = 9$ and $b^2 = 4$. Since $2abi = -12i$, or $2ab = -12$, a and b must be of opposite sign. This means that either $a = 3$ and $b = -2$ or $a = -3$ and $b = 2$. Then $\sqrt{5 - 12i} = \pm(3 - 2i)$.

[See Section 17-2.]

PROBLEM 17-19 Graph $-2i$, its negative, and its conjugate on the same coordinate system.

Solution: Figure 17-14 shows the solutions.   [See Sections 17-2 and 17-3.]

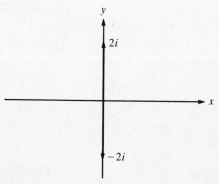

Figure 17-14

PROBLEM 17-20 Express $1 - i\sqrt{3}$ in polar form.

Solution: To express a number in polar form, you must find its magnitude and its angle with the positive x axis. The magnitude is the square root of the sum of the squares of the real and imaginary components, or $r = \sqrt{1^2 + (-\sqrt{3})^2} = \sqrt{1 + 3} = \sqrt{4} = 2$ and $\theta = \tan^{-1}(-\sqrt{3}/1) = -60°$. In polar form,

$$1 - i\sqrt{3} = r(\cos\theta + i\sin\theta)$$
$$= 2(\cos(-60°) + i\sin(-60°))$$
$$= 2(\cos 60° - i\sin 60°)$$

[See Section 17-5.]

PROBLEM 17-21   Use DeMoivre's Theorem to find $(2 + 5i)^4$.

Solution: Since $r = \sqrt{29}$ and $\theta = \tan^{-1}(\frac{5}{2})$, you can write $2 + 5i$ in polar form as $\sqrt{29}[\cos(\tan^{-1}\frac{5}{2}) + i\sin(\tan^{-1}\frac{5}{2})]$. From DeMoivre's Theorem,

$$(2 + 5i)^4 = (\sqrt{29})^4[\cos(4\tan^{-1}\tfrac{5}{2}) + i\sin(4\tan^{-1}\tfrac{5}{2})]$$
$$= 841(\cos 272.8° + i\sin 272.8°) \qquad \text{[See Section 17-5.]}$$

PROBLEM 17-22   Use DeMoivre's Theorem to find the fourth roots of -16.

Solution: You can write -16 in polar form as

$$16(\cos 180° + i\sin 180°)$$

From DeMoivre's Theorem,

$$(-16)^{1/4} = 16^{1/4}\left[\cos\frac{(180 + 360k)°}{4} + i\sin\frac{(180 + 360k)°}{4}\right]$$
$$= 2[\cos(45 + 90k)° + i\sin(45 + 90k)°], \quad \text{for } k = 0, 1, 2, 3$$

The roots are

$$2(\cos 45° + i\sin 45°) = \sqrt{2}(1 + i)$$
$$2(\cos 135° + i\sin 135°) = \sqrt{2}(-1 + i)$$
$$2(\cos 225° + i\sin 225°) = \sqrt{2}(-1 - i)$$
$$2(\cos 315° + i\sin 315°) = \sqrt{2}(1 - i) \qquad \text{[See Section 17-5.]}$$

PROBLEM 17-23   Find all complex numbers z that satisfy $z\bar{z} = 25$.

Solution: If you let $z = a + bi$, then the conjugate of z is $\bar{z} = a - bi$. If these two complex numbers satisfy the given condition, then $(a + bi)(a - bi) = a^2 - (bi)^2 = a^2 + b^2 = 25$. All complex numbers in the set $\{a + bi \mid a^2 + b^2 = 25\}$ satisfy the condition.   [See Section 17-2.]

PROBLEM 17-24   Given that $z\bar{z} = 25$ and $z = -3 + ai$, find a.

Solution: The conjugate of $z = -3 + ai$ is $\bar{z} = -3 - ai$. Then,

$$(-3 + ai)(-3 - ai) = 9 - (ai)^2 = 9 + a^2 = 25.$$

From this you get $a^2 = 16$, or $a = \pm 4$.   [See Section 17-2.]

PROBLEM 17-25   Find all solutions of the equation $z^2 = i$.

Solution: Suppose that $z = a + bi$ is any solution of the equation. Then $(a + bi)^2 = a^2 + 2abi + (bi)^2 = a^2 - b^2 + 2abi = i$. Equating real and imaginary coefficients, you get $a^2 - b^2 = 0$ and $2ab = 1$. From the first equation $a = \pm b$. From $2ab = 1$, you can see that a and b must have the same sign. Then $2b^2 = 1$, or $b = \pm\sqrt{2}/2$. The solutions are $\pm(\sqrt{2}/2 + i\sqrt{2}/2)$.
   [See Section 17-2.]

PROBLEM 17-26   Solve $z^2 = -3i$.

Solution: Let $z = a + bi$ be any solution of the equation. Then $(a + bi)^2 = a^2 + 2abi + (bi)^2 = a^2 - b^2 + 2abi = -3i$. Equating real and imaginary coefficients, you get $a^2 - b^2 = 0$ and $2ab = -3$. From the first equation $a = \pm b$. From $2ab = -3$, you can see that a and b must have opposite signs. Then $2b^2 = 3$, or $b = \pm\sqrt{3/2}$. The solutions are $\pm(\sqrt{3/2} - i\sqrt{3/2})$.
   [See Section 17-2.]

For Problems 17-27 and 17-28, find the reciprocals of the complex number and express in the form $a + bi$.   [See Section 17-2.]

PROBLEM 17-27 i

Solution: The reciprocal of i is $1/i = (1/i)(i/i) = i/i^2 = i/-1 = -i$.

PROBLEM 17-28 $4 - 3i$

Solution: The reciprocal of $4 - 3i$ is

$$\frac{1}{4 - 3i} = \frac{1}{4 - 3i}\left(\frac{4 + 3i}{4 + 3i}\right) = \frac{4 + 3i}{4^2 - (3i)^2} = \frac{4 + 3i}{16 + 9} = \frac{4 + 3i}{25} = \frac{4}{25} + \frac{3}{25}i$$

For Problems 17-29 through 17-34, convert from polar to rectangular notation or vice versa.

[See Section 17-5.]

PROBLEM 17-29 $3(\cos 45° + i \sin 45°)$

Solution: Use $\cos 45° = \sin 45° = \sqrt{2}/2$:

$$3(\cos 45° + i \sin 45°) = 3\left(\frac{\sqrt{2}}{2} + \frac{i\sqrt{2}}{2}\right) = \frac{3\sqrt{2}}{2} + \frac{3\sqrt{2}}{2}i$$

This complex number is represented by the ordered pair $\left(\frac{3\sqrt{2}}{2}, \frac{3\sqrt{2}}{2}\right)$.

PROBLEM 17-30 $5 \operatorname{cis}(-30°)$

Solution: Use $\cos 30° = \frac{\sqrt{3}}{2}$ and $\sin 30° = \frac{1}{2}$:

$$5 \operatorname{cis}(-30°) = 5(\cos(-30°) + i \sin(-30°))$$

$$= 5(\cos 30° - i \sin 30°) = \frac{5\sqrt{3}}{2} - \frac{5}{2}i$$

PROBLEM 17-31 $1 + i$

Solution: The magnitude r of $1 + i$ is $r = \sqrt{1^2 + 1^2} = \sqrt{2}$, and the angle θ is $\theta = \tan^{-1}\left(\frac{1}{1}\right) = 45°$. Thus $1 + i = r \operatorname{cis} \theta = \sqrt{2} \operatorname{cis} 45°$.

PROBLEM 17-32 -4

Solution: The magnitude of -4 is 4 and the angle is $180°$. Thus $-4 = r \operatorname{cis} \theta = 4 \operatorname{cis} 180°$.

PROBLEM 17-33 $i - 2$

Solution: The magnitude r of $i - 2$ is $r = \sqrt{(-2)^2 + 1^2} = \sqrt{5}$. The angle corresponding to $i - 2$ is a second-quadrant angle; $\tan^{-1}(1/-2) = -26.6°$, or, expressed as a second quadrant angle, $153.4°$. Then $i - 2 = r \operatorname{cis} \theta = \sqrt{5} \operatorname{cis} 153.4°$.

PROBLEM 17-34 $(\cos \theta + i \sin \theta)^{-1}$

Solution: Since this number is in polar form with $r = 1$ and angle θ, from DeMoivre's Theorem $(\cos \theta + i \sin \theta)^{-1} = \cos(-\theta) + i \sin(-\theta) = \cos \theta - i \sin \theta$.

PROBLEM 17-35   Convert to polar notation and find $(1 + i)/(2 + 2i)$.

Solution: In dividing or multiplying two complex numbers, a first step is to express each complex number in polar notation:

$$1 + i = \sqrt{2}(\cos 45° + i \sin 45°)$$
$$2 + 2i = \sqrt{8}(\cos 45° + i \sin 45°)$$

From DeMoivre's Theorem,

$$\frac{1 + i}{2 + 2i} = \frac{\sqrt{2}}{\sqrt{8}}\left[\cos(45 - 45)° + i \sin(45 - 45)°\right] = \frac{\sqrt{2}}{\sqrt{8}} = \frac{1}{2}$$

You can get the same answer by observing that $2 + 2i$ factors as $2(1 + i)$.   [See Section 17-5.]

PROBLEM 17-36   Convert to polar notation and find $(1 - i)/(1 + i)$.

Solution: First, express each complex number in polar notation:

$$1 - i = \sqrt{2}[\cos(-45°) + i\sin(-45°)]$$

$$1 + i = \sqrt{2}(\cos 45° + i\sin 45°)$$

From DeMoivre's Theorem,

$$\frac{1 - i}{1 + i} = \frac{\sqrt{2}}{\sqrt{2}}[\cos(-45 - 45)° + i\sin(-45 - 45)°] = \cos 90° - i\sin 90° = -i \qquad \text{[See Section 17-5.]}$$

For Problems 17-37 through 17-39, raise the number expressed in polar coordinates to the power shown and write the answer in both polar and standard form. [See Section 17-5.]

PROBLEM 17-37 $(\operatorname{cis}\tfrac{1}{4}\pi)^4$

Solution: From DeMoivre's Theorem, $(\operatorname{cis}\tfrac{1}{4}\pi)^4 = \operatorname{cis}\pi = \cos\pi + i\sin\pi$; in standard form, $(\operatorname{cis}\tfrac{1}{4}\pi)^4 = -1$.

PROBLEM 17-38 $(3\operatorname{cis}\tfrac{1}{2}\pi)^6$

Solution: From DeMoivre's Theorem, $(3\operatorname{cis}\tfrac{1}{2}\pi)^6 = 3^6\operatorname{cis}\tfrac{6}{2}\pi = 3^6\operatorname{cis}3\pi = 3^6\operatorname{cis}\pi$; in standard form, $(3\operatorname{cis}\tfrac{1}{2}\pi)^6 = 729(\cos\pi + i\sin\pi) = -729$.

PROBLEM 17-39 $(5\operatorname{cis}\tfrac{1}{6}\pi)^3$

Solution: From DeMoivre's Theorem, $(5\operatorname{cis}\tfrac{1}{6}\pi)^3 = 5^3\operatorname{cis}\tfrac{3}{6}\pi = 5^3(\cos\tfrac{1}{2}\pi + i\sin\tfrac{1}{2}\pi)$; in standard form, $(5\operatorname{cis}\tfrac{1}{6}\pi)^3 = 125i$.

PROBLEM 17-40 Find and graph the third roots of i.

Solution: In polar notation, i is $\operatorname{cis}\tfrac{1}{2}\pi$. The third roots of i equal $(\operatorname{cis}\tfrac{1}{2}\pi)^{1/3}$, and, by DeMoivre's Theorem, they can then be written $\operatorname{cis}(\tfrac{1}{6}\pi + \tfrac{2}{3}k\pi)$, for $k = 0, 1, 2$. The third roots are

$$\operatorname{cis}\frac{\pi}{6} = \cos\frac{\pi}{6} + i\sin\frac{\pi}{6} = \frac{\sqrt{3}}{2} + \frac{1}{2}i, \quad \text{for } k = 0$$

$$\operatorname{cis}\left(\frac{\pi}{6} + \frac{2\pi}{3}\right) = \cos\left(\frac{\pi}{6} + \frac{2\pi}{3}\right) + i\sin\left(\frac{\pi}{6} + \frac{2\pi}{3}\right)$$

$$= \cos\frac{5\pi}{6} + i\sin\frac{5\pi}{6} = \frac{-\sqrt{3}}{2} + \frac{1}{2}i, \quad \text{for } k = 1$$

$$\operatorname{cis}\left(\frac{\pi}{6} + \frac{4\pi}{3}\right) = \cos\left(\frac{\pi}{6} + \frac{4\pi}{3}\right) + i\sin\left(\frac{\pi}{6} + \frac{4\pi}{3}\right)$$

$$= \cos\frac{3\pi}{2} + i\sin\frac{3\pi}{2} = -i, \quad \text{for } k = 2$$

See Figure 17-15 for the graph. [See Section 17-5.]

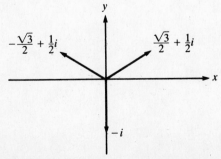

Figure 17-15

PROBLEM 17-41 Find and graph the fourth roots of $-2i$.

Solution: In polar notation, $-2i$ is $2\operatorname{cis}\frac{3}{2}\pi$. The fourth roots of $-2i$ equal $(2\operatorname{cis}\frac{3}{2}\pi)^{1/4}$, and, by DeMoivre's Theorem, they can then be written $2^{1/4}\operatorname{cis}(\frac{3}{8}\pi + \frac{1}{2}k\pi)$, for $k = 0, 1, 2, 3$. The fourth roots are

$$2^{1/4}\operatorname{cis}\frac{3\pi}{8} = 2^{1/4}\left(\cos\frac{3\pi}{8} + i\sin\frac{3\pi}{8}\right), \quad \text{for } k = 0$$

$$2^{1/4}\operatorname{cis}\left(\frac{3\pi}{8} + \frac{4\pi}{8}\right) = 2^{1/4}\left(\cos\frac{7\pi}{8} + i\sin\frac{7\pi}{8}\right), \quad \text{for } k = 1$$

$$2^{1/4}\operatorname{cis}\left(\frac{3\pi}{8} + \frac{8\pi}{8}\right) = 2^{1/4}\left(\cos\frac{11\pi}{8} + i\sin\frac{11\pi}{8}\right), \quad \text{for } k = 2$$

$$2^{1/4}\operatorname{cis}\left(\frac{3\pi}{8} + \frac{12\pi}{8}\right) = 2^{1/4}\left(\cos\frac{15\pi}{8} + i\sin\frac{15\pi}{8}\right), \quad \text{for } k = 3$$

The graph is shown in Figure 17-16.   [See Sections 17-4 and 17-5.]

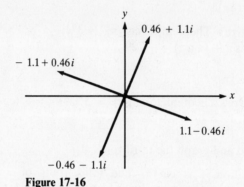

Figure 17-16

Supplementary Exercises

Perform the indicated operations and simplify.

PROBLEM 17-42 $(2 + i) - (2 - i)$

PROBLEM 17-43 $(2 + i)/(2 - i)$

PROBLEM 17-44 $i(3 + i)$

PROBLEM 17-45 $i/2i$

PROBLEM 17-46 $(5 - 3i)^3$

PROBLEM 17-47 $2i/(3 + 2i)$

PROBLEM 17-48 $i/(2 + i\sqrt{5})$

PROBLEM 17-49 $(2i)^8$

PROBLEM 17-50 $(-3i)^7$

PROBLEM 17-51 $3i(4 - i)^2$

PROBLEM 17-52 $(2 - 3i)/(2 + 3i)$

PROBLEM 17-53 $(1 - i)/(1 + i)^2$

PROBLEM 17-54 $\dfrac{(2 + i)(3 + 2i)}{2 - i}$

Perform the indicated operations both graphically and algebraically.

PROBLEM 17-55 $(3 - i) + (2 - i)$

PROBLEM 17-56 $(2 + i) - i$

PROBLEM 17-57 $(10 + i) - (10 - i)$

PROBLEM 17-58 $(10 + i) + (10 - i)$

PROBLEM 17-59 $1 + (5 - i)$

Perform the indicated operations.

PROBLEM 17-60 $(2 + 3i)(1 - i)(3 - 2i)$

PROBLEM 17-61 $\dfrac{(1 - 4i)(2 + i)}{1 + 2i}$

PROBLEM 17-62 On the same coordinate system show $-3 + 2i$, its negative, and its conjugate.

PROBLEM 17-63 Express $1 + i\sqrt{3}$ in polar form.

PROBLEM 17-64 Express $1 + i$ in polar form.

PROBLEM 17-65 Find $(1 + i\sqrt{3})(1 + i)$ and $(1 + i\sqrt{3})/(1 + i)$ in polar form.

PROBLEM 17-66 Let $w = 3 + 2i$ and $z = 5 - i$. Find the following: (a) $z + w$; (b) $2z - 3w$; (c) $\bar{z}$; (d) zw; (e) $\bar{z}\bar{w}$; (f) z/w.

PROBLEM 17-67 Find $z\bar{z}$ for $z = 4 - 3i$.

PROBLEM 17-68 Find all solutions to $x^4 = 16$.

PROBLEM 17-69 Let $z = 2 + 5i$. Find (a) $z\bar{z}$; (b) iz; (c) $|iz|$; (d) z/i.

PROBLEM 17-70 Find $(1 + i\sqrt{3})^7$ using DeMoivre's Theorem.

PROBLEM 17-71 Find all solutions of the equation $x^2 + 8 = 0$.

PROBLEM 17-72 Find all solutions of the equation $x^3 = -1$.

Express the following in terms of i.

PROBLEM 17-73 $\sqrt{-15}$

PROBLEM 17-74 $-\sqrt{-22}$

PROBLEM 17-75 $\sqrt{-16}$

Simplify the following and leave answers in terms of i.

PROBLEM 17-76 $\sqrt{5}/\sqrt{-3}$

PROBLEM 17-77 $\sqrt{-24}/\sqrt{12}$

PROBLEM 17-78 $\sqrt{-5}/\sqrt{-3}$

PROBLEM 17-79 $\sqrt{-8}/(-\sqrt{-2})$

PROBLEM 17-80 Solve $z^2/3 = 2i$.

Find the reciprocals of the following numbers and express in the form $a + bi$.

PROBLEM 17-81 $2 - i$

PROBLEM 17-82 $3i$

Express the following in rectangular notation.

PROBLEM 17-83 $8 \operatorname{cis} \pi$

PROBLEM 17-84 $\sqrt{2}(\cos \frac{1}{3}\pi + i \sin \frac{1}{3}\pi)$

PROBLEM 17-85 $3 \operatorname{cis} \frac{5}{4}\pi$

Express the following in polar notation.

PROBLEM 17-86 $\sqrt{2} + i$

PROBLEM 17-87 $-3i$

PROBLEM 17-88 Convert to polar notation and find $(1 + i)(2 + 2i)$.

PROBLEM 17-89 Convert to polar notation and find $(1 - i)/(\sqrt{3} - i)$.

Raise the following numbers expressed in polar or rectangular notation to the power shown and write the answer in both polar and standard form.

PROBLEM 17-90 $(3 \operatorname{cis} \frac{2}{3}\pi)^3$

PROBLEM 17-91 $(1 - i)^2$

PROBLEM 17-92 $[4(\cos \frac{1}{5}\pi + i \sin \frac{1}{5}\pi)]^5$

PROBLEM 17-93 $(2 \operatorname{cis} \frac{3}{2}\pi)^4$

PROBLEM 17-94 $[3 \operatorname{cis}(-\frac{1}{2}\pi)]^6$

PROBLEM 17-95 $(1 + i)^7$

PROBLEM 17-96 $[4(\cos \frac{2}{5}\pi + i \sin \frac{2}{5}\pi)]^5$

PROBLEM 17-97 Find and graph the fourth roots of 8.

PROBLEM 17-98 Find and graph the fifth roots of -1.

Answers to Supplementary Exercises

17-42: $2i$

17-43: $\frac{3}{5} + \frac{4}{5}i$

17-44: $-1 + 3i$

17-45: $\frac{1}{2}$

17-46: $-10 - 198i$

17-47: $\frac{4}{13} + \frac{6}{13}i$

17-48: $\frac{\sqrt{5}}{9} + \frac{2}{9}i$

17-49: $2^8 = 256$

17-50: $2187i$

17-51: $24 + 45i$

17-52: $-\frac{5}{13} - \frac{12}{13}i$

17-53: $-\frac{1}{2} - \frac{1}{2}i$

17-54: $\frac{1}{5} + \frac{18}{5}i$

17-55: $5 - 2i$

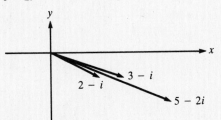

Figure 17-17

17-56: 2

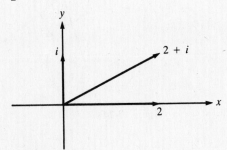

Figure 17-18

17-57: $2i$

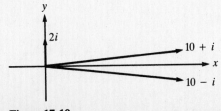

Figure 17-19

17-58: 20

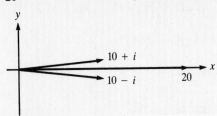

Figure 17-20

17-59: $6 - i$

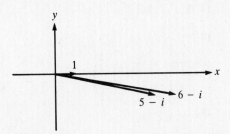

Figure 17-21

17-60: $17 - 7i$

17-61: $-\frac{8}{5} - \frac{19}{5}i$

17-62:

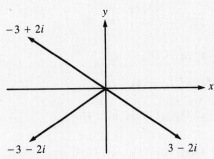

Figure 17-22

17-63: $2(\cos 60° + i \sin 60°)$

17-64: $\sqrt{2}(\cos 45° + i \sin 45°)$

17-65: product $= 2\sqrt{2}(\cos 105° + i \sin 105°)$
and quotient $= \sqrt{2}(\cos 15° + i \sin 15°)$

17-66: (a) $8 + i$; (b) $1 - 8i$; (c) $5 + i$; (d) $17 + 7i$;
(e) $17 - 7i$; (f) $1 - i$

17-67: 25

17-68: $2i, -2i, 2, -2$

17-69: (a) 29; (b) $-5 + 2i$; (c) $\sqrt{29}$; (d) $5 - 2i$

17-70: $2^7 \operatorname{cis}(7 \cdot 60°) = 64 + 64i\sqrt{3}$

17-71: $\pm 2i\sqrt{2}$

17-72: $-1, \frac{1}{2} \pm \frac{i\sqrt{3}}{2}$

17-73: $i\sqrt{15}$

17-74: $-i\sqrt{22}$

17-75: $4i$

17-76: $\dfrac{-i\sqrt{15}}{3}$

17-77: $i\sqrt{2}$

17-78: $\sqrt{15}/3$

17-79: -2

17-80: $\pm\left(\dfrac{\sqrt{3}}{2} + \dfrac{i\sqrt{3}}{2}\right)$

17-81: $\frac{2}{5} + \frac{1}{5}i$

17-82: $\frac{-i}{3}$

17-83: -8

17-84: $\dfrac{\sqrt{2}}{2} + \dfrac{i\sqrt{6}}{2}$

17-85: $\dfrac{-3\sqrt{2}}{2} - \dfrac{3i\sqrt{2}}{2}$

17-86: $\sqrt{3}$ cis 0.62

17-87: 3 cis $\frac{3}{2}\pi$

17-88: $\sqrt{16}$ cis $\frac{1}{2}\pi$, or $4i$

17-89: $\dfrac{\sqrt{2}}{2}$ cis $\left(-\dfrac{\pi}{12}\right)$

17-90: 3^3 cis 2π, or 27

17-91: 2 cis$(-\frac{1}{2}\pi)$, or $-2i$

17-92: 4^5 cis π, or -4^5

17-93: 2^4 cis 6π, or 2^4

17-94: 3^6 cis(-3π), or -3^6

17-95: $8\sqrt{2}$ cis $\frac{7}{4}\pi$, or $8 - 8i$

17-96: 4^5 cis 2π, or 4^5

17-97: $8^{1/4}$ cis 0, or $8^{1/4}$
$8^{1/4}$ cis $\frac{1}{2}\pi$, or $8^{1/4}i$
$8^{1/4}$ cis π, or $-8^{1/4}$
$8^{1/4}$ cis $\frac{3}{2}\pi$, or $-8^{1/4}i$

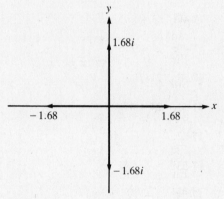

Figure 17-23

17-98: cis$\left(\dfrac{\pi}{5} + \dfrac{2\pi k}{5}\right)$ for $k = 0, 1, 2, 3, 4$

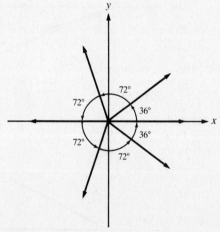

Figure 17-24

FINAL EXAM 1

1. Find the amplitude, period, horizontal axis, and phase shift of the function $y = 0.5 + 3\cos(2x + \pi)$.

2. A triangle has angles of 34° and 16°. The side opposite the 34° angle is 3 units long. Find the lengths of the other sides.

3. Convert $-338°$ to radian measure.

4. What is the size of the angle (in degrees) subtended by an arc 40 meters long on a circle of radius 1000 cm?

5. Find the value of $\cos(\sin^{-1}(-\frac{1}{2}))$.

6. Solve $\cos 2x \sin x - \sin x = 0$ for x in $[0, 2\pi)$.

7. Simplify $\dfrac{3 + 2i}{4 - 3i}$.

8. Find the fourth roots of -16. Express your answers in polar form.

9. Find the sum of the vector with magnitude 2 and direction 30° and the vector with magnitude 3 and direction 60°. Express your answer in the form $a\mathbf{i} + b\mathbf{j}$.

10. Write the polar form of the equation $x^2 + 2x + y^2 = 4$.

11. Prove the identity $\cot 2\theta = \dfrac{\cot^2\theta - 1}{2\cot\theta}$.

12. Use your calculator to find csc 1.3865.

13. If $\sin\theta = \frac{2}{3}$ and the terminal side of θ lies in quadrant II, find $\sec\theta$.

14. The angle of elevation from a point on the ground to the top of a tall building is 40°. If the angle of elevation from a second point 300 yards farther away is 28°, what is the height of the building?

15. Write the equation of the circle centered at the origin with radius 6.5 feet.

16. Find the number of degrees in 15.678 radians. Express your answer in terms of degrees, minutes, and seconds.

17. Express $\sin 30° - \sin 70°$ as a product of sines and cosines.

18. Find the exact value of $\tan 15°$.

19. What is the domain, range, and period of $\tan x$; $\sec x$?

20. Find the distance between points (x, y) and $(-x, -y)$.

ANSWERS TO FINAL EXAM 1

1. The amplitude is 3, the period is π, the axis is the line $y = 0.5$, and the phase shift is $|\frac{1}{2}\pi|$.

2. The side opposite the 16° angle is 1.48 units; the side opposite the 130° angle is 4.11 units.

3. -5.899 rad

4. $229.18°$

5. $+\sqrt{3}/2$

6. The solution set is $\{0, \pi\}$.

7. $\dfrac{6}{25} + \dfrac{17}{25}i$

8. $2(\cos\frac{1}{4}\pi + i\sin\frac{1}{4}\pi)$
   $2(\cos\frac{3}{4}\pi + i\sin\frac{3}{4}\pi)$
   $2(\cos\frac{5}{4}\pi + i\sin\frac{5}{4}\pi)$
   $2(\cos\frac{7}{4}\pi + i\sin\frac{7}{4}\pi)$

9. $\dfrac{2\sqrt{3}+3}{2}\,\mathbf{i} + \dfrac{3\sqrt{3}+2}{2}\,\mathbf{j}$

10. $r^2 + 2r\cos\theta = 4$

11. $\cot 2\theta = \dfrac{1}{\tan 2\theta} = \dfrac{1 - \tan^2\theta}{2\tan\theta}$

    $= \dfrac{(1 - \tan^2\theta)/\tan^2\theta}{2\tan\theta/\tan^2\theta}$

    $= \dfrac{(1/\tan^2\theta) - 1}{2/\tan\theta}$

    $= \dfrac{\cot^2\theta - 1}{2\cot\theta}$

12. 1.0172

13. $\sec\theta = -3\sqrt{5}/5$

14. 435.43 yds or 1306.3 ft

15. $x^2 + y^2 = (6.5)^2$

16. $898° \, 17'$

17. $-2\cos 50° \sin 20°$

18. $2 - \sqrt{3}$

19.

	$\tan x$	$\sec x$
Domain:	$\left\{x \middle| \begin{array}{l} x \text{ is real and} \\ x \neq \frac{1}{2}\pi + n\pi, \\ n \text{ any integer} \end{array}\right\}$	$\left\{x \middle| \begin{array}{l} x \text{ is real and} \\ x \neq \frac{1}{2}\pi + n\pi, \\ n \text{ any integer} \end{array}\right\}$
Range:	$\{y | y \text{ is real}\}$	$\{y | y \text{ is real and } y \geqslant 1 \text{ or } y \leqslant -1\}$
Period:	π	2π

20. $2\sqrt{x^2 + y^2}$

FINAL EXAM 2

1. For each of the following functions, find an inverse function, if it exists:
   (a) $f(x) = 3x - 2$; (b) $g(x) = 3 - x^2$.

2. Find the domain and range of (a) $y = (x - 1)/(x + 1)$; (b) $y = (1/x) + 1$;
   (c) $y = (x + 1)/(x^4 + 1)$.

3. Express 38.34° in (a) degrees, minutes, and seconds; (b) radians.

4. Write the equation for the graph of the sine function with a period of $\frac{1}{4}\pi$, an amplitude of 0.25, and a horizontal axis at $y = 1$.

5. Find the value of $\sin(\tan^{-1}(-2))$.

6. Find the solution set of $3 \sec 2x = 6$ for $0 \leqslant x < 2\pi$.

7. Simplify $(3 - 7i)(1 + 2i)$ and find its complex conjugate.

8. Express $3 + i$ in polar form; evaluate $(3 + i)^4$ and express your answer in both polar and standard form.

9. For $\mathbf{v} = \mathbf{i} - \mathbf{j}$ and $\mathbf{w} = 2\mathbf{i} - 8\mathbf{j}$, find the magnitude and direction of $\mathbf{v} + \mathbf{w}$.

10. Convert $2r \cos \theta = 3$ to Cartesian form and describe its graph.

11. Prove that $\sec^2 x - \csc^2 x = \tan^2 x - \cot^2 x$.

12. For $\sec x = 3.2468$ and $0 \leqslant x < 360°$, find x.

13. For $\tan \theta = -\frac{2}{3}$ and the terminal side of θ in quadrant IV, find $\csc \theta$.

14. The angle of elevation from a point on the ground to the top of a building is 40°. If the angle of elevation from a second point 300 yards away is 28°, how far is the first point from the base of the building?

15. Prove that $A(3, 2)$, $B(4, 7)$, and $C(8, 10)$ do not lie on a straight line.

16. The sides of a triangle measure 2, 3, and 4 feet in length. Find the angle opposite the 3-foot side.

17. Convert $\sin 30° \cos 40°$ to a sum of sines and/or cosines.

18. Find the exact value of $\sec 120°$.

19. Express $\sin x + 2 \cos x$ as the sine of a sum.

20. For $6 \sin x + \csc x - 5 = 0$ and $0 \leqslant x < 2\pi$, find x.

ANSWERS TO FINAL EXAM 2

1. (a) $f^{-1}(x) = (x + 2)/3$; (b) $g^{-1}(x)$ doesn't exist

2.

	Domain	Range
(a) $y = \dfrac{x - 1}{x + 1}$	$\{x \mid x \text{ is real, } x \neq -1\}$	$\{y \mid y \text{ is real, } y \neq 1\}$
(b) $y = \dfrac{1}{x} + 1$	$\{x \mid x \text{ is real, } x \neq 0\}$	$\{y \mid y \text{ is real, } y \neq 1\}$

	Domain	Range
(c) $y = \dfrac{x+1}{x^4+1}$	$\{x \mid x \text{ is real}\}$	$\{y \mid y \text{ is real}\}$

3. $38°20'24'' = 0.67$ rad

4. $y = 1 + 0.25 \sin 8x$

5. $-2\sqrt{5}/5$

6. $x = 30°, 150°, 210°, 330°$

7. $17 - i$; $17 + i$ (conjugate)

8. $\sqrt{10}$ cis $18.43°$; 100 cis $73.72° = 28.03 + 95.99i$

9. $\mathbf{v} + \mathbf{w} = 3\mathbf{i} - 9\mathbf{j}$; $|\mathbf{v} + \mathbf{w}| = 3\sqrt{10}$; $\arg(\mathbf{v} + \mathbf{w}) = 288.43°$

10. The vertical line $x = \frac{3}{2}$.

11.
$$\tan^2 x - \cot^2 x = \frac{\sin^2 x}{\cos^2 x} - \frac{\cos^2 x}{\sin^2 x}$$
$$= \frac{\sin^4 x - \cos^4 x}{\sin^2 x \cos^2 x}$$
$$= \frac{(\sin^2 x + \cos^2 x)(\sin^2 x - \cos^2 x)}{\sin^2 x \cos^2 x}$$
$$= \frac{\sin^2 x - \cos^2 x}{\sin^2 x \cos^2 x}$$
$$= \frac{1}{\cos^2 x} - \frac{1}{\sin^2 x}$$
$$= \sec^2 x - \csc^2 x$$

12. $72.06°$ and $287.94°$

13. $\csc \theta = -\sqrt{13}/2$

14. 518.93 yds or 1556.79 ft

15.
$$d_{AB} + d_{BC} \neq d_{AC}$$
$$\sqrt{(3-4)^2 + (2-7)^2} + \sqrt{(4-8)^2 + (7-10)^2} \neq \sqrt{(3-8)^2 + (2-10)^2}$$
$$\sqrt{1+25} + \sqrt{16+9} \neq \sqrt{25+64}$$
$$5 + \sqrt{26} \neq \sqrt{89}$$
$$5 + 5.1 \neq 9.43$$

16. $46.57°$

17. $\frac{1}{2}(\sin 70° - \sin 10°)$

18. $\sec 120° = \dfrac{1}{\cos 2(60°)} = \dfrac{1}{2\cos^2 60° - 1} = \dfrac{1}{\frac{1}{2} - 1} = -2$

19. $\sqrt{5} \sin(x + 63.43°)$

20. 0.34 rad, 2.80 rad, $\frac{1}{6}\pi$ rad, $\frac{5}{6}\pi$ rad

FINAL EXAM 3

1. Express $48.56°$ in (a) degrees, minutes, and seconds; (b) radians.

2. An arc of a circle of radius 53.4 cm is 14.82 cm in length. Find the measure of the central angle subtended by this arc in degree measure correct to the nearest minute.

3. For angle x in the third quadrant with $\cot x = \frac{4}{3}$, find $\sec x$.

4. Show $\dfrac{\tan x + \sec x}{\sin x \cot x} = \dfrac{1 + \sin x}{\cos^2 x}$.

5. A man stands 50 feet from the base of a building. If the angle of elevation to the top of the building from his position is $73°$, find the height of the building.

6. Find the $\cos(\sin^{-1}(\cos \frac{1}{3}\pi))$.

7. For $0 \leqslant x < 2\pi$, find all values of x that satisfy $2 \sin^2 x + \sin x - 1 = 0$.

8. For $y = 1 + \tan x$, find the period, domain, range, and axis.

9. Evaluate $(5 - 7i)^2$.

10. Find the fifth roots of $\frac{4}{5} - \frac{3}{5}i$.

11. What is the domain and range of $y = \cot x$?

12. Find the distance between $A(7, -2)$ and $B(-3, 4)$.

13. For $0 \leqslant x \leqslant 2\pi$, find all values of x that satisfy $2 \cos x + 1 = 0$.

14. Sketch the graph of $r = \sin 2\theta$ in polar form.

15. Find the sum of the vector with a magnitude of 5 and a direction of $45°$ and the vector with a magnitude of 6 and a direction of $30°$. Express your answer in terms of unit vectors.

16. Two observation points on level ground are 600 feet apart. A weather balloon on a line between the two points is secured to the ground by a wire. If the angles of elevation from the observation points to the balloon are $68.4°$ and $23.9°$, and the wire is vertical, find the length of the wire.

17. Two cars leave the same point at the same time on roads that form an angle of $83°12'$. One car travels at 75 km/h, the other at 60 km/h. Find the distance between the cars at the end of one hour.

18. Express $\sin 15° + \sin 75°$ as a product. (*Hint:* Use sum and difference identities.)

19. For $\cos x = -\sqrt{3}/2$ and $\tan x > 0$, find $\sin 2x$.

20. Find the exact value of $\sin \frac{5}{12}\pi$.

ANSWERS TO FINAL EXAM 3

1. (a) $48°33'36''$; (b) 0.848 rad

2. $15°54'$

3. $-\frac{5}{4}$

4. $\dfrac{\tan x + \sec x}{\sin x \cot x} = \dfrac{\dfrac{\sin x}{\cos x} + \dfrac{1}{\cos x}}{\sin x\left(\dfrac{\cos x}{\sin x}\right)}$

$= \dfrac{\sin x + 1}{\cos x}\left(\dfrac{1}{\cos x}\right)$

$= \dfrac{\sin x + 1}{\cos^2 x}$

5. 163.54 ft

6. $\sqrt{3}/2$

7. $\frac{1}{6}\pi,\ \frac{5}{6}\pi,\ \frac{3}{2}\pi$

8. The period is π; the domain is $\{x \mid x$ is real and $x \neq \frac{1}{2}\pi + n\pi,\ n$ any integer$\}$; the range is $\{y \mid y$ is real$\}$, and the axis is the line $y = 1$.

9. $-24 - 70i$

10. $\text{cis}\left(-0.1287 + k\,\dfrac{2\pi}{5}\right),\ k = 0,\ 1,\ 2,\ 3,\ 4$

11. The domain is $\{x \mid x$ is real and $x \neq n\pi,\ n$ any integer$\}$; the range is $\{y \mid y$ is real$\}$.

12. $d_{AB} = 2\sqrt{34}$

13. $\frac{2}{3}\pi,\ \frac{4}{3}\pi$

14.

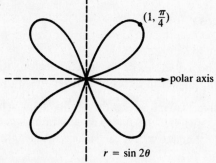

$(1, \frac{\pi}{4})$

polar axis

$r = \sin 2\theta$

15. $\left(\dfrac{5\sqrt{2}}{2} + 3\sqrt{3}\right)\mathbf{i} + \left(\dfrac{5\sqrt{2}}{2} + 3\right)\mathbf{j}$

16. 226.20 ft

17. 90.33 km

18. $2\sin 45° \cos 30°$

19. $-\sqrt{3}/2$

20. $\dfrac{\sqrt{2} + \sqrt{6}}{4}$

INDEX

Page numbers in italics indicate solved problems.